*Edited by*
*Raz Jelinek*

**Lipids and Cellular Membranes in Amyloid Diseases**

## Further Reading

Gravanis, A. G., Mellon, S. H. (eds.)

**Hormones in Neurodegeneration, Neuroprotection, and Neurogenesis**

2011

ISBN: 978-3-527-32627-3

Ramirez-Alvarado, M., Kelly, J. W., Dobson, C. M. (eds.)

**Protein Misfolding Diseases**

Current and Emerging Principles and Therapies

2010

ISBN: 978-0-471-79928-3

Sipe, J. D. (ed.)

**Amyloid Proteins**

The Beta Sheet Conformation and Disease

2005

ISBN: 978-3-527-31072-2

Fielding, C. J. (ed.)

**Lipid Rafts and Caveolae**

From Membrane Biophysics to Cell Biology

2006

ISBN: 978-3-527-31261-0

Miller, L. W. (ed.)

**Probes and Tags to Study Biomolecular Function**

for Proteins, RNA, and Membranes

2008

ISBN: 978-3-527-31566-6

Pebay-Peyroula, E. (ed.)

**Biophysical Analysis of Membrane Proteins**

Investigating Structure and Function

2008

ISBN: 978-3-527-31677-9

Pignataro, B. (ed.)

**Ideas in Chemistry and Molecular Sciences**

Where Chemistry Meets Life

2010

ISBN: 978-3-527-32541-2

*Edited by*
*Raz Jelinek*

# Lipids and Cellular Membranes in Amyloid Diseases

WILEY-VCH Verlag GmbH & Co. KGaA

**The Editor**

*Prof. Raz Jelinek*
Ben Gurion Univ. of the Negev
Department of Chemistry
Staedler Minerva Center
84105 Beer Sheva
Israel

All books published by **Wiley-VCH** are carefully produced. Nevertheless, authors, editors, and publisher do not warrant the information contained in these books, including this book, to be free of errors. Readers are advised to keep in mind that statements, data, illustrations, procedural details or other items may inadvertently be inaccurate.

**Library of Congress Card No.:** applied for

**British Library Cataloguing-in-Publication Data**
A catalogue record for this book is available from the British Library.

**Bibliographic information published by the Deutsche Nationalbibliothek**
The Deutsche Nationalbibliothek lists this publication in the Deutsche Nationalbibliografie; detailed bibliographic data are available on the Internet at http://dnb.d-nb.de.

© 2011 WILEY-VCH Verlag & Co. KGaA, Boschstr. 12, 69469 Weinheim, Germany

All rights reserved (including those of translation into other languages). No part of this book may be reproduced in any form – by photoprinting, microfilm, or any other means – nor transmitted or translated into a machine language without written permission from the publishers. Registered names, trademarks, etc. used in this book, even when not specifically marked as such, are not to be considered unprotected by law.

**Cover Design**   Adam-Design, Weinheim
**Typesetting**   Thomson Digital, Noida, India
**Printing and Binding**   betz-druck GmbH, Darmstadt

Printed in the Federal Republic of Germany
Printed on acid-free paper

**ISBN:** 978-3-527-32860-4

**ISBN: oBook:** 978-3-527-63432-3
**ISBN ePDF:** 978-3-527-63434-7
**ISBN ePub:** 978-3-527-63433-0
**ISBN Mobi:** 978-3-527-63435-4

## Contents

**Preface** *XI*
**List of Contributors** *XIII*

**1** **Interactions of α-Synuclein with Lipids and Artificial Membranes Monitored by ESIPT Probes** *1*
*Volodymyr V. Shvadchak, Lisandro J. Falomir-Lockhart, Dmytro A. Yushchenko, and Thomas M. Jovin*
1.1 Introduction to Parkinson's Disease and α-Synuclein *1*
1.2 Structural Biology of α-Synuclein *4*
1.3 Methods for Studying AS–Lipid Interactions *6*
1.4 AS–Lipid Interactions *13*
1.5 Interactions of Monomeric AS with Artificial Membranes Monitored with ESIPT Probes *15*
1.5.1 Influence of Membrane Charge *16*
1.5.2 Influence of Membrane Curvature *16*
1.5.3 Influence of Membrane Phase *17*
1.5.4 Influence of Acyl Chains *18*
1.5.5 Influence of Cholesterol *19*
1.5.6 Binding Kinetics *19*
1.6 Aggregation of AS and the Effects of Fatty Acids Monitored with ESIPT Probes *21*
1.7 Concluding Remarks *23*
References *23*

**2** **Structural and Functional Insights into α-Synuclein–Lipid Interactions** *33*
*Martin Stöckl, Bart D. van Rooijen, Mireille M.A.E Claessens, and Vinod Subramaniam*
2.1 Introduction *33*
2.2 Interaction of α-Synuclein with Model Membrane Systems *35*
2.2.1 Binding of α-Synuclein Species to Giant Unilamellar Vesicles *35*
2.2.2 Model Membrane Permeabilization by α-Synuclein Oligomers *39*
2.2.3 Structural Features of α-Synuclein Oligomers *41*

| | | |
|---|---|---|
| 2.3 | Biological Significance | 45 |
| 2.3.1 | Interaction Sites | 45 |
| 2.3.2 | Membrane Penetration | 47 |
| | References | 49 |

## 3 Surfactants and Alcohols as Inducers of Protein Amyloid: Aggregation Chaperones or Membrane Simulators? 57
*Daniel E. Otzen*

| | | |
|---|---|---|
| 3.1 | Introduction | 57 |
| 3.2 | Aggregation in the Presence of Surfactants | 58 |
| 3.2.1 | General Aspects of Protein–Surfactant Interactions | 58 |
| 3.2.2 | Effect of Surfactants on Protein Structure | 60 |
| 3.2.3 | Stoichiometry of SDS Binding | 61 |
| 3.2.4 | Aggregation of Proteins by SDS | 61 |
| 3.2.4.1 | A$\beta$ | 63 |
| 3.2.4.2 | $\beta_2$-Microglobulin and $\beta_2$-Glycoprotein I | 66 |
| 3.2.4.3 | Tau Protein | 67 |
| 3.2.4.4 | Prion Protein | 67 |
| 3.2.4.5 | Acyl CoA Binding Protein (ACBP) | 68 |
| 3.2.4.6 | $\alpha$-Synuclein ($\alpha$SN) | 69 |
| 3.3 | Palimpsests of Future Functions: Cytotoxic Protein–Lipid Complexes | 72 |
| 3.4 | Aggregation in Fluorinated Organic Solvents | 74 |
| 3.4.1 | Protein Examples | 76 |
| 3.4.1.1 | Acyl Phosphatase | 76 |
| 3.4.1.2 | $\beta_2$-Microglobulin | 76 |
| 3.4.1.3 | $\alpha$-Chymotrypsin | 77 |
| 3.4.1.4 | Alteration of Fibril Structure by TFE | 77 |
| 3.4.1.5 | Other Proteins | 78 |
| 3.5 | From Mimetics to the Real Thing: Aggregation on Lipids | 78 |
| 3.5.1 | Binding Surfaces and High Local Concentrations | 78 |
| 3.5.2 | Conformational Changes Associated with Binding | 79 |
| 3.5.3 | Chemical Variability of the Lipid Environment | 80 |
| 3.6 | Summary | 81 |
| | References | 82 |

## 4 Interaction of hIAPP and Its Precursors with Model and Biological Membranes 93
*Katrin Weise, Rajesh Mishra, Suman Jha, Daniel Sellin, Diana Radovan, Andrea Gohlke, Christoph Jeworrek, Janine Seeliger, Simone Möbitz, and Roland Winter*

| | | |
|---|---|---|
| 4.1 | Introduction | 93 |
| 4.2 | Results | 95 |
| 4.2.1 | The Conformations of Native proIAPP and hIAPP in Bulk Solution | 95 |

| | | |
|---|---|---|
| 4.2.2 | Fibrillation Kinetics and Conformational Changes of hIAPP and proIAPP in the Presence of Anionic Lipid Bilayers *96* | |
| 4.2.3 | Effect of the Membrane-Mimicking Anionic Surfactant SDS on the Amyloidogenic Propensity of hIAPP and proIAPP *102* | |
| 4.2.4 | hIAPP and proIAPP Aggregation and Fibrillation at Neutral Lipid Bilayers and Heterogeneous Model Raft Mixtures *105* | |
| 4.2.5 | Comparison with Insulin–Membrane Interaction Studies *111* | |
| 4.2.6 | Cytotoxicity of hIAPP *112* | |
| 4.3 | Conclusions *115* | |
| | References *117* | |
| | | |
| **5** | **Amyloid Polymorphisms: Structural Basis and Significance in Biology and Molecular Medicine** *121* | |
| | *Massimo Stefani* | |
| 5.1 | Introduction *121* | |
| 5.2 | Only Generic Data Are Currently Available on the Structural Features of Amyloid Oligomers *124* | |
| 5.3 | The Plasma Membrane Can Be a Primary Site of Amyloid Oligomer Generation and Interaction *127* | |
| 5.4 | Oligomer/Fibril Polymorphism Can Underlie Amyloid Cytotoxicity *129* | |
| 5.5 | Amyloid Oligomers Grown Under Different Conditions Can Display Variable Cytotoxicity by Interacting in Different Ways with the Cell Membranes *132* | |
| 5.6 | Conclusions *135* | |
| | References *136* | |
| | | |
| **6** | **Intracellular Amyloid β: a Modification to the Amyloid Hypothesis in Alzheimer's Disease** *143* | |
| | *Yan Zhang* | |
| 6.1 | Introduction *143* | |
| 6.2 | Evidence for the Presence of Intracellular Amyloid *144* | |
| 6.2.1 | Detection of Intracellular Amyloid *144* | |
| 6.2.2 | Neurotoxicty of Intracellular Amyloid *145* | |
| 6.2.3 | Possible Mechanisms of Intracellular Amyloid Toxicity *146* | |
| 6.3 | Sources of Intracellular Amyloid *146* | |
| 6.4 | Relationship Between Intracellular and Extracelluar Amyloid *148* | |
| 6.5 | Prevention of Intracellular Amyloid Toxicity *149* | |
| 6.6 | Concluding Remarks *149* | |
| 6.7 | Disclosure Statement *150* | |
| | References *150* | |

| | | |
|---|---|---|
| **7** | | **Lipid Rafts Play a Crucial Role in Protein Interactions and Intracellular Signaling Involved in Neuronal Preservation Against Alzheimer's Disease** *159* |
| | | *Raquel Marin* |
| | 7.1 | Lipid Rafts: Keys to Signaling Platforms in Neurons *159* |
| | 7.2 | Estrogen Receptors Are Part of Signaling Platforms in Neuronal Rafts *163* |
| | 7.3 | Role of Lipid Raft ERα–VDAC Interactions in Neuronal Preservation Against Aβ Toxicity *164* |
| | 7.4 | Disruption of ERα–VDAC Complex in AD Brains *167* |
| | 7.5 | Future Studies *169* |
| | | References *169* |
| | | |
| **8** | | **Alzheimer's Disease as a Membrane-Associated Enzymopathy of β-Amyloid Precursor Protein (APP) Secretases** *177* |
| | | *Saori Hata, Yuhki Saito, and Toshiharu Suzuki* |
| | 8.1 | Introduction *177* |
| | 8.1.1 | Cholesterol and Alzheimer's Disease Pathogenesis *180* |
| | 8.1.2 | ApoE, Lipoprotein Receptors and Alzheimer's Disease *181* |
| | 8.1.2.1 | LRP1 and LRP1B *182* |
| | 8.1.2.2 | SorLA/LR11 *182* |
| | 8.1.2.3 | ApoER2/LRP8 *183* |
| | 8.1.3 | Lipid Rafts and Alzheimer's Disease *183* |
| | 8.1.3.1 | APP-Cleaving Enzyme and Lipid Rafts *183* |
| | 8.1.3.2 | APP, X11 Family Proteins, and Lipid Rafts *184* |
| | 8.2 | Intramembrane-Cleaving Enzyme of Type I Membrane Proteins *186* |
| | 8.2.1 | γ-Secretase *186* |
| | 8.2.2 | γ-Secretase and Cholesterol *186* |
| | 8.3 | Alcadein Processing by γ-Secretase in Alzheimer's Disease *187* |
| | 8.3.1 | Alcadein as a γ-Secretase Substrate in Neurons *187* |
| | 8.3.2 | Alcadein Processing and γ-Secretase Dysfunction *188* |
| | | References *190* |
| | | |
| **9** | | **Impaired Regulation of Glutamate Receptor Channels and Signaling Molecules by β-Amyloid in Alzheimer's Disease** *195* |
| | | *Zhen Yan* |
| | 9.1 | Introduction *195* |
| | 9.2 | AMPAR-Mediated Synaptic Transmission and Ionic Current are Impaired by Aβ *195* |
| | 9.3 | CaMKII is Causally Involved in Aβ Impairment of AMPAR Trafficking and Function *196* |
| | 9.4 | PIP2 Regulation of NMDAR Currents is Lost by Aβ *197* |
| | 9.5 | The Effect of AChE Inhibitor on NMDAR Response is Impaired in APP Transgenic Mice *199* |

| | | |
|---|---|---|
| 9.6 | Aβ Impairs PKC-Dependent Signaling and Functions | 201 |
| 9.7 | Conclusion | 202 |
| | References | 203 |
| | | |
| **10** | **Membrane Changes in BSE and Scrapie** | **207** |
| | *Cecilie Ersdal, Gillian McGovern, and Martin Jeffrey* | |
| 10.1 | Prion Diseases | 207 |
| 10.2 | The Cellular Prion Protein (PrP$^c$) and Conversion to Disease-Associated Prion Protein (PrP$^d$) | 207 |
| 10.3 | PrP$^d$ Accumulation in the Central Nervous System and Lymphatic Tissues | 210 |
| 10.4 | Aberrant Endocytosis and Trafficking of PrP$^d$ in Neurons and Tingible Body Macrophages | 213 |
| 10.5 | Abnormal Maturation Cycle and Immune Complex Trapping of Follicular Dendritic Cells in Lymphoid Germinal Centers | 217 |
| 10.6 | Molecular Changes of Plasma Membranes Associated with PrP$^d$ Accumulation | 219 |
| 10.7 | Transfer of PrP$^d$ Between Cells | 220 |
| 10.8 | Extracellular Amyloid Form of PrP$^d$ | 221 |
| 10.9 | Strain-Directed Effects of Prion Infection | 222 |
| 10.10 | Conclusion and Perspectives | 222 |
| 10.11 | Summary | 223 |
| | References | 223 |
| | | |
| **11** | **Interaction of Alzheimer Amyloid Peptide with Cell Surfaces and Artificial Membranes** | **231** |
| | *David A. Bateman and Avijit Chakrabartty* | |
| 11.1 | Introduction | 231 |
| 11.2 | Comparison of the Neurotoxicity of Oligomeric and Fibrillar Alzheimer Amyloid Peptides | 232 |
| 11.3 | Aβ Oligomerization at the Cell Surface | 233 |
| 11.4 | Catalysis of Aβ Oligomerization by the Cell Surface | 234 |
| 11.5 | Type of Aβ Complexes that Form on the Cell Surface | 234 |
| 11.6 | Association of Alzheimer Amyloid Peptides with Lipid Particles | 238 |
| 11.7 | Future Directions | 238 |
| | References | 239 |
| | | |
| **12** | **Experimental Approaches and Technical Challenges for Studying Amyloid–Membrane Interactions** | **245** |
| | *Raz Jelinek and Tania Sheynis* | |
| 12.1 | Introduction | 245 |
| 12.2 | Unilamellar Vesicles and Micelles | 246 |
| 12.2.1 | Fluorescence spectroscopy | 246 |
| 12.2.2 | Fluorescence microscopy | 248 |
| 12.2.3 | Nuclear magnetic resonance | 249 |

| | | |
|---|---|---|
| 12.2.4 | Electron paramagnetic resonance | *250* |
| 12.2.5 | Other experimental techniques | *251* |
| 12.3 | Black Lipid Membranes | *252* |
| 12.4 | Langmuir Monolayers | *253* |
| 12.5 | Solid-Supported Bilayers | *255* |
| 12.6 | Other Techniques | *259* |
| 12.7 | Challenges and Future Work | *261* |
| | References | *263* |

**Index** *271*

## Preface

Amyloid diseases affect millions of people. However, despite the devastating health consequences and economic cost of these pathologies, we still struggle to understand their causes and develop effective therapies. While aggregation of amyloidogenic proteins is considered a hallmark of amyloid diseases, the significance and physiological impact of protein aggregation phenomena are still not clear. Cellular membranes and their lipid constituents appear to exhibit significant, perhaps even central, roles in affecting the biological properties and toxic effects of amyloid proteins. Studying the relationships between amyloid proteins and cellular membranes has been, until fairly recently, an "uncharted frontier." However, the interplay between membranes and amyloid aggregation pathways, and deciphering the physiological implications of amyloid–peptide/membrane interactions is currently attracting considerable interest both from a basic science point of view and also towards exploring new therapeutic avenues. This book is designed to highlight important aspects of this expanding topic, emphasizing the diversity of peptide systems, membrane assemblies, and biological phenomena investigated in the past few years.

This comprehensive volume reviews research work which utilized both model membranes and living cells, aimed at characterizing and better understanding membrane interactions of amyloid proteins. Chapter 1 (Jovin) describes applications of artificial membrane surfaces and fluorescent probes for studying lipid interactions of α-synuclein, the amyloidogenic peptide primarily associated with Parkinson's disease. α-Synuclein is also the focus of Chapter 2 (Subramaniam), which depicts structural and functional analyses of the peptide in different oligomeric forms, using various model membrane systems. Applications of artificial bilayer systems for studying membrane interactions are also discussed in the context of *amylin*, the amyloidogenic protein identified in type II diabetes (Chapter 4, Winter), and the amyloid-β peptide associated with Alzheimer's disease (Chapter 11, Chakrabartty).

Several chapters are devoted to investigations of amyloid–membrane interactions and their consequences within the complex environment of the cell, rather than model membrane systems. Intracellular interactions of amyloid-β as an underlining toxic factor are discussed in Chapter 6 (Zhang), while the contribution of lipid rafts within the cellular membrane as targets for amyloid-β and consequent toxicity through intracellular signaling is presented in Chapter 7 (Marin). The possible central roles of membrane-associated enzyme catalysis (Chapter 8, Suzuki) and membrane-embedded receptors (Chapter 9, Yan) in Alzheimer's disease are also analyzed. The neuropathology of *prion diseases* and putative correlations with membrane modulations are discussed in Chapter 10 (Ersdal).

Additional more general contributions complement the above chapters (which focus on specific amyloid proteins and/or cell and membrane models). A critical overview of investigations of amyloid proteins interactions with simple surfactants and alcohols perceived as membrane mimetics is provided in Chapter 3 (Otzen). The significance of *amyloid polymorphism* as a prominent factor affecting membranes and toxicity is highlighted in Chapter 5 (Stefani). Finally, we summarize the plethora of experimental approaches employed for investigating membrane interactions of amyloid peptides (Chapter 12, Jelinek).

January 2011  
Beersheva, Israel

*Raz Jelinek*

# List of Contributors

**David A. Bateman**
National Institutes of Health
National Institute of Diabetes, Digestive and Kidney Diseases
Laboratory of Biochemistry and Genetics
Bethesda, MD 20892
USA

**Avijit Chakrabartty**
University of Toronto
Department of Medical Biophysics
Toronto, ON, M5G 2M9

and

Canada
University of Toronto
Department of Biochemistry
Toronto, ON, M5S 1A8
Canada

**Mireille M.A.E. Claessens**
University of Twente
MESA+ Institute for Nanotechnology
Nanobiophysics
Drienerlolaan 5
7522 NB Enschede
The Netherlands

**Cecilie Ersdal**
Norwegian School of Veterinary Science
Department of Basic Sciences and Aquatic Medicine
Ullevålsveien 72
0033 Oslo
Norway

**Lisandro J. Falomir-Lockhart**
Max Planck Institute for Biophysical Chemistry
Laboratory of Cellular Dynamics
Am Fassberg 11
37077 Göttingen
Germany

**Andrea Gohlke**
Technical University of Dortmund
Faculty of Chemistry
Physical Chemistry I – Biophysical Chemistry
Otto-Hahn-Strasse 6
44227 Dortmund
Germany

**Saori Hata**
Hokkaido University
Graduate School of Pharmaceutical Sciences
Laboratory of Neuroscience
Kita 12 Nishi 6, Kita-ku
Sapporo 060-0812
Japan

**Martin Jeffrey**
Veterinary Laboratories
Agency – Lasswade
Pentland Science Park
Bush Loan, Penicuik
Midlothian EH26 0PZ
UK

**Raz Jelinek**
Ben Gurion University
Department of Chemistry
1 Ben Gurion Ave
Beer Sheva 84105
Israel

**Christoph Jeworrek**
Technical University of Dortmund
Faculty of Chemistry
Physical Chemistry I – Biophysical Chemistry
Otto-Hahn-Strasse 6
44227 Dortmund
Germany

**Suman Jha**
Technical University of Dortmund
Faculty of Chemistry
Physical Chemistry I – Biophysical Chemistry
Otto-Hahn-Strasse 6
44227 Dortmund
Germany

**Thomas M. Jovin**
Max Planck Institute for Biophysical Chemistry
Laboratory of Cellular Dynamics
Am Fassberg 11
37077 Göttingen
Germany

**Raquel Marin**
Laboratory of Cellular Neurobiology
Department of Physiology
School of Medicine
Institute of Biomedical Technologies
La Laguna University
38320 La Laguna
Spain

**Gillian McGovern**
Veterinary Laboratories
Agency – Lasswade
Pentland Science Park
Bush Loan, Penicuik
Midlothian EH26 0PZ
UK

**Rajesh Mishra**
Technical University of Dortmund
Faculty of Chemistry
Physical Chemistry I – Biophysical Chemistry
Otto-Hahn-Strasse 6
44227 Dortmund
Germany

**Simone Möbitz**
Technical University of Dortmund
Faculty of Chemistry
Physical Chemistry I – Biophysical Chemistry
Otto-Hahn-Strasse 6
44227 Dortmund
Germany

**Daniel E. Otzen**
Aarhus University
Interdisciplinary Nanoscience Center (iNANO)
Department of Molecular Biology
Gustav Wieds Vej 10C
8000 Aarhus C
Denmark

**Diana Radovan**
Technical University of Dortmund
Faculty of Chemistry
Physical Chemistry I – Biophysical Chemistry
Otto-Hahn-Strasse 6
44227 Dortmund
Germany

**Yuhki Saito**
Hokkaido University
Graduate School of Pharmaceutical Sciences
Laboratory of Neuroscience
Kita 12 Nishi 6, Kita-ku
Sapporo 060-0812
Japan

**Janine Seeliger**
Technical University of Dortmund
Faculty of Chemistry
Physical Chemistry I – Biophysical Chemistry
Otto-Hahn-Strasse 6
44227 Dortmund
Germany

**Daniel Sellin**
Technical University of Dortmund
Faculty of Chemistry
Physical Chemistry I – Biophysical Chemistry
Otto-Hahn-Strasse 6
44227 Dortmund
Germany

**Tania Sheynis**
Ben Gurion University
Department of Chemistry
1 Ben Gurion Ave
Beer Sheva 84105
Israel

**Volodymyr V. Shvadchak**
Max Planck Institute for Biophysical Chemistry
Laboratory of Cellular Dynamics
Am Fassberg 11
37077 Göttingen
Germany

**Massimo Stefani**
University of Florence
Department of Biochemical Sciences and Research Centre on the Molecular Basis of Neurodegeneration
Viale Morgagni 50
50134 Florence
Italy

**Martin Stöckl**
University of Twente
MESA+ Institute for Nanotechnology
Nanobiophysics
Drienerlolaan 5
7522 NB Enschede
The Netherlands

**Vinod Subramaniam**
University of Twente
MESA+ Institute for Nanotechnology and MIRA Institute for Biomedical Technology and Technical Medicine
Nanobiophysics
Drienerlolaan 5
7522 NB Enschede
The Netherlands

**Toshiharu Suzuki**
Hokkaido University
Graduate School of Pharmaceutical Sciences
Laboratory of Neuroscience
Kita 12 Nishi 6, Kita-ku
Sapporo 060-0812
Japan

**Bart D. van Rooijen**
University of Twente
MESA+ Institute for Nanotechnology
Nanobiophysics
Drienerlolaan 5
7522 NB Enschede
The Netherlands

**Katrin Weise**
Technical University of Dortmund
Faculty of Chemistry
Physical Chemistry I – Biophysical Chemistry
Otto-Hahn-Strasse 6
44227 Dortmund
Germany

**Roland Winter**
Technical University of Dortmund
Faculty of Chemistry
Physical Chemistry I – Biophysical Chemistry
Otto-Hahn-Strasse 6
44227 Dortmund
Germany

**Zhen Yan**
State University of New York at Buffalo
School of Medicine and Biomedical Sciences
Department of Physiology & Biophysics
Buffalo, NY 14214
USA

**Dmytro A. Yushchenko**
Max Planck Institute for Biophysical Chemistry
Laboratory of Cellular Dynamics
Am Fassberg 11
37077 Göttingen
Germany

**Yan Zhang**
Peking University
College of Life Sciences
State Key Laboratory of Biomembrane and Membrane Biotechnology
Laboratory of Neurobiology
5 Yiheyuan Road, Haidian District
Beijing 100871
China

# 1
## Interactions of α-Synuclein with Lipids and Artificial Membranes Monitored by ESIPT Probes

*Volodymyr V. Shvadchak, Lisandro J. Falomir-Lockhart, Dmytro A. Yushchenko, and Thomas M. Jovin*

### 1.1
#### Introduction to Parkinson's Disease and α-Synuclein

Protein conformational disorders are a group of diseases associated with misfolding and pathological aggregation of one or more proteins, leading to impairment of functionality and sequestration into insoluble bodies [1]. In the case of the neurodegenerative diseases, the dominant character of certain genetic alterations indicates that a gain of toxic function in one or more stages of the aggregation process is required for clinical progression. Parkinson's disease (PD), the second most frequent neurodegenerative syndrome affecting >1% of the population above 60 years in age, is characterized by debilitating, progressive, and irreversible neuromotor degeneration and impaired cognitive functions. The classical symptoms include muscle rigidity, bradykinesia, akinesia, postural instability, and resting tremor arising from reduced dopamine secretion, secondary to the loss of dopaminergic neurons of the substantia nigra and other nuclei of the midbrain. As PD progresses, components of the autonomic, limbic, and somatomotor systems are affected, finally extending to the neocortex with development of the full-blown clinical picture [2–4].

Although drug-based therapies serve to palliate the symptomatology of PD, a specific curative treatment is as yet unavailable [2, 3, 5, 6]. PD remains incurable primarily because its etiology remains obscure [6]. Genetic factors, exposure to certain herbicides, and oxidative stress have been established as strong etiological components; other agents and factors are undoubtedly involved. Due to the multifactorial nature of PD, it is difficult, if not impossible, to ascribe specific functional alterations to a single protein, mechanism, or neuronal type. The "amyloid hypothesis" originally proposed that aberrant protein fibrillar aggregation compromises cellular functionality, ultimately leading to neurodegeneration [7]. However, intermediates ("oligomers") of the aggregation pathway are currently considered as the more likely culprits in cellular toxicity [8], dysfunction, and death, rather than the large amyloid fibrillar aggregates, which instead may function as a protective sequestration mechanism. Among the genetic factors involved in PD,

*Lipids and Cellular Membranes in Amyloid Diseases*, First Edition. Edited by Raz Jelinek.
© 2011 Wiley-VCH Verlag GmbH & Co. KGaA. Published 2011 by Wiley-VCH Verlag GmbH & Co. KGaA.

the (over)expression/mutation of α-synuclein (AS) is most prominently implicated in cytopathology [9]. Several genetic alterations in the SNCA gene (located in the PARK1 locus) are associated with the early onset familial dominant form of PD, and AS has been identified as the major component of Lewy bodies (LBs) and Lewy neurites (LNs), amyloid aggregates found in the neuronal cell bodies and axons, respectively, of patients suffering from PD. LBs and LNs also include lower concentrations of distinctive lipids and other proteins. These amyloid deposits constitute the pathological hallmark in the brains of patients with synucleinopathies: PD, dementia with Lewy bodies (DLB) and multiple system atrophy (MSA).

Little is known about the biology of AS and its physiological functions remain obscure. It is found in high concentrations in the presynaptic nerve terminals associated with secretory vesicles, and also in glia, and it has been suggested that AS is involved in neurotransmitter secretion and reabsorption and in synaptic plasticity [10–15]. Nevertheless, SNCA knockout mice are viable, suggesting that AS is not essential either for neuronal development or for survival.

Point mutations (A30P, E46K, and A53T) and the more frequent gene duplication events are strongly correlated with early onset familial PD. Most affected individuals suffer from spontaneous PD, again indicating that a combination of genetic predisposition [16] and environmental factors is involved. Thus, at the molecular level, the physiological and aberrant interactions of AS with other biomolecules and organelles (e.g., catecholamines, membranes, mitochondria) and important influences of other external factors are at the focus of attention, with oxidative stress being of manifestly central relevance.

Lipids play a central role in PD pathology as important modulators of AS biology and protein–membrane interactions [3, 17, 18]. AS associates with secretory vesicles and the mitochondrial inner membrane [3, 17, 19]. The high affinity of AS for membranes implies that the latter not only mediate physiological protein localization but also may serve to initiate and/or modulate aberrant interactions and the rate of protein aggregation in the pathological context. Depending on the membrane composition (polar headgroups, acyl chain lengths and saturation, cholesterol content), lipids either accelerate [20–23] or inhibit [24, 25] AS fibrillation. Hence it is highly probable that preferential AS aggregation in the proximity of specific organelle membranes may account for the perturbation and dysfunction of certain cellular mechanisms (Figure 1.1).

It has been proposed that AS oligomers or low molecular weight aggregates are able to disrupt membranes and/or induce the formation of ion channels ([26–28]; see also Chapter 2), possibly leading to cellular and mitochondrial membrane depolarization [29]. Increased levels of AS can also block vesicular trafficking between the endoplasmic reticulum (ER) and the Golgi complex [30], thereby affecting many cellular processes, including autophagic recycling of damaged mitochondria. Mitochondrial dysfunction leads to oxidative stress and thus to augmented AS aggregation promoted by protein oxidation. This process may constitute a cyclic mechanism for progressive toxicity [17] that is further exacerbated by the presence of dopamine (DA) and related metabolites in specialized neuronal cells (Figure 1.1).

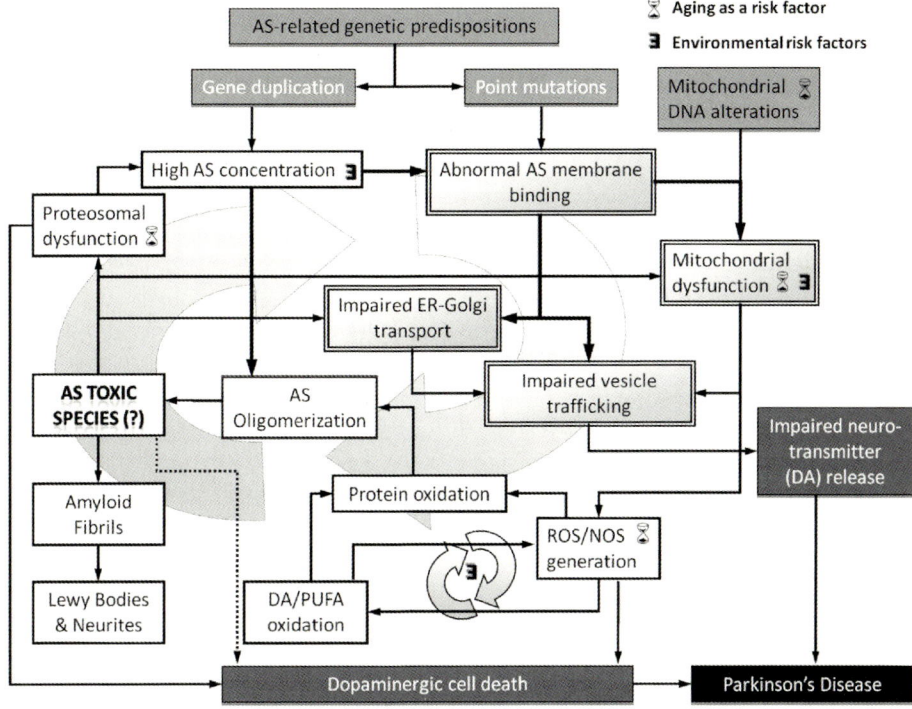

**Figure 1.1** Molecular and cellular processes related to AS and PD. PD is a multifactorial disease and neuronal dysfunction arises by the concurrent action of environmental factors stressing the functions of mitochondria, endoplasmic reticulum, and proteasomes. Genetic predisposition (gene duplication and missense mutations of SNCA gene, mutations in other genes) is further exacerbated in dopaminergic cells by the susceptibility to the promoters of reactive oxygen (ROS) and nitrogen (NOS) species. Boxes with double borders indicate membrane participation in the indicated process.

AS binds fatty acids (FAs), presumably via the hydrophobic core of the NAC (non-Aβ component) region, and the bound FAs modulate aggregation of the protein [31–33]. Particularly active are the long- and very long-chain polyunsaturated fatty acids (PUFAs) that are abundant in the central nervous system [34]. Arachidonic acid, for example, facilitates AS oligomerization when present in membrane vesicles or free in solution [22], and FAs, such as oleic acid, trigger AS aggregation *in vitro* [35]. Whereas PUFAs stabilize some kinds of soluble oligomeric and toxic species, saturated FAs inhibit their formation [36, 37]. Furthermore, the high sensitivity of PUFA- and DA-related metabolites to oxidative stress supports the notion that a combination of multifactorial alterations may converge in dopaminergic cells to trigger and promote AS toxicity, even in the absence of genetic predisposition factors. AS has also been implicated in the regulation of inflammatory response in the brain by sequestration of precursors (arachidonic acid) for prostanoids and reduction of proinflammatory cytokine secretion [38]. These factors are also known to be important for PD progression and probably for its onset.

Due to the lack of an effective PD therapy, basic research on the biology and the role of lipids in synucleinopathies is essential for understanding the etiology of neurodegenerative diseases and for the identification of potential drug targets. The classical functions attributed to lipids are energy storage and definition of physical barriers. Hence they serve as substrates for organizing the structure of biological membranes, both defining and isolating internal compartments and also providing the interface of the cell with its external milieu. Lipids function in a majority of cellular processes, ranging from protein targeting to signaling by hormones and second messengers. These phenomena are at the focus of this chapter, particularly in relation to the pathological processes involving amyloidogenic proteins.

## 1.2
### Structural Biology of α-Synuclein

AS is a 140 amino acid polypeptide of 14.5 kDa with three distinguishable domains: the N-terminal, the NAC, and the C-terminal regions (Figure 1.2). AS migrates as a 20–30 kDa protein in size-exclusion columns due to the lack of a classical secondary structure when free in solution [39, 40]. As such, AS is assigned to the family of intrinsically disordered proteins (IDPs), characterized by an unconventional energy landscape lacking the classical "folding funnel" and occupying instead a broad conformational space with no clear minimum (native folding structure) [41, 42]. However, it has been proposed that monomeric AS in solution adopts transient and

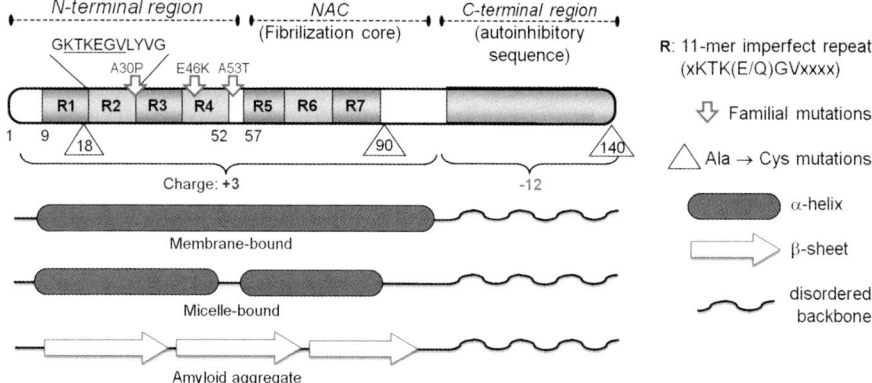

Figure 1.2 AS primary and secondary structures. AS contains seven imperfect repeats of 11-mer residues in the first two-thirds of its sequence, which are thought to be important in forming the lipid-bound α-helical and for the amyloid cross-β-sheet structures, whereas the highly acidic C-terminal region remains in a disordered state. Triangles and hollow arrows indicate the positions of Cys and familial mutants, respectively. The β-sheet regions are shown schematically; for a more detailed view, see [53].

variable conformations, for example, ring-shaped, that can rationalize some of the experimental structural data [40, 43, 44]. According to this model, the C-terminal acidic region is folded over the partially positive N-terminal region and stabilized primarily by long-range electrostatic interactions. This orientation shields the NAC region, which is rich in hydrophobic residues and constitutes the core for protein aggregation [45]. Compounds such as polyamines and cations that bind to the C-terminus and thereby disrupt the fold-back structure promote AS aggregation [43, 46].

Under given circumstances, AS can adopt two stable conformations. The first is a presumably functionally relevant α-helical conformation induced by interactions with membranes, particularly those containing acidic phospholipids, and also with detergent micelles [47–49]. The N-terminal and NAC regions include the imperfect repeats, "xKTK(E/Q)GVxxxx", highly conserved in orthologous genes (Figure 1.2). These repeats can adopt an amphiphilic helical structure, a classical motif for membrane-interacting proteins. The second, alternative, stable conformation of AS is characterized by a cross-β-sheet conformation centered in the NAC region and manifested primarily in the amyloid fibrils pathognomonic of the amyloidoses. These insoluble aggregates redissolve in high concentrations of guanidinium chloride. In both cases the C-terminal region remains disordered, possibly due to its negative net charge, leading to the suggestion that it could function as a particular target or a scavenger for protein interactions and post-translational modifications; three of the four Tyr residues and also Ser129 are within this region and serve as loci for phosphorylation. Alterations in phosphorylation, acylation, and acetylation patterns have been related in PD pathology, either by modification of AS [50] or other cellular components [51, 52], yet the relevance of these modifications to the alteration of AS function, localization, and interaction with other cellular components, particularly with membranes, remains to be established.

The conformational changes of AS have been studied extensively *in vitro* by a diverse range of physicochemical techniques, extending from fluorescence to nuclear magnetic resonance (NMR) and electron paramagnetic resonance (EPR) spectroscopy, and to molecular computation. These studies provide evidence for the ring-shaped autoinhibited monomer conformation alluded to earlier [43], the membrane/micelle-bound α-helical structures [24, 54, 55], and the stacked cross-β-sheets in fibrillar structures [56–58]. However, there is little concrete evidence about the (ultra)structures of AS *that actually form in cells* and about their role in the cytotoxic mechanisms underlying mitochondrial and proteasomal dysfunction, or their relationship to oxidative stress. Membranes are almost certainly involved in the cytotoxicity of AS. Anionic membranes promote the formation of α-helical structure in the N-terminal region of AS [24], extending as far as residue Gln99 according to recent NMR investigations [59]. Similar results have been obtained with amphiphilic micelles [54, 55]. We have focused on the analysis of AS binding to phospholipid membranes of different compositions, aiming for a broader, more comprehensive understanding of the factors that modulate such complexes by employing novel, very sensitive fluorescence multiparametric/ratiometric probes. The aggregation of AS and the influence of bound free FAs has also been explored with these new reagents and will be described below.

## 1.3
### Methods for Studying AS–Lipid Interactions

The wide variety of biophysical techniques applied in the study of protein–membrane interactions provides insight into key structural, thermodynamic, and kinetic parameters [60]. A detailed description of these methods is beyond the scope of this chapter (see [61]), and we restrict ourselves to those utilized in studies of AS. A brief summary is presented in Table 1.1.

We will consider fluorescence in more detail because of its versatility and extensive application, including in the studies featured in this chapter. Fluorescence is an inherently very sensitive method (high signal-to-noise ratio) and requires small amounts of material with concentrations on the biological scale (pM–μM). There are numerous types of fluorescence techniques useful for the investigation of AS–membrane interactions [62, 63]: steady-state fluorescence (emission and excitation spectra), time-resolved fluorescence, anisotropy, Förster or fluorescence resonance energy transfer (FRET), fluorescence correlation spectroscopy (FCS), and other correlation techniques, and far- and near-field microscopy. The reported applications of these methods in AS–membrane studies are summarized in Table 1.2. For a comprehensive, current review of small molecular fluorescent probes of amyloid aggregation, see [64].

For analyzing AS interactions with lipids by fluorescence, either the protein and/or the membrane must be fluorescently tagged. Tryptophan (Trp) and tyrosine (Tyr) are the natural intrinsic fluorophores of proteins. Trp fluorescence is more convenient than that of Tyr due to its higher quantum yield and an emission spectrum sensitive to the polarity of its environment. Inasmuch as AS lacks Trp, many groups have resorted to mutation of given residues to Trp for probing different membrane interactive domains [73, 91]. Other strategies for labeling AS are based on covalent reactions of lysine and cysteine (Cys) residues. Cysteine is likewise absent from native AS and is therefore introduced as a replacement for alanine. The Cys mutants are labeled with maleimide derivatives of fluorophores [63, 92], and double Cys mutants of AS have been used for labeling AS with two fluorophores, for example in FRET experiments [93, 94]. Genetically encoded tags of AS for studies of living cells have involved fusion with a visible fluorescent protein [87–89, 95] or with the tetracysteine tag specific for fluorogenic biarsenicals [96].

There are three categories of fluorescent probes for the selective labeling of membranes:

1) phospholipids with a headgroup or acyl chain modified with a fluorophore, such as NBD-PE, NBD-PC, and BODIPY-PC [100]
2) fluorophores containing specific anchor groups that selectively stain membranes, such as DiO, DiI [101], and, most recently, NR12S [102]
3) lipophilic fluorescent probes that bind to lipidic structures in accordance with their partition coefficient, such as DPH, DHE, and Laurdan [100].

Fluorophore-modified lipids are suitable for the preparation of fluorescent model membranes, but not always for selectively staining cellular plasma membranes,

Table 1.1 Biophysical methods used in AS–membrane investigations.

| Method[a] | Output information | Advantages | Limitations | Refs.[b] |
|---|---|---|---|---|
| NMR | Structural features: multiple conformational states of residues and interactions with membranes/micelles | Full atomistic model | Requires large quantities (solution, solid state). Low signal from large complexes. No information about unstructured and flexible regions. Slow | R1 |
| EPR | Structural features: relative spatial localization of residues | Inter-residue distances for conformational models | Requires large quantities. Limited structure information in solution | R2 |
| CD | Secondary protein structure | Fast, label free | Mean parameters. No structural details, unless combined with molecular biology | R3 |
| FTIR | Secondary protein structure | Fast, label free | See CD | R4 |
| ITC | Thermodynamic characterization | Fast, label free | See CD | R5 |
| AFM | Visualization of protein assemblies on membrane | Structures and topography of aggregates and surfaces with nanometer resolution | Experienced operator required | R6 |
| EM | Visualization of monomeric AS and aggregates in the cell or tissue | Determination of subcellular localization | Does not provide atomic-scale structure | R7 |
| Electrophysiology | Membrane conductance | Live cell studies | No structural information | R8 |
| Fluorescence | Structural and thermodynamic characterization | Wide range of probes and techniques. Nanoscale information about environment | Specific labeling techniques. Protein modification. Probe properties | R9 |

a) EPR, electron paramagnetic resonance; CD, circular dichroism; FTIR, Fourier transform infrared spectroscopy; ITC, isothermal titration calorimetry; AFM, atomic force microscopy; EM, electron microscopy.
b) R1, [59, 65–68]; R2, [69–72, 79]; R3, [26, 68, 73–75]; R4, [74, 76, 77]; R5, [78–80]; R6, [81–83]; R7,[84, 85]; R8, [86]; R9, [87–99]; see also Chapter 2.

Table 1.2 Fluorescence methods used in AS–membrane studies.

| Method | Output information | Advantages | Limitations | Refs.[a] |
|---|---|---|---|---|
| Steady-state, environment-sensitive probes | Affinity for membrane, interacting residue(s), membrane localization, kinetics of interaction | Distinguishes type of membrane, fast | Provides limited details about spatial structure of the protein | R1 |
| FRET, quenching | Structural details, spatial localization of residues | Inter-residue distances | Generally requires two (donor and acceptor) chromophores | R2 |
| Anisotropy | Affinity for membrane | Concentration-independent | Provides limited details about spatial structure | R3 |
| FCS | Affinity for membrane, size, kinetics of interaction | Single molecule technique | Provides limited details about spatial structure | R4 |
| Microscopy | Affinity for membrane, distribution | Visualization of interaction areas on the membrane | Requires genetically encoded probes | R5 |

a) R1, [73, 79, 90, 91]; R2, [93, 94]; R3, [75, 79]; R4, [94, 97, 98]; R5, [95, 99].

especially with phospholipid-like probes. In this respect, fluorophores with specific anchor groups (e.g., NR12S [102]) are advantageous, since they can be applied for labeling both model and cellular membranes. Due to their hydrophobic nature, lipid probes such as diphenylhexatriene (DPH) bind selectively to bilayers, locating in the hydrophobic core without an anchor. Nevertheless, the application of such dyes is restricted by their spectroscopic properties or the difficulty in achieving high enough densities, as is often required in FRET studies.

Irrespective of the labeled component, when dealing with soluble proteins interacting with membranes, the presence of two phases (aqueous and hydrophobic) in the system renders *environment-sensitive probes* very useful tools for analyzing by fluorescence different aspects and properties of the components under study. There are three types of environment-sensitive probes: intensiometric, band-shifting, and multiparametric ratiometric [103]:

1) *Intensiometric* dyes are mostly represented by "molecular rotor" fluorophores consisting of two planar moieties connected by a single bond. They exhibit strong changes in emission intensity (quantum yield) depending on microenvironment-dictated restriction to intramolecular rotation. Therefore, binding to larger structures or to media with higher viscosity decreases molecular rotation and the emission intensity increases. Dyes such as thioflavin T (ThT) and Congo Red are widely used for the monitoring of AS aggregation due to their increased emission intensity upon binding to rigid large aggregates and fibrils (Figure 1.3). ThT has a two-ring structure and is almost planar in the ground state ($\phi = 30°$). Upon excitation, the rings rotate in order to achieve the most stable conformation, such that the molecules lose planarity ($\phi = 90°$) and exhibit low fluorescence. Rigid and/or viscous environments restrain the rotation of the rings and as a result the quantum yield of ThT increases up to 500-fold due to the stabilization of its planar form by amyloid fibrils [104]. The main disadvantage of intensiometric dyes is the dependence of their response on concentration, a parameter that is generally very difficult to control in the cellular context.

2) *Band-shifting solvatochromic* probes do not suffer from the same limitation. These dyes exhibit strong changes in dipole moment upon excitation due to intramolecular charge transfer (ICT), such that their emission spectra shift in a manner dependent on the molecular microenvironment. The major parameters are polarity and hydration. Dipole–dipole interactions and also specific hydrogen-

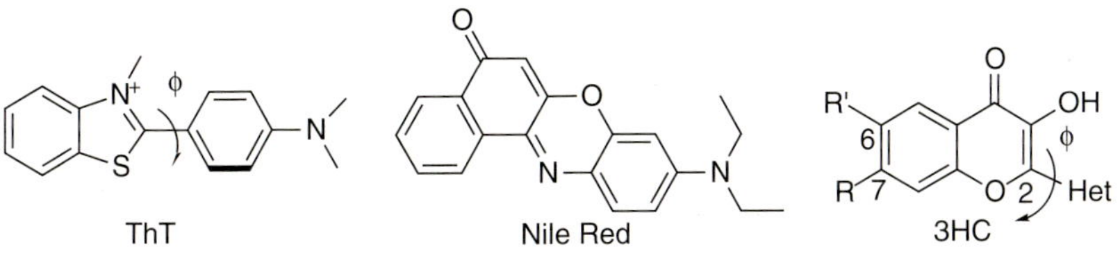

**Figure 1.3** Chemical structures of environment-sensitive dyes. Thioflavin T, Nile Red and 3-hydroxychromone (3HC) are examples of intensiometric (molecular rotor), solvatochromic, and multiparametric ratiometric dyes.

bonding interactions affect the energy of the excited state and as result lead to shifts of the excitation and emission maxima. Typical representatives of this class of probes are prodan, dialkylaminocoumarin, nitrobenzoxadiazole (NBD), and Nile Red (Figure 1.3). They find wide application in membrane studies, but their potential in investigations of AS remains to be explored fully.

3) *Multiparametric ratiometric* dyes constitute the third class of environment-sensitive probes. These fluorophores exhibit more than one emission state and as result have two or more distinct emission bands. Within this class [105, 106], the 3-hydroxychromones (3HCs) have found the widest application [103]. These probes exhibit two emission bands, the first of which, $N^*$, corresponds to the emission from the "normal" excited state. The second band, $T^*$, represents the emission from the tautomeric form of the probe formed by excited state intermolecular proton transfer (ESIPT) (Figure 1.4a). The absolute and relative intensities of the two emission bands of 3HCs are strongly dependent on the environment of the probe, sensing both polarity and hydration with high sensitivity. The ratio of the band intensities ($T^*/N^*$) is a particularly convenient and suitable indicator [107] and thus the 3HCs function as "ratiometric" probes. However, they may also report changes in the environment by systematic shifts in the positions of their emission maxima. Prominent examples are the 4′-dialkylamino-3HCs. Due to ICT from the 4′-dialkylamino to the 4-carbonyl group, the $N^*$ band position of these dyes varies significantly with polarity, as in the case of the non-ESIPT probe prodan [107].

Another feature of 3HCs is their sensitivity to hydration. In an aqueous environment these probes are hydrogen bonded to water and have excitation and emission spectra that differ from those of the nonhydrated forms. Whereas the latter exhibit two emission bands ($N^*$ and $T^*$) as described above, the hydrated form has a single emission band (H-$N^*$), particularly in protic solvents. Deconvolution of the emission spectra into three bands ($N^*$, $T^*$, and H-$N^*$) provides information about the "hydration" parameter of the microenvironment (Figure 1.4b). Due to the ability

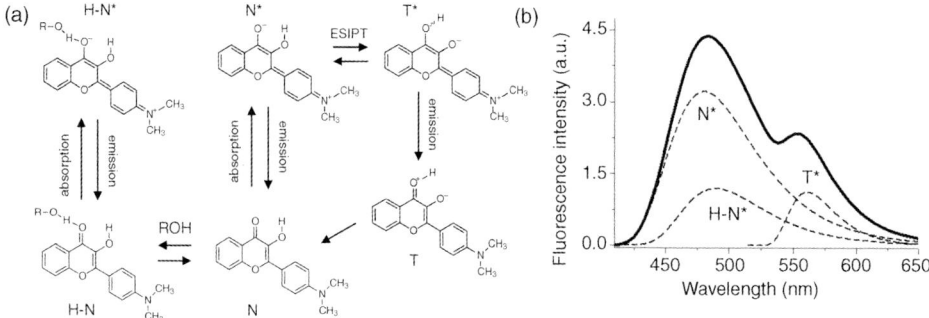

**Figure 1.4** Excited state intramolecular proton transfer (ESIPT) reaction of 3HC probe in protic media. (a) Scheme of ground and excited state transitions of 3HC probe. (b) Decomposition of 3HC spectrum into three bands. See description in the text and [108, 109].

of these probes to sense several environmental parameters, they are denoted *multiparametric*.

A great advantage of the ESIPT probes is that their ratiometric response can be optimized for the system under investigation [107]. It is possible to modify the fluorescence response of 3HCs by varying substituents at positions 2 and 7 of the chromophore heterocycle. The greater electron-donating capacity of the substituent at position 2 leads to a decrease in the $T^*:N^*$ ratio, whereas insertion of an electron-donating substituent at position 7 results in an increase in the $T^*:N^*$ ratio [110, 111]. There are different derivatives of 3HC dye with fluorescence properties optimized for various applications: probing proteins [112] and nucleotides [113], lipid bilayers and cellular membranes [102, 114], and also monitoring protein–DNA [115] and protein–protein interactions [116]. In our work, we have used derivatives of *4'-diethylamino-3-hydroxyflavone (FE)*, which has an optimized fluorescence response in hydrophobic environments, and also derivatives of 2-(2-furyl)-3-hydroxychromone (FC), which is more suitable for working in aqueous solutions (Figure 1.5).

Probes based on the FE chromophore find wide application in membrane studies. They are used for the determination of hydration levels of membranes, lipid order in bilayers, visualization of raft domains, and measurement of surface, dipole, and transmembrane potentials of membranes [110, 114, 117, 118]. In order to distinguish all these parameters of (and within) membranes, 3HC chromophores have been specifically modified with anchors to target desired elements of biological membranes (Figure 1.6). For the determination of the surface potential (electrostatic potential at the membrane–water interface) of membranes, the dyes F2N12S and F2N8 were synthesized (Figure 1.6) [114, 119]. The linkers serve to locate the chromophores of these probes at the interface, where they can report the surface potential [114, 119]. Sensing dipole and transmembrane potentials requires probes immersed in the lipid bilayer such as F8N1S or di-SFA that adopt a vertical orientation in the membrane [117, 120].

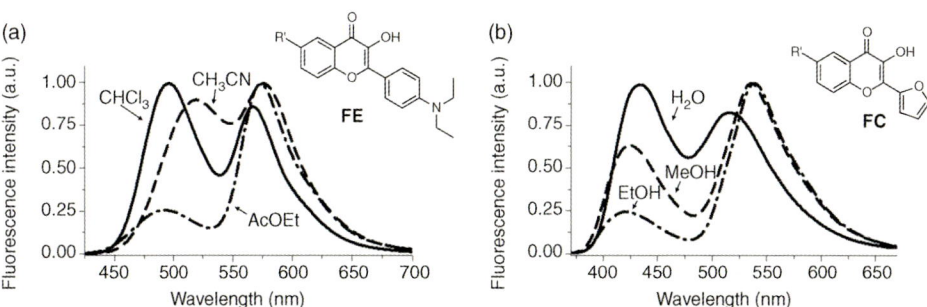

**Figure 1.5** Structures and spectral characteristics of ESIPT probes FE and FC. Multiparametric/ratiometric probes can be tuned to optimize their properties according to the molecular microenvironment. Emission spectra of (a) FE in ethyl acetate (AcOEt), acetonitrile ($CH_3CN$), and chloroform ($CHCl_3$) and (b) FC in ethanol (EtOH), methanol (MeOH) and water ($H_2O$).

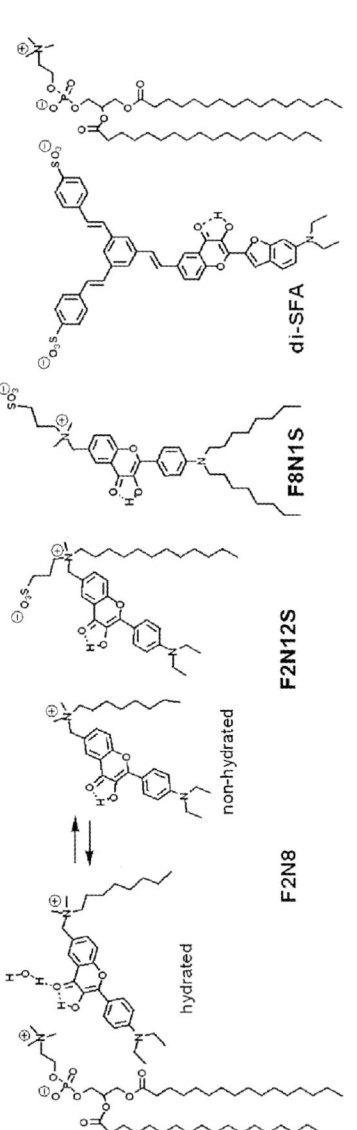

**Figure 1.6** 3HC membrane probes with different localizations in lipid bilayer. Chemical structures of several membrane probes from the family of 3HCs are targeted to position at different depths in the phospholipid bilayers, thereby sensing and reporting different membrane properties. Adapted from [103].

We have recently demonstrated the capacity of FE as an extrinsic ESIPT probe to reveal differences in the supramolecular organization of amyloid fibrils formed by wild-type (WT) AS and clinical mutants [121]. We have also shown the great utility of FE, applied in covalent form, for the investigation of AS–membrane interactions [90] and summarize the results of these investigations in the following section. The FC chromophore, a highly sensitive reporter of AS aggregation, particularly in the early stages inaccessible to ThT and other amyloid indicators (see Section 1.5) [116], is also a good reporter of AS–membrane interactions (see below).

## 1.4
## AS–Lipid Interactions

Due to the multifactorial nature of PD, the pathophysiology of AS cannot be completely understood unless it is evaluated in a relevant cellular context, which *in vitro* implies the systematic evaluation of the influence of lipids and membranes. Numerous studies [122] (see also Chapter 2) have addressed and identified the parameters that determine the thermodynamic, kinetic, and structural features of complexes of AS with lipid bilayers:

- membrane charge (ionic polar headgroups)
- membrane phase(s) (temperature, acyl chain composition, cholesterol content)
- membrane curvature (size and type of model membranes)
- nature of the polar headgroups
- supporting electrolyte (ionic strength, composition, pH)
- kinetics of protein binding, dissociation, and conformational rearrangement in the membrane interphase
- lipid–protein stoichiometry.

Most of the corresponding studies have focused on particular lipidic systems, leading to somewhat conflicting conclusions. In view of the enormous compositional diversity of natural membranes, the discrepant results and interpretations regarding membrane-associated AS conformations may reflect differences in the key experimental parameters listed above, such as protein:lipid ratio [59] and lipid (membrane) charge [123–125] (see also Chapter 2).

Early studies demonstrated that the binding of AS to extracted brain tissue lipid membranes does not require the presence of other proteins [126]. This finding led to the application of artificial model membranes in research aimed at the elucidation of the nature and functional significance of AS–lipid interactions. AS binds readily to artificial membranes [63, 71, 80, 98, 127]. Due to the positive charge of the lipophilic N-terminus, negatively charged membranes are highly preferential targets [98, 128] (see also Chapter 2). Applying microscopy with DOPC (1,2-dioleoyl-*sn*-glycero-3-phosphocholine) GUVs (giant unilamellar vesicles) [124] and circular dichroism (CD) spectroscopy with POPC (1-palmitoyl-2-oleoyl-*sn*-glycero-3-phosphocholine) vesicles [73], AS interactions with neutral membranes were not detected, whereas calorimetry with DPPC (1,2-dipalmitoyl-*sn*-glycero-3-phosphocholine) [79, 80] and fluorescence correlation spectroscopy (FCS) with POPC vesicles [98] revealed a moderate degree of protein binding. The effects of membrane phase and curvature on AS binding are complex and as yet unresolved [127]. It has been reported that AS interacts preferentially with membranes in the liquid disordered ($L_d$) phase [124, 128, 129], although there are instances in which AS demonstrates a higher affinity for raft-like domains [126, 130]. Isothermal titration calorimetry (ITC) measurements indicate that binding of AS to neutral small unilamellar vesicles (SUVs) in the gel phase ($L_o$) is stronger than to membranes in the Ld phase [80]. High membrane curvature was reported to be essential for AS binding to DPPC vesicles [80] and to increase binding to charged membranes, but this effect was not observed in POPC/POPS (1-palmitoyl-2-oleoyl-sn-glycero-3-phosphoserine) systems [98]. Such disagreement most probably reflects the use of lipids with different polar headgroups, which define the charge and curvature of the respective membranes. AS binds more strongly to membranes composed of lipids with smaller polar headgroups, for example, phosphatic acids [131], probably due to the higher negative surface energy (more and stronger membrane defects).

The conformation(s) of AS on lipid membranes are polymorphic. The CD spectra of AS complexed to artificial membranes indicate that the protein adopts an α-helical conformation [47, 132] (Figure 1.7a). High-resolution NMR shows that AS bound to small (~5 nm) SDS detergent micelles forms two antiparallel α-helices (residues 3–37 and 45–92) [54]. Discrimination between the various proposed conformations is possible by measuring the distance between the N-terminal and NAC regions of the protein by EPR (double electron electron resonance (DEER)) and single-molecule fluorescence resonance energy transfer (smFRET). For the DEER experiments, a double cysteine mutant of AS was labeled with MTSL ((1-oxyl-2,2,5,5-tetramethyl-pyrroline-3-methyl) methanethiosulfonate). Some of the EPR reports have proposed a broken helix conformation [123, 133] similar to that of micelle-bound protein [134], whereas others have indicated an extended α-helix for the sequence segment (9–90) [70, 135], an interpretation that has also been supported by smFRET measurements [94, 136] (Figure 1.7b). These determinations have been extended by

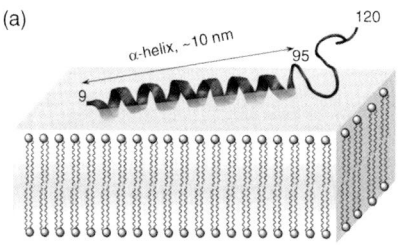
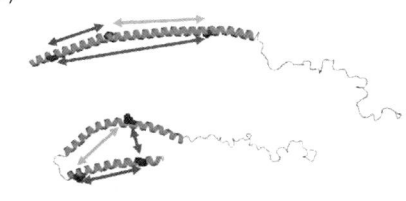

Figure 1.7 AS binds to membranes as an α-helix. (a) The α-helix is oriented parallel to the membrane–water interface. The face that interacts with the membrane is hydrophobic and is flanked by lysine residues at the nonpolar–polar interface [49]. (b) Extended (complexes with unilamellar vesicles) and broken α-helical conformations (micellar structures). From [136].

smFRET technology, leading to schemes involving linked steps of binding and conformational switching [94, 136]. One can summarize these diverse findings by proposing that AS binds to micelles as a broken helix and to more extended lipid bilayers as an extended helix, although recent reports propose that AS disrupts negatively charged vesicles with the formation of lipoprotein-like particles in which the protein adopts the broken helix conformation [69, 70]. It has also been reported that AS binds lipid membranes in the form of oligomers able to form membrane pores with α-helical [26] or β-sheet [27] conformations (see also Chapter 2).

## 1.5
### Interactions of Monomeric AS with Artificial Membranes Monitored with ESIPT Probes

We have assessed systematically the effect of different membrane properties on AS binding monitored by the ESIPT fluorescent probes described earlier, which by virtue of their inherent sensitivity permit operation under close to native conditions (concentration, temperature, etc.). Environment-sensitive dyes provide multiparametric responses characterizing the protein state that are fast enough for measurements of the interaction kinetics. For these studies we constructed a new probe based on FE [90]. The FE label was introduced into the AS molecules in each of three positions (18, 90, 140) by reaction of the maleimide derivative (MFE, Figure 1.8a) with cysteines replacing alanines at the given positions. In all three cases, attachment of the label led to an increased fluorescence emission compared with the free label in water, although the single-band profile indicated that the probes at the sites of attachment were readily accessible to water (Figure 1.8b).

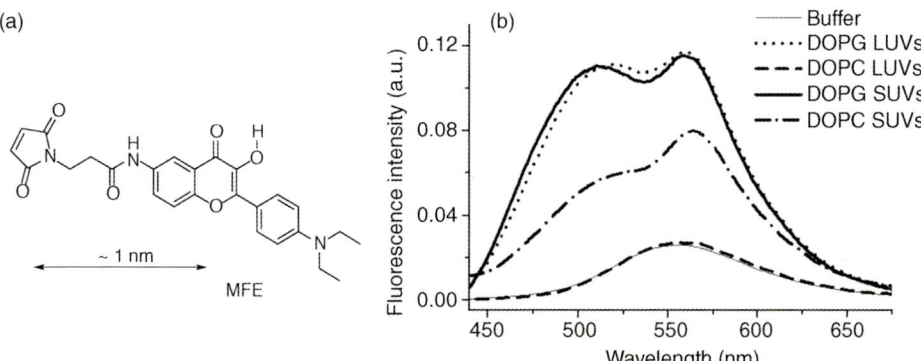

**Figure 1.8** AS binding to vesicles probed with MFE dye. (a) Chemical structure of MFE probe. (b) Emission spectra of AS18-MFE in buffer or in the presence of neutral or negatively charged SUVs and LUVs. Lipid and protein concentrations 100 μM and 100 nM, respectively.

### 1.5.1
### Influence of Membrane Charge

Titration of vesicles containing negatively charged 1,2-dioleoyl-sn-glycero-3-phosphatidylglycerol (DOPG) with AS–MFE led to a dramatic increase in the fluorescence intensity and the appearance of the two N* and T* bands characteristic of the label in an aprotic environment (Figure 1.8b). CD spectroscopy confirmed the formation of an α-helical secondary structure with all three labeled proteins, and also with the WT control. Titrations with neutral SUVs led to similar spectral changes but saturation required much higher lipid concentrations. The influence of membrane surface charge was assessed with SUVs composed of mixtures of neutral (DOPC) and negatively charged (DOPG) lipids in different ratios (Figure 1.9). The association constant of AS to neutral membranes was $\sim$1% of that observed with SUVs consisting of equimolar DOPC and DOPG (mean charge $=-$ 0.5). The binding stoichiometry of AS to neutral DOPC was $\sim$500 lipids/protein, as opposed to $\sim$30 in the case of the negatively charged DOPG (Figure 1.9a), and the binding isotherm was complex, prompting the introduction of the parameter $L_{50}$, the concentration of lipids at which 50% of the protein was bound, for comparative purposes (Figure 1.9b).

### 1.5.2
### Influence of Membrane Curvature

The interactions of AS–MFE were studied using large unilamellar vesicles (LUVs) (size $\sim$100 nm) and SUVs (size $\sim$40 nm). In addition, we compared the influence of neutral (DOPC) and charged (DOPG) lipids with the two systems. AS bound readily to SUVs composed of DOPC or DOPG; decreasing the membrane curvature (LUVs)

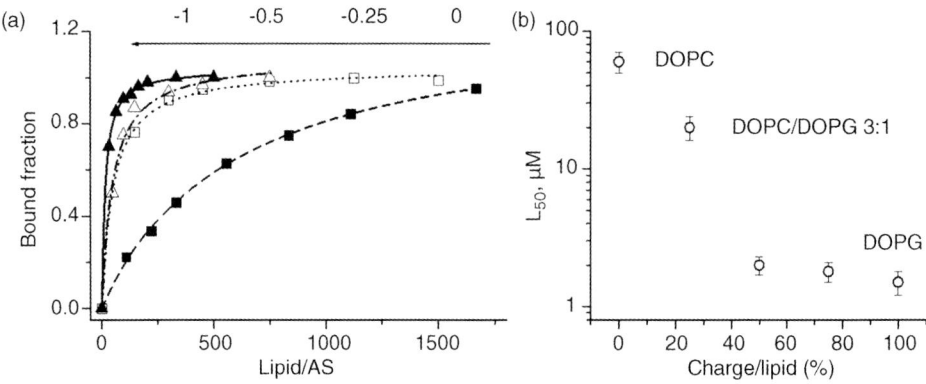

Figure 1.9 Effect of charge on the binding of AS to synthetic membranes. (a) Titration of AS18-MFE by SUVs composed of DOPG–DOPC mixtures containing 0% (■), 25% (□), 50% (Δ), and 100% (▲) of negatively charged DOPG. Lines indicate fits to the data. (b) Dependence on membrane charge of the concentration of lipids required for 50% AS binding. Experiments performed at 37 °C in 25 mM Na-PO$_4$ buffer, pH 6.5, 150 mM NaCl. Excitation wavelength, 400 nm. Data from [90].

had no perceptible effect in the case of negatively charged membranes but strongly reduced the binding to neutral vesicles (Figure 1.8b). There was no significant binding of AS to neutral 100 nm LUVs.

### 1.5.3
### Influence of Membrane Phase

The effect of lipid bilayer phase on AS binding was examined via the thermal phase transitions in membranes composed of saturated lipids. At 37 °C, the emission of AS–MFE in the presence of SUVs composed of neutral 1,2-dimyristoyl-sn-glycero-3-phosphocholine (DMPC) lipids was identical with that of the free protein in the buffer solution (Figure 1.10), indicating that AS did not bind significantly to membranes under these conditions. Cooling led to a strong increase in fluorescence and the appearance of a dual-band emission in the range 19–25 °C (the phase transition of a DMPC bilayer has a midpoint ($M_t$) of ~23 °C). The same phenomenon occurred with SUVs composed of DPPC but displaced to 37–43 °C ($M_t \approx 41$ °C). Evidently, AS has a much greater affinity for the gel phase than for the $L_d$ phase. This difference was not manifested in the case of negatively charged membranes. In fact, disordered membranes composed of DOPG lipids showed a much stronger affinity for AS than rigid 1,2-dipalmitoyl-sn-glycero-3-phosphoglycerol (DPPG) bilayers. We conclude that the membrane phase is significant only for AS binding (affinity) to uncharged membranes, although it affects the *rate* of AS binding to negatively charged membranes (much faster to the disordered phase).

Another useful ratiometric dye for the investigation of AS–membrane interactions is 6-maleimido-2-furyl-3-hydroxychromone (MFC). Binding of AS18-MFC (N-terminal region) to membranes (LUVs and SUVs) is accompanied by a pronounced increase in the T* band, analogous to the spectral alteration produced by aggregation

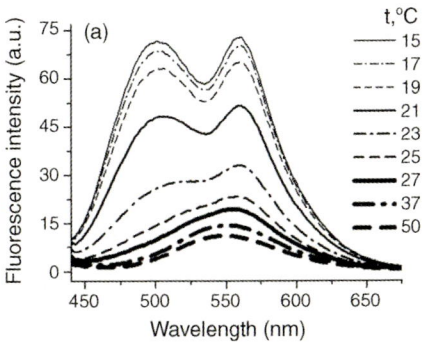
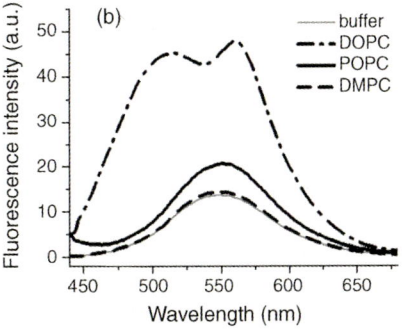
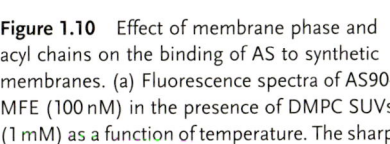

**Figure 1.10** Effect of membrane phase and acyl chains on the binding of AS to synthetic membranes. (a) Fluorescence spectra of AS90-MFE (100 nM) in the presence of DMPC SUVs (1 mM) as a function of temperature. The sharp increase corresponds to the lipid melting temperature (23 °C). (b) Fluorescence spectra of AS18-MFC (100 nM) in the presence of 1 mM DMPC, POPC, and DOPC SUVs at 37 °C. Data from [90].

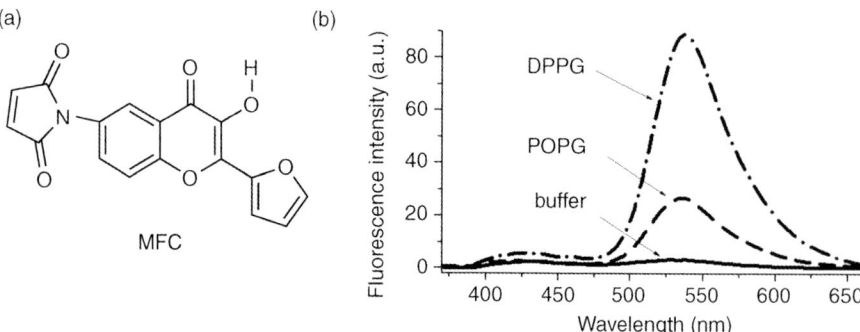

**Figure 1.11** AS binding to vesicles probed by MFC probe. (a) Chemical structure of reactive (maleimide) MFC. (b) Spectroscopic properties of AS18-MFC upon exposure to artificial membranes. Emission spectra in 25 mM Na-PO$_4$ buffer, pH 6.5, 150 mM NaCl, at 37 °C in the absence of lipids (—), in the presence of POPG SUVs (liquid phase, - - -), and in the presence of DPPG SUVs (gel phase, - · · · -).

(see below). We attribute this effect to localization of the protein-bound probe in a significantly less protic (hydrogen bonding), more hydrophobic, and more rigid environment close to or within the lipid bilayer. Interestingly, MFC is more sensitive to environmental rigidity than MFE, and may distinguish the rigid gel phase (DPPG at 25 °C) from the disordered phase (POPG at 25 °C) (Figure 1.11), whereas MFE is less sensitive to these differences yet is more sensitive to polarity.

### 1.5.4
### Influence of Acyl Chains

The presence of the double bonds in the FA acyl chains decreases membrane rigidity and the phase transition temperature. We examined the influence of acyl chain saturation on AS binding with SUVs that persists in the L$_d$ phase at 37 °C: DMPC, POPC, and DOPC, containing no, one, and two double bonds per lipid, respectively. AS interacted with SUVs composed of DOPC (requiring 600 lipids per protein at the titration midpoint), but did not bind appreciably to DMPC at 37 °C (Figure 1.10b). Since the charge, curvature, and the phase of DOPC and DMPC membranes are similar; we attribute the differences to the lower lipid packing density and reduced screening of the apolar region in the case of unsaturated lipid (DOPC). The binding of AS to DOPC SUVs (L$_d$) was much lower than that to DPPC (L$_o$), indicating a stronger influence of the lipid phase on AS binding to the neutral SUVs than of the presence of double bonds. Due to the high affinity of AS for all negatively charged membranes, an assessment of acyl chain effects based on direct titration was not feasible. In an alternative approach, AS was added to a solution containing equal amounts of SUVs composed of DOPG and 1-palmitoyl-2-oleoyl-sn-glycero-3-phosphoglycerol (POPG) (both of them in excess over protein). The resulting emission spectrum was close to that of the protein bound to DOPG, implying a greater affinity for this SUV than for that composed of DOPG.

## 1.5.5
### Influence of Cholesterol

Cholesterol is a key component of cellular membranes and is particularly abundant in neuronal cells affected in neurodegenerative diseases [137]. Cholesterol decreases membrane hydration, improves lipid packing [138], and modulates membrane raft formation. It also affects the binding of AS to membranes and the rate of AS aggregation [139]. Cholesterol (33%) in neutral SUVs led to a decrease in both AS binding stoichiometry and affinity for DOPC and DPPC, but did not affect the binding to negatively charged membranes. These results serve to emphasize the dominant role of electrostatic interactions in AS–membrane association.

## 1.5.6
### Binding Kinetics

ESIPT probes are ideal for monitoring the kinetics of binding and conformational transition by rapid chemical kinetic techniques. We measured the kinetics of AS interactions with charged and neutral membranes by the fluorescence stopped-flow method. In most cases a monoexponential course of increasing fluorescence intensity ensued after mixing vesicles with MFE-labeled AS, although some membrane compositions also exhibited at least one relatively slow component (10–100 s) (Figure 1.12a). For an initial comparison of the AS binding rates to membranes of different composition, we determined a pseudo-first-order reaction rate constant measured at 0.25 μM protein and 50 μM lipid concentrations. Negatively charged SUVs interacted with AS in the 1–20 ms range, consistent with the reported AS exchange rate in SDS micelles (10 ms) measured by $^{19}F$ NMR spectroscopy [140]. The binding to neutral SUVs was much slower, in the range of seconds, and the reaction rates and binding affinities were not linearly related.

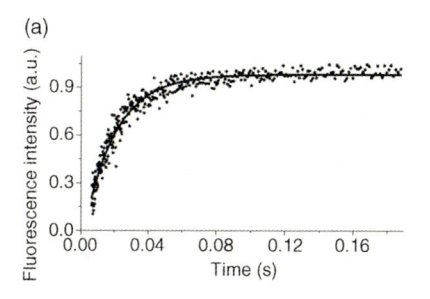

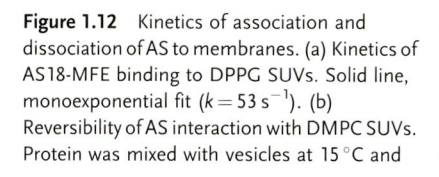

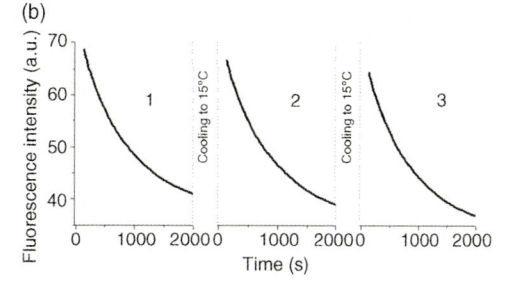

**Figure 1.12** Kinetics of association and dissociation of AS to membranes. (a) Kinetics of AS18-MFE binding to DPPG SUVs. Solid line, monoexponential fit ($k = 53\,s^{-1}$). (b) Reversibility of AS interaction with DMPC SUVs. Protein was mixed with vesicles at 15 °C and then heated to 37 °C. The protein dissociation was monitored by measuring the decrease in fluorescence intensity. After 2000 s, the sample was cooled to 15 °C for ∼5 min to restore protein binding and then again heated to 37 °C. Data from [90].

The reversibility of AS–membrane binding was explored with DMPC SUVs for which the affinity for AS changes ~100-fold across the phase transition at 23 °C. On addition of labeled protein to a solution of DMPC SUVs at 15 °C, almost complete binding ensued. Heating this solution beyond the DMPC phase transition led to a gradual dissociation of the protein from the membrane, characterized by a rate constant of ~0.001 $s^{-1}$. Successive cooling–heating cycles demonstrated the reversibility of AS association and the maintenance of membrane integrity in the process (Figure 1.12b). The dissociation of AS from membranes at constant temperature was monitored by protein migration to SUVs with a higher affinity. In the presence of DPPC SUVs and at a lipid:protein ratio of 2000:1, AS was completely bound. Addition of DOPG SUVs (which bind AS much more strongly than DPPC) led to a progressive change in fluorescence, and the final emission spectrum corresponded to that of AS–MFE bound to DOPG. Protein migration from DPPC to DOPG occurred within 10 min but was much faster from membranes in the $L_d$ phase. Migration was always to membranes exhibiting the higher affinity for AS, leading to the ranking POPC < DOPC < DPPC < DPPG < POPG < DOPG, at a fixed temperature.

The influence of all of the factors considered above on the binding of AS to membranes are summarized in Figure 1.13 and Table 1.3. Additional experiments, not discussed further here, have been directed at the influence of lipid:protein ratio [141] on the orientation of bound AS relative to the plane of the membrane. The differential signals from the ESIPT probes located at positions 18, 90, and 140 provided evidence for an N-to-C bias in AS position and a penetration into the phospholipid bilayers very dependent on the lipid:protein ratio, that is, density of bound protein.

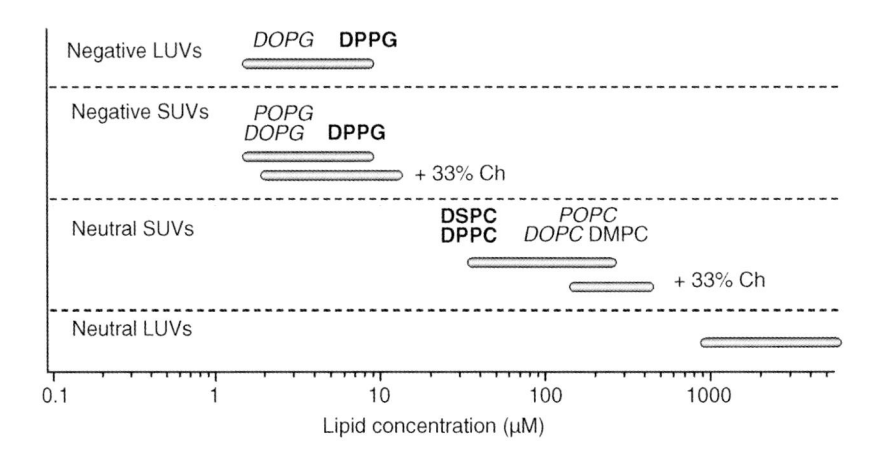

**Figure 1.13** Affinities of AS for vesicles of different composition. Affinity of AS to different membranes presented as lipid concentration at which 50% of the protein (100 nM) is bound. Text labels show affinity for particular representatives. Membranes composed of lipids containing unsaturated fatty acids are in italics. The bold type denotes the gel phase.

Table 1.3 AS binding to charged membranes.

| Factor | Neutral | Charged |
|---|---|---|
| Curvature | Increased binding | Minor effect |
| High cholesterol | Decreased binding | Minor effect |
| High rigidity (gel phase) | Increased binding | No effect or decreased binding |
| Unsaturated lipid | Increased binding | Increased binding |
| Lipid:AS | >300 | 30–100 |
| Rate of binding | ~1 s$^{-1}$ | 50–700 s$^{-1}$ |

## 1.6
### Aggregation of AS and the Effects of Fatty Acids Monitored with ESIPT Probes

As mentioned earlier, lipids affect the aggregation of AS [31–33]. We have used the ESIPT probes in preliminary investigations of AS aggregation in the presence of different categories of lipids. The aggregation of WT AS was monitored by addition of 2–3% of AS140-MFC, which exhibits significant changes in numerous emission properties as the reaction progresses [116] (Figure 1.14a). The most notable spectroscopic feature of the MFC response is a dramatic (10–15-fold) increase in the T$^*$ band intensity at an early stage of AS aggregation (Figure 1.14b). The corresponding large increase in the T$^*$:N$^*$ ratio and the quantum yield of the MFC dye reflects the transfer of the probe from the aqueous monomeric state to a significantly less protic (hydrogen bonding), more hydrophobic, and more rigid environment. It is notable that these changes precede those of the classical amyloid indicator ThT. We conclude that the T$^*$:N$^*$ ratio signal senses the formation of oligomeric and/or or other prefibrillar intermediate aggregated species of AS. The static second-order scattering signal is also recorded and reports primarily the formation of higher molecular weight aggregates such as mature amyloid fibrils [116] (Figure 1.14c).

In view of the reported different affinities of FAs for AS, we analyzed AS aggregation in the presence of saturated, monounsaturated and polyunsaturated FAs employing the ESIPT assay. The kinetic measurements were performed in 25 mM Na-PO$_4$ buffer, pH 7.4, 100 mM KCl, in the presence of a 1:4 (FA:AS) molar proportion of FAs (Figure 1.15). Under these conditions, the rise in the T$^*$:N$^*$ ratio persisted until a later increase in the N$^*$ signal led to its diminution (Figure 1.15a). Saturated FAs (16:0, 18:0, and 20:0) had little effect on the AS aggregation rate. For example, palmitic acid (16:0) showed a similar increase in the T$^*$:N$^*$ ratio as the control without FAs, as well as a small stabilization of the intermediate species. On the other hand, monounsaturated oleic acid (18:1 n-9) showed a delayed T$^*$:N$^*$ signal and an immediate decrease. The scattering signal was delayed significantly compared with those of the control and other reactions. The PUFA arachidonic acid (20:4 n-6) led to the longest stabilization of the intermediate state while the scattering curve was delayed to only a small extent. These results reveal a complex aggregation behavior of AS + lipid, yet illustrate that the ESIPT probes offer a means for devising new

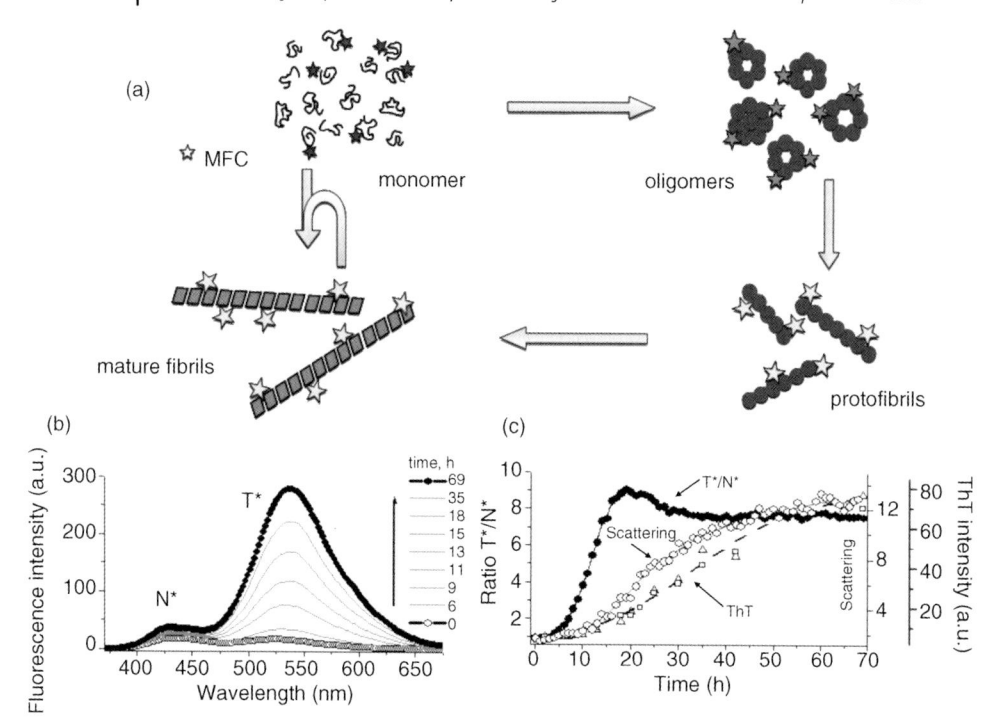

**Figure 1.14** Aggregation of AS monitored by AS140-MFC. (a) Schematic representation of different stages of aggregation probed with MFC. Typically, only 2.5% of the AS molecules are labeled (stars). (b) Representative emission spectra of AS140-MFC during coaggregation with a 40-fold excess of WT AS. (c) Comparison of MFC T*:N* ratio signal, ThT, and scattering during AS aggregation. Adapted from [116].

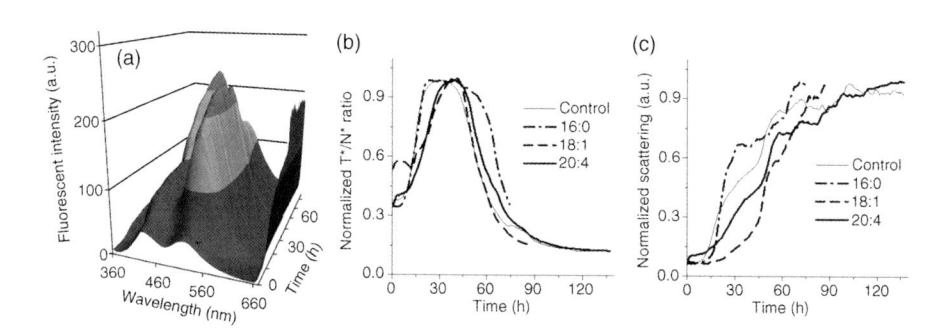

**Figure 1.15** Influence of FAs on AS aggregation. (a) Representative emission spectra of AS140-MFC during aggregation of WT AS. N*, T*, and scattering bands are indicated. The aggregation of AS in the presence of a 1:4 molar proportion of FAs was monitored by (b) MFC T*:N* ratio, and (c) light scattering, a signal that parallels that of the classical ThT indicator [116].

strategies for isolating AS aggregation intermediates and, ultimately, for screening molecules for clinically significant inhibitory properties.

## 1.7
## Concluding Remarks

The experimental approaches made possible by the ESIPT probes featured in this chapter are capable of generating data relevant to the majority of factors listed at the beginning of Section 1.4. An important practical advantage is the inherently continuous nature of their signals, providing detailed, multiparametric data for dynamic processes. The ESIPT measurements can be implemented on many instrumentation platforms and are easily automated. Appropriately modified for use in the cellular context (e.g., Figure 1.6), for example by structural modifications leading to more red-shifted spectra, ESIPT probes should serve to elucidate the roles of metabolic changes, external signals, or oxidative stress on the cytopathology underlying PD, other related neurodegenerative diseases, and functional amyloid transitions. It is virtually axiomatic that lipids and membranes will be understood as important "co-conspirators" or, at the very least, modulators of these processes.

## Acknowledgments

V.V.S and D.A.Y are recipients of Marie Curie Actions postdoctoral fellowships and L. J.F.-L of an Alexander von Humboldt Foundation Georg Förster postdoctoral fellowship. This work was funded by the Max Planck Society, the Toxic Proteins Project (Innovation Fund of the Max Planck Society), and Excellence Cluster EXC 171 – FZT103 "Microscopy at the Nanometer Range" associated with the DFG Center for the Molecular Physiology of the Brain (CMPB).

## References

1 Herczenik, E. and Gebbink, M.F. (2008) Molecular and cellular aspects of protein misfolding and disease. FASEB J., 22, 2115–2133.
2 Outeiro, T.F. and Ferreira, J. (2009) Current and future therapeutic strategies for Parkinson's disease. Curr. Pharm. Des., 15, 3968–3976.
3 Auluck, P.K., Caraveo, G., and Lindquist, S. (2010) α-Synuclein: membrane interactions and toxicity in Parkinson's disease. Annu. Rev. Cell. Dev. Biol., 26, 211–233.
4 Halliday, G.M. and McCann, H. (2010) The progression of pathology in Parkinson's disease. Ann. N. Y. Acad. Sci., 1184, 188–195.
5 Madine, J. and Middleton, D.A. (2009) Targeting α-synuclein aggregation for Parkinson's disease treatment. Drugs Future, 3, 655–663.
6 Sulzer, D. (2010) Clues to how α-synuclein damages neurons in Parkinson's disease. Mov. Disord., 25 (Suppl. 1), S27–S31.

7. Hardy, J. and Selkoe, D.J. (2002) The amyloid hypothesis of Alzheimer's disease: progress and problems on the road to therapeutics. *Science*, **297**, 353–356.
8. Lashuel, H.A., Petre, B.M., Wall, J., Simon, M., Nowak, R.J., Walz, T., and Lansbury, P.T. (2002) α-Synuclein, especially the Parkinson's disease-associated mutants, forms pore-like annular and tubular protofibrils. *J. Mol. Biol.*, **322**, 1089–1102.
9. Uversky, V.N. and Eliezer, D. (2009) Biophysics of Parkinson's disease: structure and aggregation of α-synuclein. *Curr. Protein Pept. Sci.*, **10**, 483–499.
10. Chandra, S., Gallardo, G., Fernandez-Chacon, R., Schluter, O.M., and Sudhof, T.C. (2005) α-Synuclein cooperates with CSPα in preventing neurodegeneration. *Cell*, **123**, 383–396.
11. Lotharius, J., Barg, S., Wiekop, P., Lundberg, C., Raymon, H.K., and Brundin, P. (2002) Effect of mutant α-synuclein on dopamine homeostasis in a new human mesencephalic cell line. *J. Biol. Chem.*, **277**, 38884–38894.
12. Gureviciene, I., Gurevicius, K., and Tanila, H. (2007) Role of α-synuclein in synaptic glutamate release. *Neurobiol. Dis.*, **28**, 83–89.
13. Liu, S.M., Fa, M., Ninan, I., Trinchese, F., Dauer, W., and Arancio, O. (2007) α-Synuclein involvement in hippocampal synaptic plasticity: role of NO, cGMP, cGK and CaMKII. *Eur. J. Neurosci.*, **25**, 3583–3596.
14. Cabin, D.E., Shimazu, K., Murphy, D., Cole, N.B., Gottschalk, W., McIlwain, K.L., Orrison, B., Chen, A., Ellis, C.E., Paylor, R., Lu, B., and Nussbaum, R.L. (2002) Synaptic vesicle depletion correlates with attenuated synaptic responses to prolonged repetitive stimulation in mice lacking α-synuclein. *J. Neurosci.*, **22**, 8797–8807.
15. Nemani, V.M., Lu, W., Berge, V., Nakamura, K., Onoa, B., Lee, M.K., Chaudhry, F.A., Nicoll, R.A., and Edwards, R.H. (2010) Increased expression of α-synuclein reduces neurotransmitter release by inhibiting synaptic vesicle reclustering after endocytosis. *Neuron*, **65**, 66–79.
16. Cookson, M.R. and Bandmann, O. (2010) Parkinson's disease: insights from pathways. *Hum. Mol. Genet.*, **19**, R21–R27.
17. Cookson, M.R. (2009) α-Synuclein and neuronal cell death. *Mol. Neurodegener*, **4**, 9.
18. Butterfield, S. and Lashuel, H. (2010) Amyloidogenic protein–membrane interactions: mechanistic insight from model systems. *Angew. Chem. Int. Ed.*, **49**, 5628–5654.
19. Devi, L., Raghavendran, V., Prabhu, B.M., Avadhani, N.G., and Anandatheerthavarada, H.K. (2008) Mitochondrial import and accumulation of α-synuclein impair complex I in human dopaminergic neuronal cultures and Parkinson disease brain. *J. Biol. Chem.*, **283**, 9089–9100.
20. Cole, N.B., Murphy, D.D., Grider, T., Rueter, S., Brasaemle, D., and Nussbaum, R.L. (2002) Lipid droplet binding and oligomerization properties of the Parkinson's disease protein α-synuclein. *J. Biol. Chem.*, **277**, 6344–6352.
21. Lee, H.J., Choi, C., and Lee, S.J. (2002) Membrane-bound α-synuclein has a high aggregation propensity and the ability to seed the aggregation of the cytosolic form. *J. Biol. Chem.*, **277**, 671–678.
22. Perrin, R.J., Woods, W.S., Clayton, D.F., and George, J.M. (2001) Exposure to long chain polyunsaturated fatty acids triggers rapid multimerization of synucleins. *J. Biol. Chem.*, **276**, 41958–41962.
23. Necula, M., Chirita, C.N., and Kuret, J. (2003) Rapid anionic micelle-mediated α-synuclein fibrillization in vitro. *J. Biol. Chem.*, **278**, 46674–46680.
24. Zhu, M. and Fink, A.L. (2003) Lipid binding inhibits α-synuclein fibril formation. *J. Biol. Chem.*, **278**, 16873–16877.
25. Zhu, M., Li, J., and Fink, A.L. (2003) The association of α-synuclein with membranes affects bilayer structure,

26. Zakharov, S.D., Hulleman, J.D., Dutseva, E.A., Antonenko, Y.N., Rochet, J.C., and Cramer, W.A. (2007) Helical α-synuclein forms highly conductive ion channels. *Biochemistry*, **46**, 14369–14379.

27. Kim, H.Y., Cho, M.K., Kumar, A., Maier, E., Siebenhaar, C., Becker, S., Fernandez, C.O., Lashuel, H.A., Benz, R., Lange, A., and Zweckstetter, M. (2009) Structural properties of pore-forming oligomers of α-synuclein. *J. Am. Chem. Soc.*, **131**, 17482–17489.

28. van Rooijen, B.D., Claessens, M.M., and Subramaniam, V. (2010) Membrane interactions of oligomeric α-synuclein: potential role in Parkinson's disease. *Curr. Protein Pept. Sci.*, **11**, 334–342.

29. Banerjee, K., Sinha, M., Pham, C.L.L., Jana, S., Chanda, D., Cappai, R., and Chakrabarti, S. (2010) α-Synuclein induced membrane depolarization and loss of phosphorylation capacity of isolated rat brain mitochondria: implications in Parkinson's disease. *FEBS Lett.*, **584**, 1571–1576.

30. Cooper, A.A., Gitler, A.D., Cashikar, A., Haynes, C.M., Hill, K.J., Bhullar, B., Liu, K., Xu, K., Strathearn, K.E., Liu, F., Cao, S., Caldwell, K.A., Caldwell, G.A., Marsischky, G., Kolodner, R.D., LaBaer, J., Rochet, J.-C., Bonini, N.M., and Lindquist, S. (2006) α-Synuclein blocks ER-Golgi traffic and Rab1 rescues neuron loss in Parkinson's models. *Science*, **313**, 324–328.

31. Darios, F., Ruiperez, V., Lopez, I., Villanueva, J., Gutierrez, L.M., and Davletov, B. (2010) α-Synuclein sequesters arachidonic acid to modulate SNARE-mediated exocytosis. *EMBO Rep.*, **11**, 528–533.

32. De Franceschi, G., Frare, E., Bubacco, L., Mammi, S., Fontana, A., and de Laureto, P.P. (2009) Molecular insights into the interaction between α-synuclein and docosahexaenoic acid. *J. Mol. Biol.*, **394**, 94–107.

33. Karube, H., Sakamoto, M., Arawaka, S., Hara, S., Sato, H., Ren, C.H., Goto, S., Koyama, S., Wada, M., Kawanami, T., Kurita, K., and Kato, T. (2008) N-Terminal region of α-synuclein is essential for the fatty acid-induced oligomerization of the molecules. *FEBS Lett.*, **582**, 3693–3700.

34. Youdim, K.A., Martin, A., and Joseph, J.A. (2000) Essential fatty acids and the brain: possible health implications. *Int. J. Dev. Neurosci.*, **18**, 383–399.

35. Lucke, C., Gantz, D.L., Klimtchuk, E., and Hamilton, J.A. (2006) Interactions between fatty acids and α-synuclein. *J. Lipid Res.*, **47**, 1714–1724.

36. Johansson, A.S., Garlind, A., Berglind-Dehlin, F., Karlsson, G., Edwards, K., Gellerfors, P., Ekholm-Pettersson, F., Palmblad, J., and Lannfelt, L. (2007) Docosahexaenoic acid stabilizes soluble amyloid-beta protofibrils and sustains amyloid-beta-induced neurotoxicity in vitro. *FEBS J.*, **274**, 990–1000.

37. Sharon, R., Bar-Joseph, I., Frosch, M.P., Walsh, D.M., Hamilton, J.A., and Selkoe, D.J. (2003) The formation of highly soluble oligomers of α-synuclein is regulated by fatty acids and enhanced in Parkinson's disease. *Neuron*, **37**, 583–595.

38. Golovko, M.Y., Barcelo-Coblijn, G., Castagnet, P.I., Austin, S., Combs, C.K., and Murphy, E.J. (2009) The role of α-synuclein in brain lipid metabolism: a downstream impact on brain inflammatory response. *Mol. Cell. Biochem.*, **326**, 55–66.

39. Bertoncini, C.W., Fernandez, C.O., Griesinger, C., Jovin, T.M., and Zweckstetter, M. (2005) Familial mutants of α-synuclein with increased neurotoxicity have a destabilized conformation. *J. Biol. Chem.*, **280**, 30649–30652.

40. Dedmon, M.M., Lindorff-Larsen, K., Christodoulou, J., Vendruscolo, M., and Dobson, C.M. (2005) Mapping long-range interactions in α-synuclein using spin-label NMR and ensemble molecular dynamics simulations. *J. Am. Chem. Soc.*, **127**, 476–477.

41. Uversky, V.N. and Dunker, A.K. (2010) Understanding protein non-folding. *Biochim. Biophys. Acta Proteins Proteomics*, **1804**, 1231–1264.

42 Huang, A. and Stultz, C.M. (2009) Finding order within disorder: elucidating the structure of proteins associated with neurodegenerative disease. *Future Med. Chem.*, **1**, 467–482.

43 Bertoncini, C.W., Jung, Y.S., Fernandez, C.O., Hoyer, W., Griesinger, C., Jovin, T.M., and Zweckstetter, M. (2005) Release of long-range tertiary interactions potentiates aggregation of natively unstructured α-synuclein. *Proc. Natl. Acad. Sci. USA*, **102**, 1430–1435.

44 Sung, Y.H. and Eliezer, D. (2007) Residual structure, backbone dynamics, and interactions within the synuclein family. *J. Mol. Biol.*, **372**, 689–707.

45 Giasson, B.I., Murray, I.V.J., Trojanowski, J.Q., and Lee, V.M.Y. (2001) A hydrophobic stretch of 12 amino acid residues in the middle of α-synuclein is essential for filament assembly. *J. Biol. Chem.*, **276**, 2380–2386.

46 Rasia, R.M., Bertoncini, C.W., Marsh, D., Hoyer, W., Cherny, D., Zweckstetter, M., Griesinger, C., Jovin, T.M., and Fernandez, C.O. (2005) Structural characterization of copper(II) binding to α-synuclein: insights into the bioinorganic chemistry of Parkinson's disease. *Proc. Natl. Acad. Sci. USA*, **102**, 4294–4299.

47 Davidson, W.S., Jonas, A., Clayton, D.F., and George, J.M. (1998) Stabilization of α-synuclein secondary structure upon binding to synthetic membranes. *J. Biol. Chem.*, **273**, 9443–9449.

48 Bussell, R. and Eliezer, D. (2003) A structural and functional role for 11-mer repeats in α-synuclein and other exchangeable lipid binding proteins. *J. Mol. Biol.*, **329**, 763–778.

49 Jao, C.C., Der-Sarkissian, A., Chen, J., and Langen, R. (2004) Structure of membrane-bound α-synuclein studied by site-directed spin labeling. *Proc. Natl. Acad. Sci. USA*, **101**, 8331–8336.

50 McFarland, M.A., Ellis, C.E., Markey, S.P., and Nussbaum, R.L. (2008) Proteomics analysis identifies phosphorylation-dependent α-synuclein protein interactions. *Mol. Cell. Proteomics*, **7**, 2123–2137.

51 Outeiro, T.F., Kontopoulos, E., Altmann, S.M., Kufareva, I., Strathearn, K.E., Amore, A.M., Volk, C.B., Maxwell, M.M., Rochet, J.C., McLean, P.J., Young, A.B., Abagyan, R., Feany, M.B., Hyman, B.T., and Kazantsev, A.G. (2007) Sirtuin 2 inhibitors rescue α-synuclein-mediated toxicity in models of Parkinson's disease. *Science*, **317**, 516–519.

52 Fukata, Y. and Fukata, M. (2010) Protein palmitoylation in neuronal development and synaptic plasticity. *Nat. Rev. Neurosci.*, **11**, 161–175.

53 Heise, H., Celej, M.S., Becker, S., Riedel, D., Pelah, A., Kumar, A., Jovin, T.M., and Baldus, M. (2008) Solid-state NMR reveals structural differences between fibrils of wild-type and disease-related A53T mutant of α-synuclein. *J. Mol. Biol.*, **380**, 444–450.

54 Ulmer, T.S., Bax, A., Cole, N.B., and Nussbaum, R.L. (2005) Structure and dynamics of micelle-bound human α-synuclein. *J. Biol. Chem.*, **280**, 9595–9603.

55 Rao, J.N., Kim, Y.E., Park, L.S., and Ulmer, T.S. (2009) Effect of pseudorepeat rearrangement on α-synuclein misfolding, vesicle binding, and micelle binding. *J. Mol. Biol.*, **390**, 516–529.

56 Heise, H., Hoyer, W., Becker, S., Andronesi, O.C., Riedel, D., and Baldus, M. (2005) Molecular-level secondary structure, polymorphism, and dynamics of full-length α-synuclein fibrils studied by solid-state NMR. *Proc. Natl. Acad. Sci. USA*, **102**, 15871–15876.

57 Vilar, M., Chou, H.T., Luhrs, T., Maji, S.K., Riek-Loher, D., Verel, R., Manning, G., Stahlberg, H., and Riek, R. (2008) The fold of α-synuclein fibrils. *Proc. Natl. Acad. Sci. USA*, **105**, 8637–8642.

58 Chen, M., Margittai, M., Chen, J., and Langen, R. (2007) Investigation of α-synuclein fibril structure by site-directed spin labeling. *J. Biol. Chem.*, **282**, 24970–24979.

59 Bodner, C.R., Dobson, C.M., and Bax, A. (2009) Multiple tight phospholipid-binding modes of α-synuclein revealed by solution NMR spectroscopy. *J. Mol. Biol.*, **390**, 775–790.

60 Serdyuk, I.N., Zaccai, N.R., and Zaccai, J. (2007) *Methods in Molecular Biophysics*, 1st edn, Cambridge University Press, Cambridge.
61 Jelinek, R. and Sheynis, T. (2010) Amyloid–membrane interactions: experimental approaches and techniques. *Curr. Protein Pept. Sci.*, **11**, 372–384.
62 Gorbenko, G.P. (2010) Fluorescence spectroscopy of protein oligomerization in membranes. *J. Fluoresc.*, **6**, 6.
63 Munishkina, L.A. and Fink, A.L. (2007) Fluorescence as a method to reveal structures and membrane interactions of amyloidogenic proteins. *Biochim. Biophys. Acta*, **1768**, 1862–1885.
64 Bertoncini, C.W. and Celej, M.S. (2011) Small molecule fluorescent probes for the detection of amyloid self-assembly *in vitro* and *in vivo*. *Curr. Protein Pept. Sci.*, in press.
65 Madine, J., Hughes, E., Doig, A.J., and Middleton, D.A. (2008) The effects of α-synuclein on phospholipid vesicle integrity: a study using $^{31}$P NMR and electron microscopy. *Mol. Membr. Biol.*, **25**, 518–527.
66 Wang, G. (2008) NMR of membrane-associated peptides and proteins. *Curr. Protein Pept. Sci.*, **9**, 50–69.
67 Woods, W.S., Boettcher, J.M., Zhou, D.H., Kloepper, K.D., Hartman, K.L., Ladror, D.T., Qi, Z., Rienstra, C.M., and George, J.M. (2007) Conformation-specific binding of α-synuclein to novel protein partners detected by phage display and NMR spectroscopy. *J. Biol. Chem.*, **282**, 34555–34567.
68 de Laureto, P.P., Tosatto, L., Frare, E., Marin, O., Uversky, V.N., and Fontana, A. (2006) Conformational properties of the SDS-bound state of α-synuclein probed by limited proteolysis: unexpected rigidity of the acidic C-terminal tail. *Biochemistry*, **45**, 11523–11531.
69 Drescher, M., van Rooijen, B.D., Veldhuis, G., Subramaniam, V., and Huber, M. (2010) A stable lipid-induced aggregate of α-synuclein. *J. Am. Chem. Soc.*, **132**, 4080–4081.
70 Jao, C.C., Hegde, B.G., Chen, J., Haworth, I.S., and Langen, R. (2008) Structure of membrane-bound α-synuclein from site-directed spin labeling and computational refinement. *Proc. Natl. Acad. Sci. USA*, **105**, 19666–19671.
71 Drescher, M., Veldhuis, G., van Rooijen, B.D., Milikisyants, S., Subramaniam, V., and Huber, M. (2008) Antiparallel arrangement of the helices of vesicle-bound α-synuclein. *J. Am. Chem. Soc.*, **130**, 7796–7797.
72 Margittai, M. and Langen, R. (2006) Spin labeling analysis of amyloids and other protein aggregates. *Methods Enzymol.*, **413**, 122–139.
73 Pfefferkorn, C.M. and Lee, J.C. (2010) Tryptophan probes at the α-synuclein and membrane interface. *J. Phys. Chem. B*, **114**, 4615–4622.
74 Hong, D.P., Fink, A.L., and Uversky, V.N. (2008) Structural characteristics of α-synuclein oligomers stabilized by the flavonoid baicalein. *J. Mol. Biol.*, **383**, 214–223.
75 Smith, D.P., Tew, D.J., Hill, A.F., Bottomley, S.P., Masters, C.L., Barnham, K.J., and Cappai, R. (2008) Formation of a high affinity lipid-binding intermediate during the early aggregation phase of α-synuclein. *Biochemistry*, **47**, 1425–1434.
76 Cambrea, L.R., Haque, F., Schieler, J.L., Rochet, J.C., and Hovis, J.S. (2007) Effect of ions on the organization of phosphatidylcholine/phosphatidic acid bilayers. *Biophys. J.*, **93**, 1630–1638.
77 Munishkina, L.A., Phelan, C., Uversky, V.N., and Fink, A.L. (2003) Conformational behavior and aggregation of α-synuclein in organic solvents: modeling the effects of membranes. *Biochemistry*, **42**, 2720–2730.
78 Ferreon, A.C. and Deniz, A.A. (2007) α-Synuclein multistate folding thermodynamics: implications for protein misfolding and aggregation. *Biochemistry*, **46**, 4499–4509.
79 Kamp, F. and Beyer, K. (2006) Binding of α-synuclein affects the lipid packing in bilayers of small vesicles. *J. Biol. Chem.*, **281**, 9251–9259.

80 Nuscher, B., Kamp, F., Mehnert, T., Odoy, S., Haass, C., Kahle, P.J., and Beyer, K. (2004) α-Synuclein has a high affinity for packing defects in a bilayer membrane – a thermodynamics study. *J. Biol. Chem.*, **279**, 21966–21975.

81 Giannakis, E., Pacifico, J., Smith, D.P., Hung, L.W., Masters, C.L., Cappai, R., Wade, J.D., and Barnham, K.J. (2008) Dimeric structures of α-synuclein bind preferentially to lipid membranes. *Biochim. Biophys. Acta*, **1778**, 1112–1119.

82 Ding, T.T., Lee, S.J., Rochet, J.C., and Lansbury, P.T. Jr. (2002) Annular α-synuclein protofibrils are produced when spherical protofibrils are incubated in solution or bound to brain-derived membranes. *Biochemistry*, **41**, 10209–10217.

83 Jo, E., McLaurin, J., Yip, C.M., St. George-Hyslop, P., and Fraser, P.E. (2000) α-Synuclein membrane interactions and lipid specificity. *J. Biol. Chem.*, **275**, 34328–34334.

84 Scott, D.A., Tabarean, I., Tang, Y., Cartier, A., Masliah, E., and Roy, S. (2010) A pathologic cascade leading to synaptic dysfunction in α-synuclein-induced neurodegeneration. *J. Neurosci.*, **30**, 8083–8095.

85 Parihar, M.S., Parihar, A., Fujita, M., Hashimoto, M., and Ghafourifar, P. (2009) α-Synuclein overexpression and aggregation exacerbates impairment of mitochondrial functions by augmenting oxidative stress in human neuroblastoma cells. *Int. J. Biochem. Cell Biol.*, **41**, 2015–2024.

86 Feng, L.R., Federoff, H.J., Vicini, S., and Maguire-Zeiss, K.A. (2010) α-Synuclein mediates alterations in membrane conductance: a potential role for α-synuclein oligomers in cell vulnerability. *Eur. J. Neurosci.*, **32**, 10–17.

87 Schwach, G., Tschemmernegg, M., Pfragner, R., Ingolic, E., Schreiner, E., and Windisch, M. (2010) Establishment of stably transfected rat neuronal cell lines expressing α-synuclein GFP fusion proteins. *J. Mol. Neurosci.*, **41**, 80–88.

88 van Ham, T.J., Esposito, A., Kumita, J.R., Hsu, S.T.D., Schierle, G.S.K., Kaminsk, C.F., Dobson, C.M., Nollen, E.A.A., and Bertoncini, C.W. (2010) Towards multiparametric fluorescent imaging of amyloid formation: studies of a YFP model of α-synuclein aggregation. *J. Mol. Biol.*, **395**, 627–642.

89 McLean, P.J., Kawamata, H., and Hyman, B.T. (2001) α-Synuclein-enhanced green fluorescent protein fusion proteins form proteasome sensitive inclusions in primary neurons. *Neuroscience*, **104**, 901–912.

90 Shvadchak V. V., Yushchenko D. A., Falomir-Lockhart L. J., and Jovin T. M. (2011) Specificity and kinetics of alpha-synuclein binding to model membranes determined with fluorescent ESIPT probe. *J. Biol. Chem. Under revision*.

91 van Rooijen, B.D., van Leijenhorst-Groener, K.A., Claessens, M.M., and Subramaniam, V. (2009) Tryptophan fluorescence reveals structural features of α-synuclein oligomers. *J. Mol. Biol.*, **394**, 826–833.

92 Thirunavukkuarasu, S., Jares-Erijman, E.A., and Jovin, T.M. (2008) Multiparametric fluorescence detection of early stages in the amyloid protein aggregation of pyrene-labeled α-synuclein. *J. Mol. Biol.*, **378**, 1064–1073.

93 Veldhuis, G., Segers-Nolten, I., Ferlemann, E., and Subramaniam, V. (2009) Single-molecule FRET reveals structural heterogeneity of SDS-bound α-synuclein. *ChemBioChem*, **10**, 436–439.

94 Ferreon, A.C., Gambin, Y., Lemke, E.A., and Deniz, A.A. (2009) Interplay of α-synuclein binding and conformational switching probed by single-molecule fluorescence. *Proc. Natl. Acad. Sci. USA*, **106**, 5645–5650.

95 Outeiro, T.F., Putcha, P., Tetzlaff, J.E., Spoelgen, R., Koker, M., Carvalho, F., Hyman, B.T., and McLean, P.J. (2008) Formation of toxic oligomeric α-synuclein species in living cells. *PLoS ONE*, **3**, e1867.

96 Roberti, M.J., Bertoncini, C.W., Klement, R., Jares-Erijman, E.A., and Jovin, T.M. (2007) Fluorescence imaging of amyloid formation in living cells by a functional, tetracysteine-tagged α-synuclein. *Nat. Methods*, **4**, 345–351.

97 Nath, S., Meuvis, J., Hendrix, J., Carl, S.A., and Engelborghs, Y. (2010) Early aggregation steps in α-synuclein as measured by FCS and FRET: evidence for a contagious conformational change. *Biophys. J.*, **98**, 1302–1311.

98 Rhoades, E., Ramlall, T.F., Webb, W.W., and Eliezer, D. (2006) Quantification of α-synuclein binding to lipid vesicles using fluorescence correlation spectroscopy. *Biophys. J.*, **90**, 4692–4700.

99 Outeiro, T.F. and Lindquist, S. (2003) Yeast cells provide insight into α-synuclein biology and pathobiology. *Science*, **302**, 1772–1775.

100 Cairo, C.W., Key, J.A., and Sadek, C.M. (2010) Fluorescent small-molecule probes of biochemistry at the plasma membrane. *Curr. Opin. Chem. Biol.*, **14**, 57–63.

101 Hougland, R.P. (2002) Probes for lipids and membranes, in *Handbook of Fluorescent Probes and Research Products*, 9th edn., Molecular Probes, Eugene, OR, 503–542.

102 Kucherak, O.A., Oncul, S., Darwich, Z., Yushchenko, D.A., Arntz, Y., Didier, P., Mely, Y., and Klymchenko, A.S. (2010) Switchable Nile Red-based probe for cholesterol and lipid order at the outer leaflet of biomembranes. *J. Am. Chem. Soc.*, **132**, 4907–4916.

103 Demchenko, A.P., Mely, Y., Duportail, G., and Klymchenko, A.S. (2009) Monitoring biophysical properties of lipid membranes by environment-sensitive fluorescent probes. *Biophys. J.*, **96**, 3461–3470.

104 Biancalana, M., Makabe, K., Koide, A., and Koide, S. (2009) Molecular mechanism of Thioflavin-T binding to the surface of β-rich peptide self-assemblies. *J. Mol. Biol.*, **385**, 1052–1063.

105 Yushchenko, D.A., Shvadchak, V.V., Bilokin, M.D., Klymchenko, A.S., Duportail, G., Mely, Y., and Pivovarenko, V.G. (2006) Modulation of dual fluorescence in a 3-hydroxyquinolone dye by perturbation of its intramolecular proton transfer with solvent polarity and basicity. *Photochem. Photobiol. Sci.*, **5**, 1038–1044.

106 Yushchenko, D.A., Shvadchak, V.V., Klymchenko, A.S., Duportail, G., Pivovarenko, V.G., and Mely, Y. (2007) Steric control of the excited-state intramolecular proton transfer in 3-hydroxyquinolones: steady-state and time-resolved fluorescence study. *J. Phys. Chem. A*, **111**, 8986–8992.

107 Klymchenko, A.S. and Demchenko, A.P. (2004) 3-Hydroxychromone dyes exhibiting excited-state intramolecular proton transfer in water with efficient two-band fluorescence. *New J. Chem.*, **28**, 687–692.

108 Caarls, W., Soledad Celej, M., Demchenko, A.P., and Jovin, T.M. (2009) Characterization of coupled ground state and excited state equilibria by fluorescence spectral deconvolution. *J. Fluoresc.*, **20**, 181–190.

109 M'Baye, G., Shynkar, V.V., Klymchenko, A.S., Mely, Y., and Duportail, G. (2006) Membrane dipole potential as measured by ratiometric 3-hydroxyflavone fluorescence probes: accounting for hydration effects. *J. Fluoresc.*, **16**, 35–42.

110 M'Baye, G., Klymchenko, A.S., Yushchenko, D.A., Shvadchak, V.V., Ozturk, T., Mely, Y., and Duportail, G. (2007) Fluorescent dyes undergoing intramolecular proton transfer with improved sensitivity to surface charge in lipid bilayers. *Photochem. Photobiol. Sci.*, **6**, 71–76.

111 Klymchenko, A.S., Yushchenko, D.A., and Mely, Y. (2007) Tuning excited state intramolecular proton transfer in 3-hydroxyflavone derivative by reaction of its isothiocyanate group with an amine. *J. Photochem. Photobiol. A*, **192**, 93–97.

112 Avilov, S.V., Bode, C., Tolgyesi, F.G., Klymchenko, A.S., Fidy, J., and Demchenko, A.P. (2005) Heat perturbation of bovine eye lens α-crystallin probed by covalently attached ratiometric fluorescent dye 4′-diethylamino-3-hydroxyflavone. *Biopolymers*, **78**, 340–348.

113 Yushchenko, D.A., Vadzyuk, O.B., Kosterin, S.O., Duportail, G., Mely, Y., and Pivovarenko, V.G. (2007) Sensing of

adenosine-5′-triphosphate anion in aqueous solutions and mitochondria by a fluorescent 3-hydroxyflavone dye. *Anal. Biochem.*, **369**, 218–225.

114 Shynkar, V.V., Klymchenko, A.S., Kunzelmann, C., Duportail, G., Muller, C.D., Demchenko, A.P., Freyssinet, J.M., and Mely, Y. (2007) Fluorescent biomembrane probe for ratiometric detection of apoptosis. *J. Am. Chem. Soc.*, **129**, 2187–2193.

115 Klymchenko, A.S., Shvadchak, V.V., Yushchenko, D.A., Jain, N., and Mely, Y. (2008) Excited-state intramolecular proton transfer distinguishes microenvironments in single- and double-stranded DNA. *J. Phys. Chem. B*, **112**, 12050–12055.

116 Yushchenko, D.A., Fauerbach, J.A., Thirunavukkuarasu, S., Jares-Erijman, E.A., and Jovin, T.M. (2010) Fluorescent ratiometric MFC probe sensitive to early stages of α-synuclein aggregation. *J. Am. Chem. Soc.*, **132**, 7860–7861.

117 Klymchenko, A.S., Duportail, G., Mely, Y., and Demchenko, A.P. (2003) Ultrasensitive two-color fluorescence probes for dipole potential in phospholipid membranes. *Proc. Natl. Acad. Sci. USA*, **100**, 11219–11224.

118 Klymchenko, A.S., Mely, Y., Demchenko, A.P., and Duportail, G. (2004) Simultaneous probing of hydration and polarity of lipid bilayers with 3-hydroxyflavone fluorescent dyes. *Biochim. Biophys. Acta*, **1665**, 6–19.

119 Klymchenko, A.S., Duportail, G., Ozturk, T., Pivovarenko, V.G., Mely, Y., and Demchenko, A.P. (2002) Novel two-band ratiometric fluorescence probes with different location and orientation in phospholipid membranes. *Chem. Biol.*, **9**, 1199–1208.

120 Klymchenko, A.S., Stoeckel, H., Takeda, K., and Mely, Y. (2006) Fluorescent probe based on intramolecular proton transfer for fast ratiometric measurement of cellular transmembrane potential. *J. Phys. Chem. B*, **110**, 13624–13632.

121 Celej, M.S., Caarls, W., Demchenko, A.P., and Jovin, T.M. (2009) A triple-emission fluorescent probe reveals distinctive amyloid fibrillar polymorphism of wild-type α-synuclein and its familial Parkinson's disease mutants. *Biochemistry*, **48**, 7465–7472.

122 Ruiperez, V., Darios, F., and Davletov, B. (2010) α-Synuclein, lipids and Parkinson's disease. *Prog. Lipid. Res.*, **23, 49**, 420–428.

123 Drescher, M., Godschalk, F., Veldhuis, G., van Rooijen, B.D., Subramaniam, V., and Huber, M. (2008) Spin-label EPR on α-synuclein reveals differences in the membrane binding affinity of the two antiparallel helices. *ChemBioChem*, **9**, 2411–2416.

124 Stockl, M., Fischer, P., Wanker, E., and Herrmann, A. (2008) α-Synuclein selectively binds to anionic phospholipids embedded in liquid-disordered domains. *J. Mol. Biol.*, **375**, 1394–1404.

125 Haque, F., Pandey, A.P., Cambrea, L.R., Rochet, J.C., and Hovis, J.S. (2010) Adsorption of α-synuclein on lipid bilayers: modulating the structure and stability of protein assemblies. *J. Phys. Chem. B*, **114**, 4070–4081.

126 Kubo, S., Nemani, V.M., Chalkley, R.J., Anthony, M.D., Hattori, N., Mizuno, Y., Edwards, R.H., and Fortin, D.L. (2005) A combinatorial code for the interaction of α-synuclein with membranes. *J. Biol. Chem.*, **280**, 31664–31672.

127 Bisaglia, M., Mammi, S., and Bubacco, L. (2009) Structural insights on physiological functions and pathological effects of α-synuclein. *FASEB J.*, **23**, 329–340.

128 van Rooijen, B.D., Claessens, M.M., and Subramaniam, V. (2008) Membrane binding of oligomeric α-synuclein depends on bilayer charge and packing. *FEBS Lett.*, **582**, 3788–3792.

129 Kjaer, L., Giehm, L., Heimburg, T., and Otzen, D. (2009) The influence of vesicle size and composition on α-synuclein structure and stability. *Biophys. J.*, **96**, 2857–2870.

130 Fortin, D.L., Troyer, M.D., Nakamura, K., Kubo, S., Anthony, M.D., and Edwards, R.H. (2004) Lipid rafts mediate the synaptic localization of α-synuclein. *J. Neurosci.*, **24**, 6715–6723.

131 van Rooijen, B.D., Claessens, M.M., and Subramaniam, V. (2009) Lipid bilayer

disruption by oligomeric α-synuclein depends on bilayer charge and accessibility of the hydrophobic core. *Biochim. Biophys. Acta*, **1788**, 1271–1278.

132  Perrin, R.J., Woods, W.S., Clayton, D.F., and George, J.M. (2000) Interaction of human α-synuclein and Parkinson's disease variants with phospholipids. Structural analysis using site-directed mutagenesis. *J. Biol. Chem.*, **275**, 34393–34398.

133  Bortolus, M., Tombolato, F., Tessari, I., Bisaglia, M., Mammi, S., Bubacco, L., Ferrarini, A., and Maniero, A.L. (2008) Broken helix in vesicle and micelle-bound α-synuclein: insights from site-directed spin labeling-EPR experiments and MD simulations. *J. Am. Chem. Soc.*, **130**, 6690–6691.

134  Rao, J.N., Jao, C.C., Hegde, B.G., Langen, R., and Ulmer, T.S. (2010) A combinatorial NMR and EPR approach for evaluating the structural ensemble of partially folded proteins. *J. Am. Chem. Soc.*, **132**, 8657–8668.

135  Georgieva, E.R., Ramlall, T.F., Borbat, P.P., Freed, J.H., and Eliezer, D. (2008) Membrane-bound α-synuclein forms an extended helix: long-distance pulsed ESR measurements using vesicles, bicelles, and rodlike micelles. *J. Am. Chem. Soc.*, **130**, 12856–12857.

136  Trexler, A.J. and Rhoades, E. (2009) α-Synuclein binds large unilamellar vesicles as an extended helix. *Biochemistry*, **48**, 2304–2306.

137  Liu, J.P., Tang, Y., Zhou, S.F., Toh, B.H., McLean, C., and Li, H. (2010) Cholesterol involvement in the pathogenesis of neurodegenerative diseases. *Mol. Cell. Neurosci.*, **43**, 33–42.

138  Mills, T.T., Toombes, G.E., Tristram-Nagle, S., Smilgies, D.M., Feigenson, G.W., and Nagle, J.F. (2008) Order parameters and areas in fluid-phase oriented lipid membranes using wide angle X-ray scattering. *Biophys. J.*, **95**, 669–681.

139  Bar-On, P., Rockenstein, E., Adame, A., Ho, G., Hashimoto, M., and Masliah, E. (2006) Effects of the cholesterol-lowering compound methyl-β-cyclodextrin in models of α-synucleinopathy. *J. Neurochem.*, **98**, 1032–1045.

140  Li, C., Lutz, E.A., Slade, K.M., Ruf, R.A., Wang, G.F., and Pielak, G.J. (2009) $^{19}$F NMR studies of α-synuclein conformation and fibrillation. *Biochemistry*, **48**, 8578–8584.

141  Pandey, A.P., Haque, F., Rochet, J.C., and Hovis, J.S. (2009) Clustering of α-synuclein on supported lipid bilayers: role of anionic lipid, protein, and divalent ion concentration. *Biophys. J.*, **96**, 540–551.

# 2
# Structural and Functional Insights into α-Synuclein–Lipid Interactions

*Martin Stöckl, Bart D. van Rooijen, Mireille M.A.E. Claessens, and Vinod Subramaniam*

## 2.1
### Introduction

With the current increase in the lifespan of the population, neurodegenerative diseases caused by amyloid aggregates are leading to dementia in an increasing number of patients. Apart from Alzheimer's disease, one of the most prevalent neurodegenerative diseases is Parkinson's disease (PD), which affects about 1–2% of the population above the age of 65 years [1]. On the physiological level, the cause of the disease is a severe loss of neuromelanin-containing dopaminergic neurons, especially in the substantia nigra pars compacta [2–4]. Although most cases of PD are sporadic, there are several genetic factors associated with the disease. Among these, a key player is the protein α-synuclein, as several point mutations (A30P, A53T, and E46K) and also gene duplication lead to the development of early onset familiar forms of PD [5–9]. Additionally, the loss of neurons in both sporadic and familial forms of PD is accompanied by the formation of fibrillar cytoplasmic inclusions, the so-called "Lewy bodies" and "Lewy neurites" both of which are rich in α-synuclein fibrils [10, 11]. It has therefore been suggested that PD development is associated with the formation of misfolded α-synuclein species that trigger the formation of amyloid fibrils and finally fibrillar deposits, which potentially interfere with several cellular structures and processes. Especially the interaction of early α-synuclein aggregates with cellular membranes is thought to play an important role in the development of PD.

α-Synuclein belongs to the class of intrinsically disordered proteins lacking a well-folded structure under physiological conditions. Whereas in circular dichroism (CD) spectroscopy α-synuclein shows a spectrum indicative of random coil structure, the measured hydrodynamic radius of the monomer deviates from that predicted for a pure random coil protein of equal size. This suggests the existence of residual structure and long-range interactions between different parts of α-synuclein, and has been shown using various approaches [12–16]. Despite the lack of a well-defined secondary structure, the protein can be broadly divided into three different "domains" with distinct properties and functions. The C-terminal part of α-synuclein (approximately amino acids 100–140), which is highly negatively charged, shows no defined structure in solution or when bound to lipid membranes. However, some studies report long-range electrostatic interactions with other parts of the protein, which may inhibit the aggregation of the protein [15, 17–22]. The central part of the α-synuclein protein is formed by a hydrophobic stretch, the so-called NAC (non-Aβ component, amino acids 61–95), which shows a high propensity for aggregation and is critical for the formation of oligomers and fibrillar aggregates [23–26]. The sequence of the N-terminal part of the protein closely resembles amphipathic helix regions also found in lipoproteins. Indeed, approximately the first 100 amino acids form amphipathic helices upon membrane binding [27–29]. The ability to interact with membranes may be related to the physiological function of the protein. Since α-synuclein has been found to be associated with presynaptic termini, a potential physiological function of the protein may be the stabilization of the synaptic vesicle reserve pool [30–35].

Although loss of function due to aggregation may play a role in toxicity, the oligomeric and prefibrillar forms of the protein may also act actively as toxic species with respect to neuronal death during PD progression [36–41]. However, the exact nature of the oligomers that may be responsible for cell death in PD is still elusive. Studies on cellular models show that cytotoxicity seems to be associated with β-sheet content, as oligomers with a high fraction of α-helical motifs appear to be benign. Interestingly, certain small molecules trigger this conversion of prefibrillar aggregates to the α-helical species [42, 43]. Hence these compounds are very interesting drug candidates for the treatment of PD.

*In vitro*, using different protocols, it is possible to synthesize and purify oligomeric species which differ in cytotoxicity, secondary structure, morphology, and whether the oligomers are on- or off-pathway towards fibrillization [44]. In addition, the nature of the oligomeric species and the aggregation process can be influenced by a variety of factors, ranging from physico-chemical parameters such as pH and molecular crowding [45–50], to alterations caused by oxidation, single point mutations or deletions [19–22, 51–63], and interactions with other compounds, such as ions, small molecules, and lipid membranes [17, 18, 42, 63–80]. *In vitro* experiments have shown that with respect to membrane binding, especially the protein to lipid ratio affects the aggregation of α-synuclein, as conditions where the protein is in excess enhance the fibrillization, whereas an excess of lipids has a stabilizing effect on the monomers [47, 48, 81]. This effect is presumably related to different binding modes of α-synuclein depending on the surface concentration [82–84]. Also in biological systems, membrane interactions may trigger the formation of aggregates, where a reduced cholesterol level may have a protective effect [65, 85, 86].

The complexity of the web of interactions between α-synuclein monomers and the possible presence of different types of oligomers makes it difficult to dissect the mechanisms by which early α-synuclein aggregates lead to the death of the dopaminergic neurons [87, 88]. On the one hand, misfolded early aggregate species may interfere with the protein degradation pathway, and the subsequent inhibition of the proteasome could cause cellular stress [89]. On the other hand, several *in vivo* and *in vitro* studies have shown that oligomeric α-synuclein species can induce permeabilization of both artificial and natural lipid membrane structures [90–93]. Based on this observation, it was postulated that α-synuclein oligomers could interfere with many cellular pathways, ranging from the impairment of mitochondria, possibly leading to energy depletion and proapoptotic signals [94], to the leakage of synaptic vesicles, which would explain the observed high cytoplasmic concentrations of catecholamines in affected dopaminergic neurons [95]. A summary of pathways posing a possible target for toxic α-synuclein species can be found in a recent review by Auluck *et al.* [30]. In addition, recent studies have reported that oligomeric α-synuclein species are able to cross the plasma membrane [96, 97]. Upon uptake, these oligomeric species possibly seed α-synuclein aggregation in "uninfected" neurons, contributing to the progression of cell death through formerly unaffected regions of the brain [98, 99]. Hence the impairment of cellular membranes presumably plays an important role in the pathophysiology of PD. However, the exact mechanism by which the α-synuclein oligomers cause the permeabilization of membranes is still under dispute. In order to gain insights into the details of the mechanisms of cellular toxicity, it is crucial to identify the underlying mechanism in well-controlled model systems.

In the following sections, we review essential insights from recent experiments into those factors having an impact on interactions between monomeric and oligomeric α-synuclein and phospholipid vesicles. In addition, functional and structural properties of both the α-synuclein oligomers and the phospholipid bilayers are discussed with regard to their impact on membrane binding and penetration. The aggregation of α-synuclein leads to a distribution of oligomeric species whose properties are dependent on the exact details of the preparation procedures. In our own work, we have established a protocol for α-synuclein oligomer preparation that leads to a stable, β-sheet-rich oligomeric species that we have characterized extensively [100–102]. The results from our laboratory presented in the following sections therefore always refer to the α-synuclein oligomers prepared according to this protocol. However, it cannot be precluded that oligomers prepared under different conditions show different behavior regarding membrane binding, permeabilization, and oligomer structure. In the future, it will therefore be necessary to characterize carefully the structures and membrane binding and permeabilization properties of different α-synuclein oligomer species to establish common principles for these processes, should they exist.

## 2.2
## Interaction of α-Synuclein with Model Membrane Systems

### 2.2.1
### Binding of α-Synuclein Species to Giant Unilamellar Vesicles

Interactions between α-synuclein and phospholipid membranes seem to play an important role in the physiological function of the monomeric protein and also in the events resulting in the development of PD, as membrane-bound α-synuclein shows a different aggregation behavior and α-synuclein oligomers are able to permeabilize phospholipid membranes.

It is known that in vesicle model systems, negatively charged lipids strongly enhance membrane binding of monomeric α-synuclein (see Chapter 1). This requirement cannot explain, however, the preferential enrichment of α-synuclein at synaptic vesicles and in the presynaptic termini [33, 103]. The forces driving the local enrichment of the protein remain elusive, although interactions with a yet unidentified protein may explain the preferential enrichment of α-synuclein at specific membrane locations. However, the protein is able to bind to phospholipid membranes by itself, adopting an amphipathic helical structure [27, 28, 104]. In doing so, membrane lipid domains in which certain lipid species are enriched could direct the α-synuclein protein to certain fractions of the membrane. However, the information on the lipids or membrane properties involved in α-synuclein membrane binding is somewhat contradictory. Although it has been reported that aggregated α-synuclein purified from cells is associated with lipids which are typically found in so-called "lipid raft" domains which are enriched in saturated lipid species and cholesterol, polyunsaturated lipids, which are normally excluded from such domains, also seem to play important role in membrane binding of monomeric α-synuclein [34, 105, 106]. Since the lipid composition of the plasma membrane is too complex to identify lipid species and membrane properties that influence the membrane binding of α-synuclein directly, a well-defined model system is required. Micrometer-sized giant unilamellar vesicles (GUVs) are well suited for such studies since visible lateral phase separation in $l_d$ (liquid disordered) and lipid raft-like $l_o$ (liquid ordered) phases can be induced if appropriate lipid mixtures are used. Consequently, using fluorescent probes which partition preferentially into distinct lipid domains in combination with fluorescently labeled proteins, the visualization of the binding of α-synuclein to the phospholipid bilayer and the identification of lipid phase-specific binding sites is straightforward using fluorescence microscopy.

This approach has helped to identify important factors contributing to and regulating the membrane binding of both monomeric and oligomeric α-synuclein. It could be corroborated that monomeric and oligomeric α-synuclein bind to GUVs that contained negatively charged lipid species (PS (phosphatidylserine), PG (phosphatidylglycerol), or PA (phosphatidic acid)). Binding of monomeric α-synuclein to GUVs seems to be mainly mediated by complementary charges irrespective of the nature of the incorporated negatively charged lipid, while no binding to GUVs containing only zwitterionic lipids (PC (phosphatidylcholine), or SM (sphingomyelin)) was detected [107]. Similar observations were made for α-synuclein oligomers [101]. In all cases, α-synuclein did not alter the observed morphology of the GUVs and in addition did not penetrate through the membrane.

The impact of the saturation grade of the acyl chains, which is mainly responsible for the enrichment of lipids in certain lipid domains, on the binding behavior of α-synuclein can be studied using GUVs prepared from lipid mixtures containing negatively charged saturated lipids. Monomeric α-synuclein binds to GUVs containing the saturated lipids distearoyl-phosphatidylserine (DSPS) or dipalmitoyl-phosphatidic acid (DPPA), whereas no binding is observed for dipalmitoyl-phosphatidylglycerol (DPPG). This result cannot be explained by differential binding efficiencies for the lipid headgroups, as PA, PS and PG with unsaturated tails all bind α-synuclein to a similar extent. This points to the involvement of the lipid phase behavior in α-synuclein binding. Depending on the nature of their acyl chains (e.g., length, degree of saturation, or branching) and also the extent of the interactions between headgroups, lipids show varying phase transition temperatures. As a result, for lipid mixtures composed of at least two lipid species with different transition temperatures, a certain temperature interval exists in which different membrane domains which are respectively enriched in one of the lipid species coexist. In this view, DSPS and DPPA form a homogeneous $l_d$ phase with DOPC (dioleoyl-phosphatidylcholine), whereas DPPG redistributes into laterally segregated lipid domains, which are most likely in the $s_o$ (solid ordered) phase [107]. Despite the high concentration of negatively charged lipids in these ordered domains, no binding of monomeric and oligomeric α-synuclein can be observed. As gel phase domains are thought to be absent in biological membranes, lipid mixtures consisting of an unsaturated PC, a saturated SM species, cholesterol, and charged lipids that show a coexistence of $l_d$ and $l_o$ domains resemble the plasma membrane more closely. Again, binding of α-synuclein to GUVs is dependent on the presence of negatively charged lipids. In this case, α-synuclein monomers bind to phase-separated GUVs containing DSPS (which has been shown to distribute equally between $l_d$ and $l_o$ domains), albeit only to those parts of the membrane that are in the $l_d$ phase (Figure 2.1a) [107]. A similar behavior is observed for oligomeric α-synuclein species. In vesicles prepared from a lipid mixture that contains only PG as a phospholipid, α-synuclein oligomers only bind to dilauroyl-phosphatidylglycerol (DLPG)-rich domains which form the $l_d$ phase, and no binding to the DPPG-rich domains ($l_o$ phase) is observed (Figure 2.1b) [100]. An overview of the results of the binding studies is given in Table 2.1.

These results clearly show that the presence of negatively charged lipids is not sufficient to trigger the membrane association of monomeric and oligomeric α-synuclein. The protein is able to sense the phase state of the membrane and therefore the binding is not regulated by specific lipid species but by the physico-chemical properties of the lipid domains present. In this respect, especially the degree of lipid packing plays an important role in protein binding [101, 107].

Quantification of the fluorescence intensities of labeled proteins at the GUV membrane allows the determination of the amount of membrane-bound α-synuclein and thereby the membrane affinity, yielding mechanistic and functional insights into the interactions between the monomeric protein and phospholipid bilayers. The sensitivity of the membrane affinity of the monomeric proteins to ionic concentrations and also the required presence of negatively charged lipids show that the interactions of the α-synuclein monomers are mediated mainly by electrostatic interactions [107, 108]. At first glance, this observation seems surprising, as at neutral pH both the lipid bilayer and the protein are net negatively charged. However, due to the formation of the amphipathic helices in the N-terminal region of α-synuclein, the lysine residues orient towards one surface of the helix, generating a positively

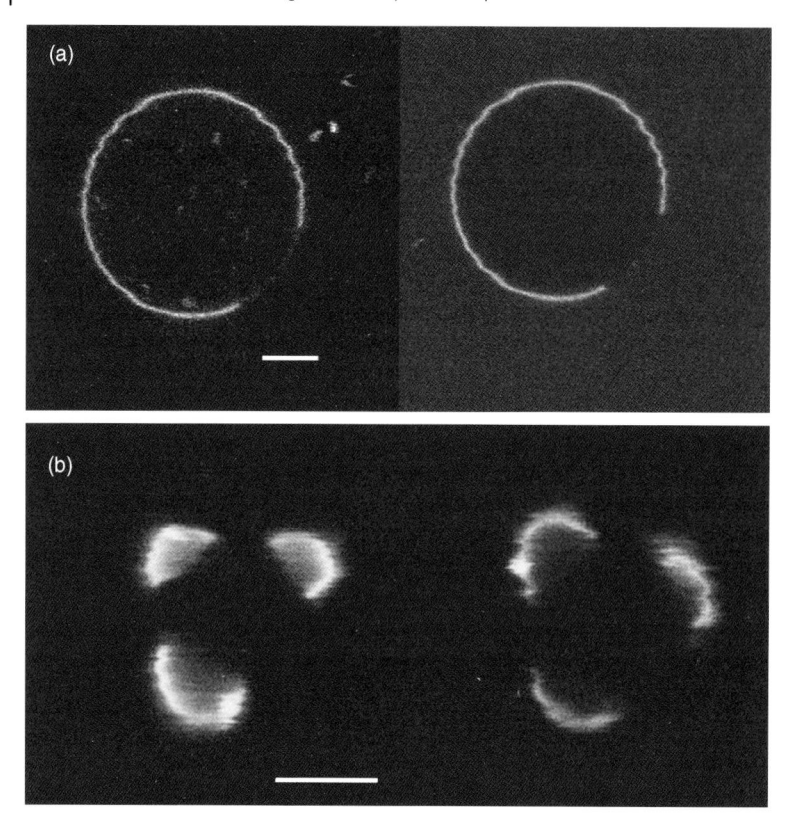

**Figure 2.1** Confocal microscopy images of (a) DOPC–18:0-SM–Chol–DSPS (25:25:25:25) and (b) DLPG–DPPG–Chol (33:33:33) GUVs. Images on the left show the fluorescence of probes labeling the $l_d$ domain and the images on the right show fluorescently labeled α-synuclein monomers (a) or oligomers (b). The scale bar indicates 5 μm. For experimental details, see [101, 107].

charged interface. In competition assays, the binding efficiencies of the pathogenic protein variants (A30P, A53T, and E46K) identified in inheritable forms of PD and the wild-type (WT) protein can be compared [107]. Whereas the E46K variant shows enhanced, and the A30P reduced, affinity towards the membrane relative to the WT α-synuclein, no changes are observed for the A53T variant, which corroborates findings on other membrane systems [34, 82, 109, 110]. This again underlines the importance of electrostatic interactions during membrane binding, as the introduction of two net positive charges in the E46K variant causes stronger membrane binding. In contrast, the proline at position 30 hinders helix formation in this part of the protein, which is a prerequisite for proper alignment of the lysine residues towards the phospholipid bilayer. Hence the membrane affinity of this α-synuclein variant is strongly reduced, especially at high ionic concentrations. In summary, the alterations in binding strength of the disease-related α-synuclein mutants could

**Table 2.1** Lipid phase specificity of lipid binding of $\alpha$-synuclein.

| Lipid composition (mol:mol) | $\alpha$-Synuclein | Lipid phases | Charged lipids | Binding of $\alpha$-synuclein[c] | |
|---|---|---|---|---|---|
| | | | | To $l_d$ | To $l_o/s_o$ |
| DOPC[a] | Monomer | $l_d$ only | N.a. | – | N.a. |
| DOPC–SSM–Chol (33:33:33)[a] (Raft) | Monomer | $l_d$ and $l_o$ | N.a. | – | – |
| DOPC–DPPA (85:15)[a] | Monomer | $l_d$ only | In $l_d$ | + | N.a. |
| DOPC–DSPS (70:30)[a] | Monomer | $l_d$ only | In $l_d$ | + | N.a. |
| DOPC–DPPG (70:30)[a] | Monomer | $l_d$ and $s_o$ | In $s_o$ | – | – |
| Raft–DOPS (75:25)[a] | Monomer | $l_d$ and $l_o$ | In $l_d$ | + | – |
| Raft–DSPS (75:25)[a] | Monomer | $l_d$ and $l_o$ | $l_d$ and $l_o$ | + | – |
| DLPG–DPPG–Chol (33:33:33)[b] | Oligomer | $l_d$ and $l_o$ | $l_d$ and $l_o$ | + | – |

a) 8.7 mM NaH$_2$PO$_4$, 210 mM glucose, 62.5 mM sucrose, pH 7.2 [107].
b) 10 mM HEPES, 150 mM NaCl, pH 7.4 [101].
c) N.a., not applicable; –, no binding observed; +, binding observed.

interfere with the physiological function of α-synuclein and have an impact on protein aggregation events. Thus, *in vivo* the altered membrane affinity could be an important factor during development of PD.

### 2.2.2
### Model Membrane Permeabilization by α-Synuclein Oligomers

In addition to binding to negatively charged $l_d$ membranes, α-synuclein oligomers are also known to permeabilize phospholipid bilayers. Membrane permeabilization by α-synuclein oligomers may play a major role in the pathological mechanisms of PD, potentially due to the impairment of membranous cellular structures such as mitochondria and synaptic vesicles [95, 111, 112]. However, the mechanisms by which oligomers cause bilayer permeabilization still remain unclear. On the one hand it has been hypothesized that α-synuclein oligomers are able to form pore-like structures mediating the permeation of small molecules [93, 113–117], whereas on the other hand membrane thinning by the interaction with the oligomeric protein, which increases the permeability of the phospholipid bilayers, has been postulated [118, 119]. Understanding which factors are decisive for membrane impairment by α-synuclein oligomers will give clues about how to counteract these processes.

A straightforward approach to quantify the extent of membrane permeabilization utilizes the self-quenching properties of calcein. To this end, calcein-filled LUVs (large unilamellar vesicles) are prepared from different lipid mixtures and incubated with α-synuclein oligomers. Leakage of calcein from the LUVs leads to dilution and reduction in the self-quenching, and a corresponding increase in the fluorescence signal is observed.

An important factor affecting the vulnerability to disruption of phospholipid bilayers is the nature of the lipid headgroup. Headgroup properties have a strong impact on lipid packing and membrane phase behavior. The incubation of LUVs prepared from POPC (1-palmitoyl-2-oleoyl-phosphatidylcholine) with α-synuclein oligomers shows no leakage of calcein, which is to be expected since the protein binds only weakly to such vesicles [108]. In contrast, strong disruption of the vesicles can be observed in the case of LUVs prepared from the negatively charged lipids DOPS (dioleoyl-phosphatidylserine), POPG (1-palmitoyl-2-oleoyl-phosphatidylglycerol), and soy PI (phosphatidylinositol) (Figure 2.2). The kinetic measurements show that membrane permeabilization occurs directly after addition of the oligomers and is complete within minutes [100]. Whereas for PG and PI complete leakage of the LUV contents is achieved for micromolar concentrations of oligomeric α-synuclein, PS LUVs are disrupted to a weaker extent (Figure 2.2). In addition to the oligomers, the impact of the monomeric and fibrillar species of α-synuclein on vesicle disruption has been determined. In this case, a 10-fold higher concentration of monomeric or fibrillar α-synuclein is needed to achieve substantial membrane permeabilization [100].

Lipid bilayers consisting of only negatively charged lipids have a very high surface charge density and do not reflect the situation present at cellular membranes. The oligomer-induced permeabilization of lipid bilayers has therefore also been investigated for lower, more relevant, surface charge densities. For LUVs containing 50 mol% of POPC and 50 mol% of negatively charged lipids, membrane permeabilization by α-synuclein oligomers is not observed [100]. GUV experiments indicate that α-synuclein oligomers do bind to GUVs containing 50 mol% PG or PA [101], which suggests that although negatively charged lipids are sufficient to trigger binding of α-synuclein to membranes in the $l_d$ phase, to cause membrane permeabilization an addition initial destabilization of the membrane is necessary. To test this hypothesis, LUVs can be prepared in which the lipid packing in

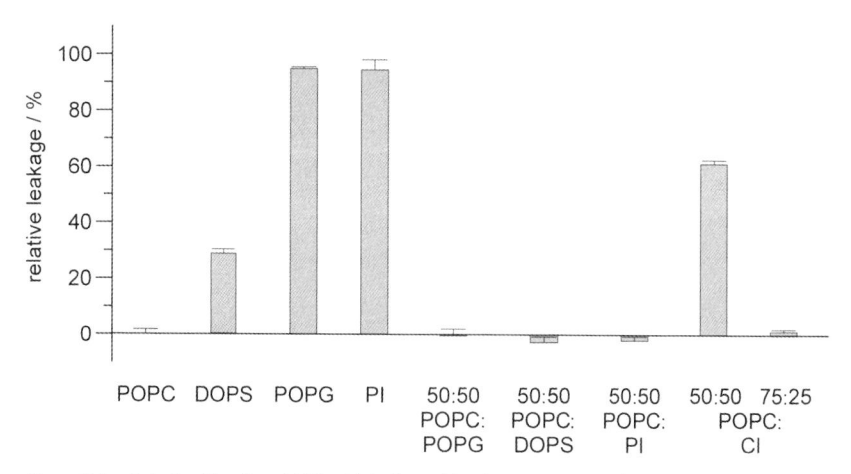

**Figure 2.2** Calcein efflux from LUVs with indicated lipid composition induced by 1 μM α-synuclein oligomers at a phospholipid concentration of 20 μM. Complete leakage upon addition of 0.5% (w/v) Triton X-100 is equal to 100%. For experimental details, see van Rooijen et al. [100].

the phospholipid bilayer is affected to some extent. To this end, lipids that have an inverted cone shape, that is, their acyl chains occupy a larger area than their headgroups (e.g., PA, cardiolipin), can be used, as these lipids induce spontaneous negative curvature and therefore defects in the headgroup region of the bilayer. α-Synuclein oligomers are able to permeabilize 50:50 (mol:mol) mixtures of bovine cardiolipin (CL) with POPC (Figure 2.2). The impact of membrane order has been corroborated using POPG LUVs containing varying amounts of lysoPG (lysophosphatidylglycerol). As lysoPG has only one acyl chain, it has a pronounced conical shape and therefore increases the molecular density of bilayers in the headgroup region. Upon incubation with α-synuclein oligomers, the membrane permeabilization is reduced in a lysoPG concentration-dependent fashion. These results show that the degree of order in the headgroup region, and thus the accessibility of the hydrophobic membrane core, has a strong impact on the ability of α-synuclein oligomers to disrupt membrane bilayers (Figure 2.3a). In order to determine the impact of the lipid packing, which also influences headgroup spacing and is dependent on, for example, the number of double bonds in the acyl chains of the lipids, LUVs from PG species with varying degree of saturation were prepared. Additionally, cholesterol can increase the degree of order in lipid bilayers. In a series of LUVs prepared from 18:2-PG (dilinoleoyl-phosphatidylglycerol), DOPG (dioleoyl-phosphatidylglycerol), POPG and POPG–Chol (cholesterol) (75:25), the degree of lipid packing increases in ascending order (Figure 2.3b). Again an increased membrane order is accompanied by decreased oligomer-induced membrane permeabilization [100]. Thus, in analogy with membrane binding, the degree of lipid packing plays an important role in the permeabilization of membranes. However, whereas less ordered regions ($l_d$ phase) may serve as binding sites for α-synuclein but highly ordered membrane regions ($l_o$ and gel phases) do not – in both cases the lipids still are well packed – the presence of lipid defects seems to be a prerequisite for the permeabilization of phospholipid bilayers.

To determine the impact of the known pathogenic α-synuclein point mutants on the ability to impair membranous structures, oligomers prepared from these α-synuclein variants were incubated with POPG LUVs. The exchange of single amino acids has no impact on the membrane disruptive effect of the oligomers. Only for the A30P variant is a slight reduction observed, which could relate to the lower membrane affinity of the monomer [34, 82, 107, 109, 110].

## 2.2.3
### Structural Features of α-Synuclein Oligomers

In order to understand the processes and interactions that underlie the permeabilization of phospholipid bilayers by α-synuclein oligomers, insights into the oligomer structure are required. In our studies, we focused on a stable oligomeric α-synuclein species prepared by shaking highly concentrated monomer solutions at room temperature followed by a short incubation at 37 °C. Although CD spectroscopy indicates a significant amount of β-sheets in the oligomers [102], the detailed structure of oligomers has not been resolved. To gain insights into the topography

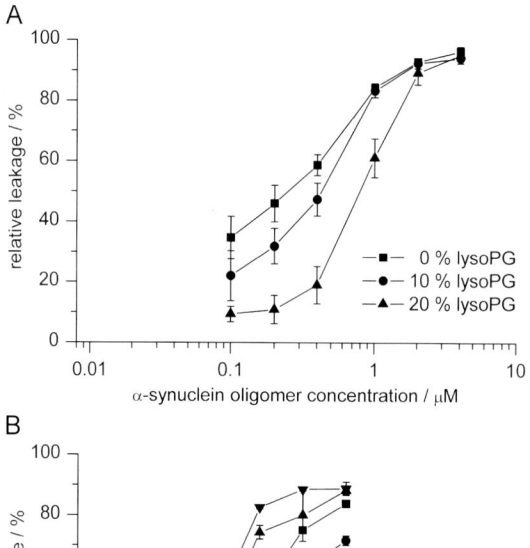

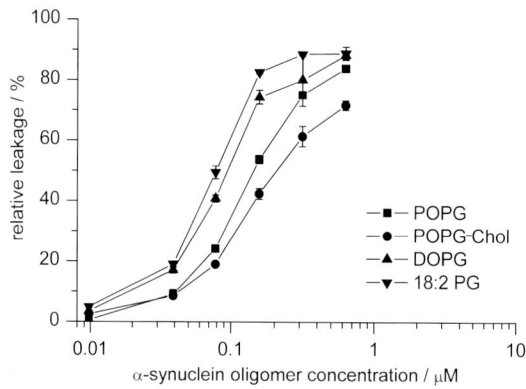

**Figure 2.3** Headgroup spacing (a) and lipid packing (b) have a strong influence on the membrane permeabilization caused by α-synuclein oligomers. (a) To determine the impact of headgroup spacing on LUV leakage, POPG LUVs containing varying amounts of lysoPG (20 μM lipids in total) were prepared. (b) To study the influence of the degree of lipid order, PG LUVs (25 μM lipids) were prepared from PG with different acyl chains: 18:2-PG, DOPG, POPG, and POPG–Chol (75:25). In all cases the vesicles were incubated with oligomeric α-synuclein and the relative amount of released calcein was plotted (0.5% (w/v) Triton X-100 is equal to 100% leakage). For experimental details, see van Rooijen et al. [100].

of the peptide chains of α-synuclein oligomers, the environment-sensitive fluorescence emission of Trp (tryptophan) has been used. This probe is especially suited to study α-synuclein as the WT protein lacks Trp residues, so that Trp probes can be introduced with ease at specific positions. Control experiments also show that the introduction of a single Trp does not change the aggregation behavior of α-synuclein [102].

The assay to determine structural features of the oligomers is based on the dependence of the peak wavelength of the fluorescence on the polarity of the surrounding medium. Thus, a blue shift of the Trp fluorescence emission indicates a transition of the probe from the aqueous buffer into a more hydrophobic environment, for example, the core of a protein or, in case of membrane binding, the hydrophobic interior of the phospholipid bilayer. The introduction of Trp residues at different positions and the measurement of the fluorescence emission spectra of α-synuclein monomers and oligomers in the presence and absence of DOPS LUVs allow the extent of shielding of the probes in free and membrane-bound α-synuclein to be determined. In the case of the monomers, the relatively red-shifted peak wavelengths of all measured Trp positions (4W, 39W, 69W, 90W, 124W, 140W) show that in general all amino acid residues are solvent exposed. However, a slight protection for the first 90 amino acids in contrast to the C-terminus can be observed, which suggests the presence of long-range intramolecular interactions (Figure 2.4). The emission spectra of α-synuclein oligomers show a substantial blue shift, which is especially pronounced for the Trp probes in the N-terminal part of the protein (4W–90W) (Figure 2.4a). These data clearly demonstrate that the core of the oligomers is formed by the N-terminal part and the NAC peptide. The shift towards shorter wavelengths is much less pronounced for the Trp probes in the negatively charged C-terminal part of α-synuclein, indicating that this region is solvent exposed on the outside of the oligomer. The slight blue shift is accompanied by spectral broadening of the fluorescence emission, demonstrating the presence of conformational heterogeneity in the C-termini of the α-synuclein molecules forming the oligomers [102]. These observations suggest that, although a full integration into the formed structure is potentially not possible due to the presence of a high fraction of negatively charged amino acids, the C-terminal region still interacts – at least transiently – with the largely hydrophobic oligomeric core.

Whereas for α-synuclein monomers the N-terminal region forms an amphipathic α-helix on membrane binding [27, 28, 104], the binding motif of the β-sheet containing oligomeric α-synuclein remains under investigation. To identify the amino acids that are closely involved in the interactions with phospholipid bilayers, the extent of additional shielding of Trp probes in α-synuclein oligomers upon incubation with DOPS LUVs was determined. Embedding of Trp residues in the hydrophobic core of the bilayer should cause an additional shift of the fluorescence emission of the respective Trp probe towards shorter wavelengths. Upon membrane binding, a slight additional shielding is observed for the probes in positions 4, 69, and 90 whereas no wavelength changes are found for positions 39, 124, and 140 (Figure 2.4a). This result shows that in the free oligomers the probes are well protected from the aqueous medium and that the interaction with the hydrophobic membrane core contributes only marginally. However, it is still possible to identify the regions of the α-synuclein oligomers that interact with the membrane. Astonishingly, for α-synuclein monomers and oligomers, the same regions of the protein are involved in membrane binding. Oligomer binding is facilitated by the amino acids which form the two membrane-binding amphipathic helices in monomeric α-synuclein; as observed in monomer binding, the short stretch of amino acids including position 39 connecting the helices does not embed in the membrane [28]. Although for the peak fluorescence emission of the Trp probes in the C-terminus upon membrane binding no shift to shorter wavelength is observed, the spectra are somewhat broadened. This observation agrees with the

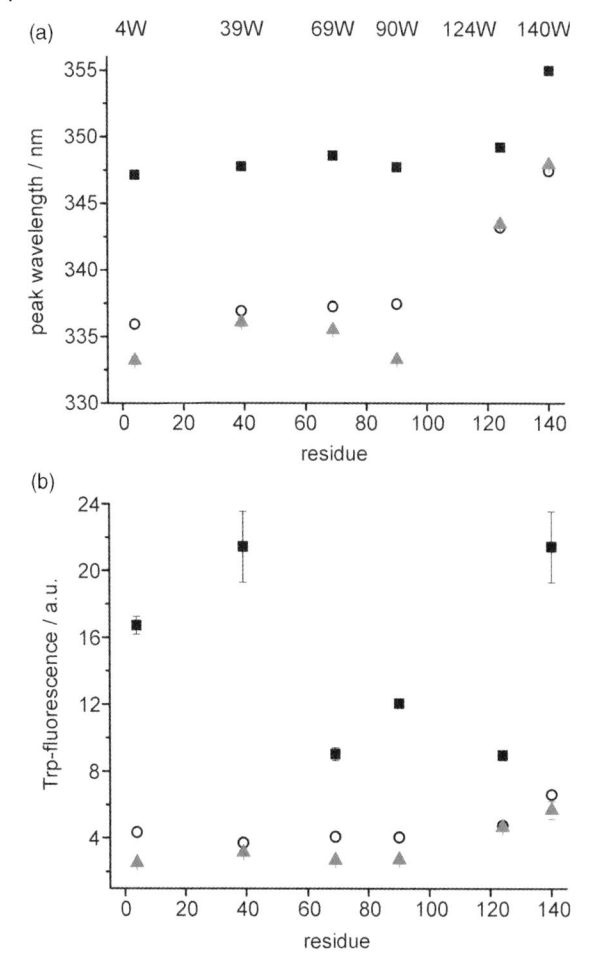

**Figure 2.4** Determination of α-synuclein oligomer structure by Trp fluorescence spectroscopy on α-synuclein mutants containing single Trp residues. Exposure of the probes in monomers (black squares), and oligomers (open circles) was measured by two independent methods: (a) the peak wavelength of the fluorescence emission is dependent of the polarity of the medium surrounding the probe; (b) upon incubation with acrylamide (0.5 M), fluorescence of accessible Trp probes is quenched. To determine which residues become protected upon membrane binding, oligomeric α-synuclein was incubated with DOPS LUVs (gray triangles, lipid: protein = 300:1). Where present, error bars indicate the standard deviation. For experimental details, see van Rooijen et al. [102].

idea that the C-terminal residues interact transiently with the core of the oligomers, and that upon oligomer binding the C-terminal part of the protein will likely have to remodel.

A complementary approach to investigate changes in oligomer structure upon membrane binding uses acrylamide to quench the Trp fluorescence in order to characterize the accessibility of the Trp probes to hydrophilic molecules [120]. As collisional quenching of fluorophores by acrylamide requires direct contact, the measured quenching rates contain information about the accessibility of the fluorophore. In monomeric α-synuclein, the Trp probes in all positions are readily accessible to the quencher molecules. However, substantial differences in the quenching ratios observed for different probe positions suggest that the N- and C-terminal residues are particularly exposed to the medium whereas the central part of the protein is partially protected from quenching, presumably by transient intramolecular interactions. For α-synuclein oligomers, the quenching rates for all tested positions are lower than minimal values measured for the monomers. The results agree very well with the approach based on the polarity dependence of the Trp fluorescence emission (Figure 2.4a and b). Whereas the N-terminal substitutions up to 90W are strongly shielded from the quencher, which suggests their incorporation into a compact structure, higher accessibility to acrylamide is found for the more solvent-exposed C-terminal part of α-synuclein. Upon binding to DOPS LUVs, a slightly reduced quenching is found for the substitutions 4W, 69W, and 90W, whereas positions 39W, 124W, and 140W are not further protected from the quencher (Figure 2.4b). The quenching data corroborate the Trp emission spectra results and provide evidence that for oligomeric α-synuclein the membrane interactions are mainly mediated by the N-terminal and central regions of the protein, whereas a patch surrounding amino acid 39 and the C-terminus do not contribute. These results are consistent with those found for monomeric α-synuclein.

Both experimental approaches confirm that in the oligomers the hydrophobic core is formed by the N-terminal amino acids, at least up to position 90, whereas the C-terminal part remains solvent exposed and associates at most transiently with the oligomer core. However, the comparison with the values (peak emission wavelength and quenching) determined for the 69W variant in α-synuclein fibrils shows that the shielding of the residue is even more pronounced in the fibrils [102]. Hence it can be assumed that in the core of the oligomers, the final compact amyloid structure, which is typical for the fibrils, is not yet present. Strikingly, the experiments characterizing the interaction between α-synuclein oligomers and the phospholipid bilayer show that presumably the same amino acids which mediate membrane binding in the α-synuclein monomer are part of the interface which becomes protected from the aqueous medium upon membrane binding of the oligomer, despite the overall structural differences of monomers (amphipathic helices) and oligomers (β-sheets).

## 2.3
## Biological Significance

### 2.3.1
### Interaction Sites

In cells, α-synuclein is found to be associated with distinct membrane regions such as the presynaptic termini [34, 121, 122]. However, the physiological function of the protein and the factors which regulate the membrane association are still unknown.

Apart from scaffolding proteins, special lipids or the different properties of membrane domains are possible factors that could define α-synuclein binding sites. Since up to now no proteins involved in the membrane association of α-synuclein have been identified, it is likely that the protein itself recognizes its interaction sites. In model systems explored, all types of negatively charged lipids were able to mediate membrane binding of α-synuclein monomers, and a specific preference for a certain type of lipid could not be identified [34, 91, 101, 106–108, 123]. However, the composition of lipids in the plasma membrane is highly complex. In contrast to model lipid systems in which few defined lipid species are used, the lipids found in cellular membranes differ, for example, in the length and saturation grade of their fatty acids. This leads to the lateral segregation of lipids and is the cause behind the formation of lipid domains (the so-called lipid "rafts") with distinct physico-chemical properties. In this context, we tested the impact of the saturation grade on the membrane binding of α-synuclein monomers. Although a direct relation between the degree of saturation and membrane binding of α-synuclein was not observed, the phase of the membrane in which the negatively charged lipids were embedded was decisive [107]. We observed that α-synuclein monomers only bound to GUVs if the negatively charged lipids were incorporated into $l_d$ domains, which demonstrates that lipid packing directly affects the membrane interactions (Figure 2.1a). This suggests that although an interaction of α-synuclein monomers with the plasma membrane should be possible, a specific enrichment of α-synuclein at well-ordered raft domains is unlikely. Binding studies using small unilamellar vesicles (SUVs), which due to their high curvature contain packing defects, have shown that α-synuclein monomers bind to those vesicles even in the absence of negatively charged lipids and partially countervails the membrane defects (see also Chapter 1) [124–126]. Although the sizes and curvatures of SUVs and synaptic vesicles are comparable, it is unlikely that synaptic vesicles contain defects that mediate binding of α-synuclein. It is more likely the specific composition of synaptic vesicles, with the presence of negatively charged lipids, that make them prime binding sites for α-synuclein monomers, where they are able to exert their presumed physiological function in stabilizing the synaptic vesicle reserve pool and in inhibiting premature fusion with the synaptic membrane [32, 35, 127]. In this view, a moderate overexpression of α-synuclein reduces the neurotransmitter release and creates defects in the regulation of the synaptic vesicle reserve pool, as its size and the vesicle reclustering upon endocytosis are reduced [128]. Thereby, α-synuclein presumably regulates the homeostasis of various proteins (e.g., SNARE proteins) maintaining the synaptic vesicle pool or controlling vesicle fusion with the synapse [31, 129–132]. Hence protein aggregation could lead to a loss of function that would impair the regulation of neurotransmitter release. This is corroborated by recent work in which the RNA-mediated silencing of α-synuclein triggers the development of a neurodegenerative phenotype [133]. Hence the careful balancing of the local α-synuclein concentration at the synaptic vesicle reserve pool and the synapse seems to be a major factor maintaining the functionality of the neurotransmitter release and preventing neurodegeneration. This hypothesis is in good agreement with the model for the physiological action of α-synuclein at the synaptic vesicles proposed by Auluck et al. [30].

In membrane binding experiments, α-synuclein oligomers have shown a behavior analogous to that of monomeric α-synuclein, as membrane binding was observed only for $l_d$ phase regions (Figure 2.1b). This suggests that *in vivo* α-synuclein monomers and oligomers share binding sites, an observation of special importance as membrane interactions influence the aggregation behavior of α-synuclein. It has been reported that a low protein to lipid ratio inhibits fibril formation whereas a high protein to lipid ratio accelerates it [47, 48]. This can be explained by assuming that α-synuclein is less aggregation prone when membrane bound, as a transition from a random coil structure in solution to a helical structure occurs, provided that the α-synuclein monomers have a higher affinity towards the phospholipid bilayer than to each other. However, in the cases that the lipid to protein ratio decreases, the aggregation of α-synuclein could be triggered either by the high local concentration of the protein due to the two-dimensional confinement on the membrane, where the formation of α-synuclein dimers could be an initial step [134], or the increased amount of unbound protein in solution with an unfolded structure. In this view, also in a recent review by Auluck *et al.* [30] it is proposed that the initial formation of α-synuclein dimers at the membrane, which are on-pathway to fibrillization, is responsible for the generation of toxic oligomeric α-synuclein species.

In this context, this model may explain the development of α-synuclein aggregates in the case of the pathogenic A30P protein variant, which is as such not more prone to aggregation [38, 135, 136]. However, the observed reduced membrane affinity of A30P results in a higher concentration of α-synuclein in solution, therefore presumably promoting aggregation propensity. On a mechanistic level, the experiments suggest that a simple model in which monomeric and oligomeric α-synuclein merely interact with the lipid headgroups by complementary charges when long-range electrostatic repulsion is sufficiently screened is not sufficient to explain our findings. Partial penetration of the lipid bilayer by the monomer and oligomer seems to be necessary for binding, a situation which is precluded in the case of well-packed lipids bilayers in $l_o$ or gel phase membranes.

### 2.3.2
**Membrane Penetration**

Apart from an impairment of the cellular protein degradation machinery causing subsequent mitochondrial damage and oxidative stress, a second suggested pathogenic mechanism involved in the death of neurons during progression of PD is the disruption of membranous structures in the cells by α-synuclein oligomers. In particular, impairment of the vesicular transport and the synaptic vesicles are suspected targets [79, 128, 137–140]. In the model system investigated, the oligomers show an affinity for negatively charged lipids if they were embedded in an $l_d$ phase (Figure 2.1b). This suggests that binding of the α-synuclein oligomers to the plasma membrane is possible. However, for LUVs, leakage was only observed for very high concentrations of negatively charged lipids, which implies the existence of intrinsic instabilities in the phospholipid bilayer (Figures 2.2 and 2.3). In contrast to the model system, due to the lower fraction of negatively charged lipids and the presence of

Chol, which is able to ameliorate defects and improve lipid packing, the presence of such defects in the plasma membrane is not to be expected. Therefore, the plasma membrane is not a likely target for membrane permeabilization by α-synuclein oligomers. However, recent research has provided evidence for the appearance of misfolded α-synuclein species also in the extracellular medium and their subsequent uptake by neighboring cells [97, 98]. Whether these proteins breach the plasma membrane on their own or are excreted by an exocytotic mechanism is still to be established. Intriguing possibilities as targets for membrane disruption by α-synuclein oligomers are synaptic vesicles. In fact, a correlation of dopamine release into the cytoplasm to α-synuclein in cellular PD models has been described [95]. In addition, dopamine and its oxidized species have a strong effect on the aggregation behavior of α-synuclein, but the exact role of these species in the development of disease is still to be determined [55, 64, 141–146].

Another common trait that is often found associated with PD models involving α-synuclein is the impairment of the mitochondria. As mitochondrial malfunction can induce the formation of reactive oxygen species, it is also a potential mode of action in the death of neurons. In fact, α-synuclein has been found to associate with the mitochondrial membrane whereas α-synuclein overexpression and acidification of the cytoplasm even enhance these interactions (for a review, see [94]). Possibly the mitochondrial lipid cardiolipin plays an important role, as it is able to trigger membrane binding of α-synuclein due to its negative charge. The shape of this lipid is a fairly broad inverted cone, which could induce strong local membrane distortions due to the spontaneous negative curvature. Such membrane defects would facilitate membrane permeabilization by α-synuclein oligomers. Admittedly, there is only a molecular fraction of 4% of cardiolipin in the outer membrane of the mitochondria, whereas in our model system for 25% cardiolipin no membrane permeabilization was observed (Figure 2.2). However, one has to keep in mind that in cells normally the lipids are not equally distributed, but can enrich strongly in confined lipid domains. In this view, a strong enrichment of cardiolipin to over 20% at contact sites between inner and outer membranes of mitochondria has been reported [147]. As these lipid patches have been isolated by swelling and centrifugation steps, one can safely assume that the local concentration of cardiolipin may even be significantly higher, as surrounding material with low cardiolipin content presumably was also isolated. Hence it is plausible that such membrane patches are vulnerable to membrane permeabilization by α-synuclein oligomers.

In summary, there is a significant body of evidence that membrane permeabilization of cellular vesicles or mitochondria is a likely mechanism leading to the death of dopaminergic neurons during progression of PD. Apart from a loss of function of the affected organelles, the liberation of reactive oxygen species or dopamine into the cytoplasm would also be toxic for the affected cells, as these compounds are able to oxidize or react with α-synuclein to influence strongly the aggregation kinetics of the protein. Therefore, further research is needed to determine what is cause and what is effect, as the observed α-synuclein aggregates could also just be symptomatic for neurons suffering from these stress factors. In addition, elucidating the underlying mechanisms by which the α-synuclein oligomers are able to penetrate the phospholipid bilayer will give valuable insights into their mode of action and presumably facilitate the development of a possible intervention strategy.

## Acknowledgments

M.S. is sponsored by a German Academic Exchange Service (DAAD) research grant (D/09/50722). B.D.v.R. was supported by the Nanotechnology Network in The Netherlands, NanoNed, project 7921.

## References

1 de Rijk, M.C. et al. (2000) Prevalence of Parkinson's disease in Europe: a collaborative study of population-based cohorts. Neurologic Diseases in the Elderly Research Group. *Neurology*, **54** (11 Suppl. 5), S21–S30.
2 Heath, J.W. (1947) Clinicopathologic aspects of parkinsonian states; review of the literature. *Arch. Neurol. Psychiatry*, **58** (4), 484–497.
3 Jellinger, K. and Jirasek, A. (1971) Neuroaxonal dystrophy in man: character and natural history. *Acta Neuropathol.*, **5** (Suppl. 5), 3–16.
4 Scott, T.R. and Netsky, M.G. (1961) The pathology of Parkinson's syndrome: a critical review. *Int. J. Neurol.*, **2**, 51–60.
5 Kruger, R. et al. (1998) Ala30Pro mutation in the gene encoding α-synuclein in Parkinson's disease. *Nat. Genet*, **18** (2), 106–108.
6 Polymeropoulos, M.H. et al. (1997) Mutation in the α-synuclein gene identified in families with Parkinson's disease. *Science*, **276** (5321), 2045–2047.
7 Ahn, T.B. et al. (2008) α-Synuclein gene duplication is present in sporadic Parkinson disease. *Neurology*, **70** (1), 43–49.
8 Zarranz, J.J. et al. (2004) The new mutation, E46K, of α-synuclein causes Parkinson and Lewy body dementia. *Ann. Neurol.*, **55** (2), 164–173.
9 Wider, C., Foroud, T., and Wszolek, Z.K. (2010) Clinical implications of gene discovery in Parkinson's disease and parkinsonism. *Mov. Disord.*, **25** (Suppl. 1), S15–S20.

10 Förster, E. and Lewy, F.H. (1912) Paralysis agitans. *Handbuch der Neurologie* (ed. M. Lewandowsky), Springer, Berlin, pp. 920–933.
11 Ohama, E. and Ikuta, F. (1976) Parkinson's disease: distribution of Lewy bodies and monoamine neuron system. *Acta Neuropathol.*, **34** (4), 311–319.
12 Frimpong, A.K. et al. (2010) Characterization of intrinsically disordered proteins with electrospray ionization mass spectrometry: conformational heterogeneity of α-synuclein. *Protein*, **78** (3), 714–722.
13 Dedmon, M.M. et al. (2005) Mapping long-range interactions in α-synuclein using spin-label NMR and ensemble molecular dynamics simulations. *J. Am. Chem. Soc.*, **127** (2), 476–477.
14 Sung, Y.H. and Eliezer, D. (2007) Residual structure, backbone dynamics, and interactions within the synuclein family. *J. Mol. Biol.*, **372** (3), 689–707.
15 Bertoncini, C.W. et al. (2005) Release of long-range tertiary interactions potentiates aggregation of natively unstructured α-synuclein. *Proc. Natl. Acad. Sci. USA*, **102** (5), 1430–1435.
16 Salmon, L. et al. (2010) NMR characterization of long-range order in intrinsically disordered proteins. *J. Am. Chem. Soc.*, **132** (24), 8407–8418.
17 Antony, T. et al. (2003) Cellular polyamines promote the aggregation of α-synuclein. *J. Biol. Chem.*, **278** (5), 3235–3240.

18 Fernandez, C.O. et al. (2004) NMR of α-synuclein–polyamine complexes elucidates the mechanism and kinetics of induced aggregation. *EMBO J.*, **23** (10), 2039–2046.

19 Hoyer, W. et al. (2004) Impact of the acidic C-terminal region comprising amino acids 109–140 on α-synuclein aggregation in vitro. *Biochemistry*, **43** (51), 16233–16242.

20 Kim, J. et al. (2010) The inhibitory effect of pyrroloquinoline quinone on the amyloid formation and cytotoxicity of truncated α-synuclein. *Mol. Neurodegener.*, **5**, 20.

21 Murray, I.V. et al. (2003) Role of α-synuclein carboxy-terminus on fibril formation in vitro. *Biochemistry*, **42** (28), 8530–8540.

22 Ulusoy, A. et al. (2010) Co-expression of C-terminal truncated α-synuclein enhances full-length α-synuclein-induced pathology. *Eur. J. Neurosci.*, **32** (3), 409–422.

23 Giasson, B.I. et al. (2001) A hydrophobic stretch of 12 amino acid residues in the middle of α-synuclein is essential for filament assembly. *J. Biol. Chem.*, **276** (4), 2380–2386.

24 Waxman, E.A., Mazzulli, J.R., and Giasson, B.I. (2009) Characterization of hydrophobic residue requirements for α-synuclein fibrillization. *Biochemistry*, **48** (40), 9427–9436.

25 Zibaee, S. et al. (2007) Sequence determinants for amyloid fibrillogenesis of human α-synuclein. *J. Mol. Biol.*, **374** (2), 454–464.

26 Koo, H.J., Lee, H.J., and Im, H. (2008) Sequence determinants regulating fibrillation of human α-synuclein. *Biochem. Biophys. Res. Commun.*, **368** (3), 772–778.

27 Eliezer, D. et al. (2001) Conformational properties of α-synuclein in its free and lipid-associated states. *J. Mol. Biol.*, **307** (4), 1061–1073.

28 Ulmer, T.S. et al. (2005) Structure and dynamics of micelle-bound human α-synuclein. *J. Biol. Chem.*, **280** (10), 9595–9603.

29 Chandra, S. et al. (2003) A broken α-helix in folded α-synuclein. *J. Biol. Chem.*, **278** (17), 15313–15318.

30 Auluck, P.K., Caraveo, G., and Lindquist, S. (2010) α-Synuclein: membrane interactions and toxicity in Parkinson's disease. *Annu. Rev. Cell Dev. Biol.*,

31 Burre, J. et al. (2010) α-Synuclein promotes SNARE-complex assembly in vivo and in vitro. *Science*. doi: 10.1126/science.1195227.

32 Cabin, D.E. et al. (2002) Synaptic vesicle depletion correlates with attenuated synaptic responses to prolonged repetitive stimulation in mice lacking α-synuclein. *J. Neurosci.*, **22** (20), 8797–8807.

33 Clayton, D.F. and George, J.M. (1998) The synucleins: a family of proteins involved in synaptic function, plasticity, neurodegeneration and disease. *Trends Neurosci.*, **21** (6), 249–254.

34 Fortin, D.L. et al. (2004) Lipid rafts mediate the synaptic localization of α-synuclein. *J. Neurosci.*, **24** (30), 6715–6723.

35 Gureviciene, I., Gurevicius, K., and Tanila, F.H. (2007) Role of α-synuclein in synaptic glutamate release. *Neurobiol. Dis.*, **28** (1), 83–89.

36 Periquet, M. et al. (2007) Aggregated α-synuclein mediates dopaminergic neurotoxicity in vivo. *J. Neurosci.*, **27** (12), 3338–3346.

37 Volles, M.J. and Lansbury, P.T. Jr. (2007) Relationships between the sequence of α-synuclein and its membrane affinity, fibrillization propensity, and yeast toxicity. *J. Mol. Biol.*, **366** (5), 1510–1522.

38 Conway, K.A. et al. (2000) Acceleration of oligomerization, not fibrillization, is a shared property of both α-synuclein mutations linked to early-onset Parkinson's disease: implications for pathogenesis and therapy. *Proc. Natl. Acad. Sci. USA*, **97** (2), 571–576.

39 Goldberg, M.S. and Lansbury, P.T. Jr. (2000) Is there a cause-and-effect relationship between α-synuclein fibrillization and Parkinson's disease? *Nat. Cell Biol.*, **2** (7), E115–E119.

40 Masliah, E. et al. (2000) Dopaminergic loss and inclusion body formation in α-synuclein mice: implications for neurodegenerative disorders. *Science*, **287** (5456), 1265–1269.

41 Emadi, S. *et al.* (2009) Detecting morphologically distinct oligomeric forms of α-synuclein. *J. Biol. Chem.*, **284** (17), 11048–11058.

42 Ehrnhoefer, D.E. *et al.* (2008) EGCG redirects amyloidogenic polypeptides into unstructured, off-pathway oligomers. *Nat. Struct. Mol. Biol.*, **15** (6), 558–566.

43 Bieschke, J. *et al.* (2010) EGCG remodels mature α-synuclein and amyloid-β fibrils and reduces cellular toxicity. *Proc. Natl. Acad. Sci. USA*, **107** (17), 7710–7715.

44 Danzer, K.M. *et al.* (2007) Different species of α-synuclein oligomers induce calcium influx and seeding. *J. Neurosci.*, **27** (34), 9220–9232.

45 Hoyer, W. *et al.* (2002) Dependence of α-synuclein aggregate morphology on solution conditions. *J. Mol. Biol.*, **322** (2), 383–393.

46 Munishkina, L.A., Fink, A.L., and Uversky, V.N. (2008) Concerted action of metals and macromolecular crowding on the fibrillation of α-synuclein. *Protein Pept. Lett.*, **15** (10), 1079–1085.

47 Zhu, M., Li, J., and Fink, A.L. (2003) The association of α-synuclein with membranes affects bilayer structure, stability, and fibril formation. *J. Biol. Chem.*, **278** (41), 40186–40197.

48 Zhu, M. and Fink, A.L. (2003) Lipid binding inhibits α-synuclein fibril formation. *J. Biol. Chem.*, **278** (19), 16873–16877.

49 McNulty, B.C., Young, G.B., and Pielak, G.J. (2006) Macromolecular crowding in the *Escherichia coli* periplasm maintains α-synuclein disorder. *J. Mol. Biol.*, **355** (5), 893–897.

50 Kjaer, L. *et al.* (2009) The influence of vesicle size and composition on α-synuclein structure and stability. *Biophys. J.*, **96** (7), 2857–2870.

51 Fredenburg, R.A. *et al.* (2007) The impact of the E46K mutation on the properties of α-synuclein in its monomeric and oligomeric states. *Biochemistry*, **46** (24), 7107–7118.

52 Glaser, C.B. *et al.* (2005) Methionine oxidation, α-synuclein and Parkinson's disease. *Biochim. Biophys. Acta*, **1703** (2), 157–169.

53 Harada, R. *et al.* (2009) The effect of amino acid substitution in the imperfect repeat sequences of α-synuclein on fibrillation. *Biochim. Biophys. Acta*, **1792** (10), 998–1003.

54 Lee, J.T. *et al.* (2008) Ubiquitination of α-synuclein by Siah-1 promotes α-synuclein aggregation and apoptotic cell death. *Hum. Mol. Genet*, **17** (6), 906–917.

55 Leong, S.L. *et al.* (2009) Modulation of α-synuclein aggregation by dopamine: a review. *Neurochem. Res.*, **34** (10), 1838–1846.

56 Li, J., Uversky, V.N., and Fink, A.L. (2001) Effect of familial Parkinson's disease point mutations A30P and A53T on the structural properties, aggregation, and fibrillation of human α-synuclein. *Biochemistry*, **40** (38), 11604–11613.

57 Pandey, N., Schmidt, R.E., and Galvin, J.E. (2006) The α-synuclein mutation E46K promotes aggregation in cultured cells. *Exp. Neurol.*, **197** (2), 515–520.

58 Rao, J.N. *et al.* (2009) Effect of pseudorepeat rearrangement on α-synuclein misfolding, vesicle binding, and micelle binding. *J. Mol. Biol.*, **390** (3), 516–529.

59 Rott, R. *et al.* (2008) Monoubiquitylation of α-synuclein by seven in absentia homolog (SIAH) promotes its aggregation in dopaminergic cells. *J. Biol. Chem.*, **283** (6), 3316–3328.

60 Shamoto-Nagai, M. *et al.* (2007) In parkinsonian substantia nigra, α-synuclein is modified by acrolein, a lipid-peroxidation product, and accumulates in the dopamine neurons with inhibition of proteasome activity. *J. Neural Transm.*, **114** (12), 1559–1567.

61 Waxman, E.A., Emmer, K.L., and Giasson, B.I. (2010) Residue Glu83 plays a major role in negatively regulating α-synuclein amyloid formation. *Biochem. Biophys. Res. Commun.*, **391** (3), 1415–1420.

62 Bodner, C.R. *et al.* (2010) Differential phospholipid binding of α-synuclein variants implicated in Parkinson's disease revealed by solution NMR spectroscopy. *Biochemistry*, **49** (5), 862–871.

63 Meng, X. et al. (2009) Molecular mechanisms underlying the flavonoid-induced inhibition of α-synuclein fibrillation. *Biochemistry*, **48** (34), 8206–8224.

64 Pham, C.L. et al. (2009) Dopamine and the dopamine oxidation product 5,6-dihydroxyindole promote distinct on-pathway and off-pathway aggregation of α-synuclein in a pH-dependent manner. *J. Mol. Biol.*, **387** (3), 771–785.

65 Lee, H.J., Choi, C., and Lee, S.J. (2002) Membrane-bound α-synuclein has a high aggregation propensity and the ability to seed the aggregation of the cytosolic form. *J. Biol. Chem.*, **277** (1), 671–678.

66 Brown, D.R. (2009) Metal binding to α-synuclein peptides and its contribution to toxicity. *Biochem. Biophys. Res. Commun.*, **380** (2), 377–381.

67 Amijee, H. et al. (2009) Inhibitors of protein aggregation and toxicity. *Biochem. Soc. Trans.*, **37** (Pt 4), 692–696.

68 Bharathi, Indi, S.S., and Rao, K.S. (2007.) Copper- and iron-induced differential fibril formation in α-synuclein: TEM study. *Neurosci. Lett*, **424** (2), 78–82.

69 Cole, N.B. et al. (2002) Lipid droplet binding and oligomerization properties of the Parkinson's disease protein α-synuclein. *J. Biol. Chem.*, **277** (8), 6344–6352.

70 Du, H.N. et al. (2006) Acceleration of α-synuclein aggregation by homologous peptides. *FEBS Lett.*, **580** (15), 3657–3664.

71 Emadi, S. et al. (2007) Isolation of a human single chain antibody fragment against oligomeric α-synuclein that inhibits aggregation and prevents α-synuclein-induced toxicity. *J. Mol. Biol.*, **368** (4), 1132–1144.

72 Grammatopoulos, T.N. et al. (2007) Angiotensin II protects against α-synuclein toxicity and reduces protein aggregation *in vitro*. *Biochem. Biophys. Res. Commun.*, **363** (3), 846–851.

73 Haque, F. et al. (2010) Adsorption of α-synuclein on lipid bilayers: modulating the structure and stability of protein assemblies. *J. Phys. Chem. B*, **114** (11), 4070–4081.

74 Latawiec, D. et al. (2010) Modulation of α-synuclein aggregation by dopamine analogs. *PLoS ONE*, **5** (2), e9234.

75 Pandey, N. et al. (2008) Curcumin inhibits aggregation of α-synuclein. *Acta Neuropathol.*,

76 Rao, J.N., Dua, V., and Ulmer, T.S. (2008) Characterization of α-synuclein interactions with selected aggregation-inhibiting small molecules. *Biochemistry*, **47** (16), 4651–4656.

77 Uversky, V.N., Li, J., and Fink, A.L. (2001) Metal-triggered structural transformations, aggregation, and fibrillation of human α-synuclein. A possible molecular NK between Parkinson's disease and heavy metal exposure. *J. Biol. Chem.*, **276** (47), 44284–44296.

78 Wang, X. et al. (2010) Copper binding regulates intracellular α-synuclein localisation, aggregation and toxicity. *J. Neurochem.*, **113** (3), 704–714.

79 Zhou, R.M. et al. (2009) Molecular interaction of α-synuclein with tubulin influences on the polymerization of microtubule *in vitro* and structure of microtubule in cells. *Mol. Biol. Rep.*,

80 Jiang, M. et al. (2010) Baicalein reduces E46K α-synuclein aggregation *in vitro* and protects cells against E46K α-synuclein toxicity in cell models of familiar Parkinsonism. *J. Neurochem.*, **114** (2), 419–429.

81 Perrin, R.J. et al. (2001) Exposure to long chain polyunsaturated fatty acids triggers rapid multimerization of synucleins. *J. Biol. Chem.*, **276** (45), 41958–41962.

82 Kim, Y.S. et al. (2006) A novel mechanism of interaction between α-synuclein and biological membranes. *J. Mol. Biol.*, **360** (2), 386–397.

83 Bodner, C.R., Dobson, C.M., and Bax, A. (2009) Multiple tight phospholipid-binding modes of α-synuclein revealed by solution NMR spectroscopy. *J. Mol. Biol.*, **390** (4), 775–790.

84 Ferreon, A.C. et al. (2009) Interplay of α-synuclein binding and conformational switching probed by single-molecule fluorescence. *Proc. Natl. Acad. Sci. USA*, **106** (14), 5645–5650.

85 Jo, E. et al. (2004) α-Synuclein–synaptosomal membrane interactions: implications for fibrillogenesis. *Eur. J. Biochem.*, **271** (15), 3180–3189.

86 Bar-On, P. et al. (2008) Statins reduce neuronal α-synuclein aggregation in *in vitro* models of Parkinson's disease. *J. Neurochem.*, **105** (5), 1656–1667.

87 Moore, D.J. et al. (2005) Molecular pathophysiology of Parkinson's disease. *Annu. Rev. Neurosci.*, **28**, 57–87.

88 Franssens, V. et al. (2010) Yeast unfolds the road map toward α-synuclein-induced cell death. *Cell Death Differ.*, **17** (5), 746–753.

89 Cuervo, A.M., Wong, E.S., and Martinez-Vicente, M. (2010) Protein degradation, aggregation, and misfolding. *Mov. Disord.*, **25** (Suppl. 1), S49–S54.

90 Madine, J., Doig, A.J., and Middleton, D.A. (2006) A study of the regional effects of α-synuclein on the organization and stability of phospholipid bilayers. *Biochemistry*, **45** (18), 5783–5792.

91 Jo, E. et al. (2000) α-Synuclein membrane interactions and lipid specificity. *J. Biol. Chem.*, **275** (44), 34328–34334.

92 Giannakis, E. et al. (2008) Dimeric structures of α-synuclein bind preferentially to lipid membranes. *Biochim. Biophys. Acta*, **1778** (4), 1112–1119.

93 Feng, L.R. et al. (2010) α-Synuclein mediates alterations in membrane conductance: a potential role for α-synuclein oligomers in cell vulnerability. *Eur. J. Neurosci.*, **32** (1), 10–17.

94 Devi, L. and Anandatheerthavarada, H.K. (2010) Mitochondrial trafficking of APP and alpha synuclein: relevance to mitochondrial dysfunction in Alzheimer's and Parkinson's diseases. *Biochim. Biophys. Acta*, **1802** (1), 11–19.

95 Mosharov, E.V. et al. (2006) α-Synuclein overexpression increases cytosolic catecholamine concentration. *J. Neurosci.*, **26** (36), 9304–9311.

96 Jang, A. et al. (2010) Non-classical exocytosis of α-synuclein is sensitive to folding states and promoted under stress conditions. *J. Neurochem.*, 1263–1274.

97 Danzer, K.M. et al. (2009) Seeding induced by α-synuclein oligomers provides evidence for spreading of α-synuclein pathology. *J. Neurochem.*, **111** (1), 192–203.

98 Park, J.Y. et al. (2009) On the mechanism of internalization of α-synuclein into microglia: roles of ganglioside GM1 and lipid raft. *J. Neurochem.*, **110** (1), 400–411.

99 Goedert, M., Clavaguera, F., and Tolnay, M. (2010) The propagation of prion-like protein inclusions in neurodegenerative diseases. *Trends Neurosci.*,

100 van Rooijen, B.D., Claessen, M.M., and Subramaniam, V. (2009) Lipid bilayer disruption by oligomeric α-synuclein depends on bilayer charge and accessibility of the hydrophobic core. *Biochim. Biophys. Acta*, **1788** (6), 1271–1278.

101 van Rooijen, B.D., Claessen, M.M., and Subramaniam, V. (2008) Membrane binding of oligomeric α-synuclein depends on bilayer charge and packing. *FEBS Lett.*, **582** (27), 3788–3792.

102 van Rooijen, B.D. et al. (2009) Tryptophan fluorescence reveals structural features of α-synuclein oligomers. *J. Mol. Biol.*, **394** (5), 826–833.

103 Maroteaux, L., Campanelli, J.T., and Scheller, R.H. (1988) Synuclein: a neuron-specific protein localized to the nucleus and presynaptic nerve terminal. *J. Neurosci.*, **8** (8), 2804–2815.

104 Jao, C.C. et al. (2004) Structure of membrane-bound α-synuclein studied by site-directed spin labeling. *Proc. Natl. Acad. Sci. USA*, **101** (22), 8331–8336.

105 Martinez, Z. et al. (2007) GM1 specifically interacts with α-synuclein and inhibits fibrillation. *Biochemistry*, **46** (7), 1868–1877.

106 Kubo, S. et al. (2005) A combinatorial code for the interaction of α-synuclein with membranes. *J. Biol. Chem.*, **280** (36), 31664–31672.

107 Stöckl, M. et al. (2008) α-Synuclein selectively binds to anionic phospholipids embedded in liquid-disordered domains. *J. Mol. Biol.*, **375** (5), 1394–1404.

108 Rhoades, E. et al. (2006) Quantification of α-synuclein binding to lipid vesicles using fluorescence correlation

108 spectroscopy. *Biophys. J.*, **90** (12), 4692–4700.
109 Perrin, R.J. et al. (2000) Interaction of human α-synuclein and Parkinson's disease variants with phospholipids. Structural analysis using site-directed mutagenesis. *J. Biol. Chem.*, **275** (44), 34393–34398.
110 Choi, W. et al. (2004) Mutation E46K increases phospholipid binding and assembly into filaments of human α-synuclein. *FEBS Lett.*, **576** (3), 363–368.
111 Sulzer, D. (2010) Clues to how α-synuclein damages neurons in Parkinson's disease. *Mov. Disord.*, **25** (Suppl. 1), S27–S31.
112 Furukawa, K. et al. (2006) Plasma membrane ion permeability induced by mutant α-synuclein contributes to the degeneration of neural cells. *J. Neurochem.*, **97** (4), 1071–1077.
113 Kim, H.Y. et al. (2009) Structural properties of pore-forming oligomers of α-synuclein. *J. Am. Chem. Soc.*, **131** (47), 17482–17489.
114 Kostka, M. et al. (2008) Single particle characterization of iron-induced pore-forming α-synuclein oligomers. *J. Biol. Chem.*, **283** (16), 10992–11003.
115 Lashuel, H.A. et al. (2002) α-Synuclein, especially the Parkinson's disease-associated mutants, forms pore-like annular and tubular protofibrils. *J. Mol. Biol.*, **322** (5), 1089–1102.
116 Volles, M.J. and Lansbury, P.T. Jr. (2002) Vesicle permeabilization by protofibrillar α-synuclein is sensitive to Parkinson's disease-linked mutations and occurs by a pore-like mechanism. *Biochemistry*, **41** (14), 4595–4602.
117 Di Pasquale, E. et al. (2010) Altered ion channel formation by the Parkinson's-disease-linked E46K mutant of α-synuclein is corrected by GM3 but not by GM1 gangliosides. *J. Mol. Biol.*, **397** (1), 202–218.
118 Kayed, R. et al. (2004) Permeabilization of lipid bilayers is a common conformation-dependent activity of soluble amyloid oligomers in protein misfolding diseases. *J. Biol. Chem.*, **279** (45), 46363–46366.
119 Sokolov, Y. et al. (2006) Soluble amyloid oligomers increase bilayer conductance by altering dielectric structure. *J. Gen. Physiol.*, **128** (6), 637–647.
120 Eftink, M.R. and Ghiron, C.A. (1976) Exposure of tryptophanyl residues in proteins. Quantitative determination by fluorescence quenching studies. *Biochemistry*, **15** (3), 672–680.
121 Iwai, A. et al. (1995) The precursor protein of non-A beta component of Alzheimer's disease amyloid is a presynaptic protein of the central nervous system. *Neuron*, **14** (2), 467–475.
122 Jakes, R., Spillantini, M.G., and Goedert, M. (1994) Identification of two distinct synucleins from human brain. *FEBS Lett.*, **345** (1), 27–32.
123 Davidson, W.S. et al. (1998) Stabilization of α-synuclein secondary structure upon binding to synthetic membranes. *J. Biol. Chem.*, **273** (16), 9443–9449.
124 Kamp, F. and Beyer, K. (2006) Binding of α-synuclein affects the lipid packing in bilayers of small vesicles. *J. Biol. Chem.*, **281** (14), 9251–9259.
125 Beyer, K. (2007) Mechanistic aspects of Parkinson's disease: α-synuclein and the biomembrane. *Cell Biochem. Biophys.*, **47** (2), 285–299.
126 Nuscher, B. et al. (2004) α-Synuclein has a high affinity for packing defects in a bilayer membrane: a thermodynamics study. *J. Biol. Chem.*, **279** (21), 21966–21975.
127 Abeliovich, A. et al. (2000) Mice lacking α-synuclein display functional deficits in the nigrostriatal dopamine system. *Neuron*, **25** (1), 239–252.
128 Nemani, V.M. et al. (2010) Increased expression of α-synuclein reduces neurotransmitter release by inhibiting synaptic vesicle reclustering after endocytosis. *Neuron*, **65** (1), 66–79.
129 Garcia-Reitbock, P. et al. (2010) SNARE protein redistribution and synaptic failure in a transgenic mouse model of Parkinson's disease. *Brain*, **133** (Part 7), 2032–2044.
130 Darios, F. et al. (2010) α-Synuclein sequesters arachidonic acid to modulate SNARE-mediated exocytosis. *EMBO Rep.*, **11** (7), 528–533.

131 Scott, D.A. *et al.* (2010) A pathologic cascade leading to synaptic dysfunction in α-synuclein-induced neurodegeneration. *J. Neurosci.*, **30** (24), 8083–8095.

132 Murphy, D.D. *et al.* (2000) Synucleins are developmentally expressed, and α-synuclein regulates the size of the presynaptic vesicular pool in primary hippocampal neurons. *J. Neurosci.*, **20** (9), 3214–3220.

133 Gorbatyuk, O.S. *et al.* (2010) In *vivo* RNAi-mediated α-synuclein silencing induces nigrostriatal degeneration. *Mol. Ther.*, **18** (8), 1450–1457.

134 Drescher, M. *et al.* (2010) A stable lipid-induced aggregate of α-synuclein. *J. Am. Chem. Soc.*, **132** (12), 4080–4082.

135 Hoyer, W. *et al.* (2004) Rapid self-assembly of α-synuclein observed by *in situ* atomic force microscopy. *J. Mol. Biol.*, **340** (1), 127–139.

136 Kamiyoshihara, T. *et al.* (2007) Observation of multiple intermediates in α-synuclein fibril formation by singular value decomposition analysis. *Biochem. Biophys. Res. Commun.*, **355** (2), 398–403.

137 Chen, L. *et al.* (2007) Oligomeric α-synuclein inhibits tubulin polymerization. *Biochem. Biophys. Res. Commun.*, **356** (3), 548–553.

138 Kim, M. *et al.* (2008) Impairment of microtubule system increases α-synuclein aggregation and toxicity. *Biochem. Biophys. Res. Commun.*, **365** (4), 628–635.

139 Cooper, A.A. *et al.* (2006) α-Synuclein blocks ER-Golgi traffic and Rab1 rescues neuron loss in Parkinson's models. *Science*, **313** (5785), 324–328.

140 Gitler, A.D. *et al.* (2008) The Parkinson's disease protein α-synuclein disrupts cellular Rab homeostasis. *Proc. Natl. Acad. Sci. USA*, **105** (1), 145–150.

141 Leong, S.L. *et al.* (2009) Formation of dopamine-mediated α-synuclein-soluble oligomers requires methionine oxidation. *Free Radic. Biol. Med.*, **46** (10), 1328–1337.

142 Bisaglia, M. *et al.* (2010) Dopamine quinones interact with α-synuclein to form unstructured adducts. *Biochem. Biophys. Res. Commun.*, **394** (2), 424–428.

143 Follmer, C. *et al.* (2007) Dopamine affects the stability, hydration, and packing of protofibrils and fibrils of the wild type and variants of α-synuclein. *Biochemistry*, **46** (2), 472–482.

144 Outeiro, T.F. *et al.* (2009) Dopamine-induced conformational changes in α-synuclein. *PLoS ONE*, **4** (9), e6906.

145 Rochet, J.C. *et al.* (2004) Interactions among α-synuclein, dopamine, and biomembranes: some clues for understanding neurodegeneration in Parkinson's disease. *J. Mol. Neurosci.*, **23** (1–2), 23–34.

146 Yamakawa, K. *et al.* (2010) Dopamine facilitates α-synuclein oligomerization in human neuroblastoma SH-SY5Y cells. *Biochem. Biophys. Res. Commun.*, **391** (1), 129–134.

147 Ardail, D. *et al.* (1990) Mitochondrial contact sites. Lipid composition and dynamics. *J. Biol. Chem.*, **265** (31), 18797–18802.

# 3
# Surfactants and Alcohols as Inducers of Protein Amyloid: Aggregation Chaperones or Membrane Simulators?
*Daniel E. Otzen*

## 3.1
### Introduction

Studies of the dynamics between aggregation-prone proteins/peptides and biological membranes have been under intense scrutiny for several decades [1], stimulated by the reports that the β peptide forms ion channels in planar bilayers [2] and the toxic species in the Aβ aggregation cascade are soluble non-fibrillated oligomers. We need to understand both the biophysical mechanisms that govern the formation and decay of these species in the context of the overall fibrillation process and also the biological consequences that follow. This chapter reviews the many biophysical attempts to recreate the aggregation process in a simple and well-controlled environment that simultaneously allows for detailed structural and biophysical analysis while *to some extent* mimicking the biological membrane. The italics emphasize that the level of mimicry should be taken *cum grano salis*. Two different environments consistently yield β-sheet-rich structures with either prefibrillar or fibril-like properties, namely anionic surfactants (ASs), predominantly sodium dodecyl sulfate (SDS, and fluorinated organic solvents (FOSs), mainly trifluoroethanol (TFE). Both of these additives strengthen protein–protein hydrogen bonds in addition to solvating hydrophobic side chains. The tug-of-war between these two forces typically leads to an intermediate AS or FOS concentration range where intermolecular hydrogen bonding between β-sheet structures dominates, whereas higher concentrations give rise to intramolecular hydrogen bonding in α-helices. It is debatable whether these conditions correspond directly to membrane conditions, but there are interesting parallels. Aggregation in ASs and FOSs also provides detailed insights into the conformational properties of proteins which emphasize the marvelous – although potentially detrimental – structural diversity provided by a given polypeptide chain. In terms of applications, these additives, particularly SDS, also lead to a much more reproducible aggregation behavior, and this makes both detailed studies and large-scale screening efforts much more feasible.

First, the effects of ASs on protein aggregation are presented and discussed, followed by a description of a detailed structural and energetic model for how surfactants modulate the aggregation process which has recently emerged from the author's group in a collaborative effort. The emerging topic of cytotoxic protein–oleic acid complexes, which provide an unexpected biological twist to an otherwise synthetic topic, is then considered, followed by a description of the aggregation of proteins in the presence of FOSs. Finally, aspects of the aggregation of proteins on membranes which are relevant in this context are summarized.

## 3.2
## Aggregation in the Presence of Surfactants

### 3.2.1
### General Aspects of Protein–Surfactant Interactions

Surfactants are surface-active compounds which consist of a hydrophilic and a hydrophobic group [3]. Based on the charge of the hydrophilic group, they are grouped as nonionic, anionic, cationic, or zwitterionic. Above a certain concentration, known as the critical micelle concentration (cmc), individual surfactant molecules will associate to form micelles by burying their hydrophobic groups in a loosely packed core [4], which leaves the hydrophilic groups in contact with water. Further increases in surfactant concentration feed into the micelle population so that the monomeric surfactant concentration remains constant around the cmc. These two concentration regimes, namely monomers below the cmc and micelles in dynamic equilibrium with monomers above the cmc, determine how surfactants interact with proteins. Micelles form because of the hydrophobic effect, that is, the entropically favorable liberation of bound water, rather than any enthalpic stabilization (due to the very loose and unspecific interactions in the micellar core). The upshot is that alternative hydrophobic interactions, for example, those that can be made with suitable binding sites on a protein, may be stable enough to induce the formation of higher order surfactant assemblies, even below the cmc. In practice, only ionic surfactants have this option available, since nonionic surfactants have high mutual affinity and generally only interact with proteins with preformed hydrophobic binding sites, for example, membrane proteins or certain semi-unfolded proteins such as α-lactalbumin [5]. Repulsion between similarly charged headgroups raises the cmc values of ionic surfactants relative to their nonionic counterparts. Thus. despite an identical alkyl chain length, the cmc values for SDS and dodecyl maltoside in water are 7.0 and 0.17 mM, respectively. However, this very repulsion facilitates strong interactions with proteins; binding to oppositely charged side chains will not only provide binding energy but also reduce charge repulsion. Already in 1945, SDS was shown to bind serum albumin (that stalwart of protein–surfactant studies) through strong electrostatic interactions, so that the binding stoichiometry is determined by the number of cationic protein groups [6]. A simple increase in the number of Lys residues at the expense of Glu residues greatly improves a peptide's SDS binding affinity [7]. The geometry of neighboring hydrophobic patches on the protein, which provide binding sites for the alkyl chain, will then determine the precise position of the bound SDS molecules [8].

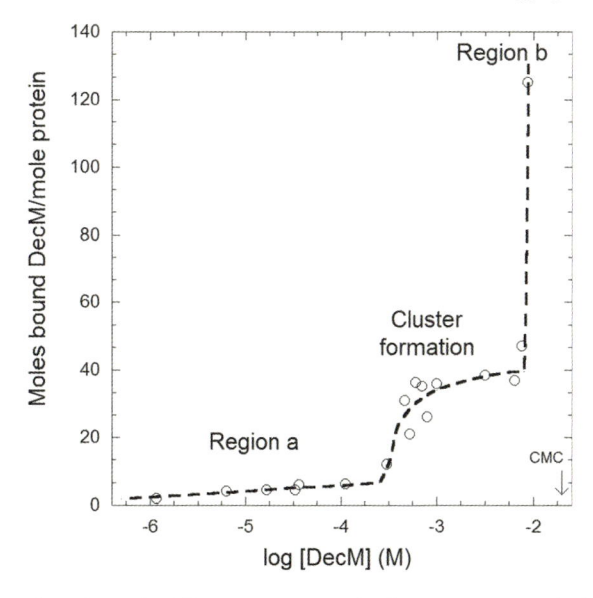

**Figure 3.1** The different stages in SDS binding to serum albumin. Initially, individual SDS molecules bind in a non-cooperative fashion (region a), followed by a more cooperative process in which clusters are formed, before massive uptake of SDS into large micellar structures takes place (region b). Adapted from [3].

Tanford and others demonstrated early on that SDS binding is a multi-step process, starting with binding of a small number of individual SDS molecules, followed by a more cooperative binding of SDS molecules before complete denaturation and bulk micelle formation (Figure 3.1). As more SDS binds, greater structural changes occur on the protein and this in turn provides more binding sites for SDS to bind. Reynolds and Tanford established, in their pioneering studies with equilibrium dialysis experiments [9], that complexes of around 0.4 g of SDS per gram of protein (complex A) form around 0.5–0.8 mM SDS, whereas higher SDS concentrations lead to the well-known ratio of 1.4 g of SDS to 1 g of protein or about one SDS per two peptide residues (complex B). The ability of SDS either to stimulate or to inhibit protein aggregation is based on these two different types of complexes. As we shall see in Section 3.2.4.5–3.2.4.6, the small number of SDS molecules bound in complex A may drive protein aggregation by combining SDS from different protein molecules to form a shared micelle, effectively exploiting micellization to drive protein aggregation.

Cluster formation below the cmc can be studied with the fluorescent probe pyrene, whose fluorescence spectrum is very different in an aqueous and a hydrophobic environment, such as that provided by surfactant clusters or bulk micelles [10]. Pyrene fluorescence typically shows a relatively sharp transition as the surfactant reaches micellar conditions. Such a transition will occur at much lower SDS concentrations (and may occur in several steps) if cluster-inducing proteins are present [11–13]. Pyrene titrations should be performed routinely whenever proteins are shown to interact with SDS below the bulk cmc in order to test whether such clusters are formed or other mechanisms are at work.

## 3.2.2
### Effect of Surfactants on Protein Structure

Although sub- and super-cmc concentration regimes have significantly different effects on protein structure, they are influenced by the same dual effect. First, the hydrophobic groups of SDS can interact with hydrophobic groups on the protein, thus solvating the protein efficiently. Second, these groups also displace water from the protein surface, removing potential hydrogen bonding partners. Hydrophobic groups are very poor hydrogen bonding partners by themselves, and this drives protein hydrogen bonding groups (carbonyl oxygen and amide nitrogen) to form inter- or intra-protein hydrogen bonds, either as α-helices or as β-sheets. The same phenomenon stabilizes hydrogen bonds buried in a globular protein's hydrophobic core [14] and formed within membrane proteins immersed in the membrane bilayer [15]. Whether one or the other secondary structure prevails depends on the balance between these effects and also intrinsic properties of the protein. A lucid exposition of this was provided by Goto and co-workers using the K3 peptide (a fragment of $\beta_2$-microglobulin) [16]. K3 aggregation shows a bell-shaped dependence on SDS concentration, which reflects a delicate balance between decreasing inter-protein hydrophobic interactions and increasing intra-protein polar interactions.

Micellar SDS generally stabilizes α-helical structure [17–20]. Here local hydrogen bonding dominates because the high amount of SDS efficiently solvates and dilutes out the peptide, thus destabilizing intermolecular β-sheet contacts. Over the years, many different models for the structural arrangement of proteins in micellar SDS have been proposed: a necklace model with micelle-like clusters on the polypeptide chain based on electrophoretic mobility measurements [21], a rod-like particle or "prolate ellipsoid" model based on viscosity measurements [22], and a deformable prolate ellipsoid model based on transient electrical birefringence [23]. The last model has been modified further by taking into account the enhanced tendency of Lys, Arg, and His residues to become α-helical in the presence of detergent. This would rationalize the large variations in the degree of different proteins' α-helicity. This has led to a model where the SDS–protein complex is flexible rather than a rigid rod and has an overall shape of a random coil but contains α-helical structure [24].

Below the cmc, in contrast, SDS is present in limiting amounts and cannot coat the protein comprehensively. Here SDS is able to induce β-sheet structure, stabilized by global hydrogen bonding (i.e., between residues far apart in the sequence), in a wide variety of peptides including the Alzheimer peptide Aβ [25–28]. Whether hydrogen bonding is inter- or intramolecular will depend on the conformational properties of the protein in question; the same protein can pass through a number of rearrangements over longer time scales. Below the cmc, many all-α-helix proteins such as acyl coenzyme A binding protein (ACBP) [12] and myoglobin [29] remain stably α-helical below the cmc, whereas intrinsically disordered proteins such as α-synuclein (αSN) can initially possess α-helix structures in the monomer but subsequently collapse to a β-sheet-rich aggregate over several hours [30].

### 3.2.3
### Stoichiometry of SDS Binding

Before discussing how SDS is able to induce protein aggregation, we need to clarify how to refer to SDS concentrations in a protein–SDS system. In a solution containing only SDS, water, and buffer, the only species present are monomeric and micellar SDS. The distribution of SDS between the two states is governed by the cmc value, which in turn declines sharply with increase in ionic strength [31] due to reduced headgroup repulsion. Changing the ionic strength therefore changes the concentration range of monomeric SDS, that is, its "window of opportunity" for interactions with proteins. When we also introduce proteins into the solution, SDS is able to occupy a third state, namely in complex with protein. Since this reduces the concentration of free (unbound) monomeric SDS, it will effectively increase the cmc. If we start to titrate SDS into a protein solution, the binding relationship may be simply quantified at any given SDS concentration $[SDS]_{total}$ by the following equation [32]:

$$[SDS]_{total} = [SDS]_{free} + n_{SDS}[P] \tag{3.1}$$

where $n_{SDS}$ is the number of SDS molecules bound per protein molecule and [P] is the concentration of protein. Hence we can determine $[SDS]_{free}$ and $n_{SDS}$ by monitoring the distinct stages of binding of SDS to protein at different protein concentrations, identifying key transition points, and plotting $[SDS]_{total}$ versus [P] at each of these points. Isothermal titration calorimetry (ITC) is the method of choice to measure the different stages of binding, since it monitors all binding reactions irrespective of structural changes, provided that there is an associated heat flow (Figure 3.2). This makes it superior to Trp fluorescence [29] and circular dichroism (CD) spectroscopy. This linear relationship described in Eq. (3.1) has been reported for a significant number of proteins, making it a general phenomenon [11–13, 29, 30, 32]. The corollary is that protein concentrations higher than a few micromolar will therefore significantly affect the concentration of free SDS and therefore increase the apparent cmc. Claims that a protein aggregates "close to the cmc" should be judged in this light.

### 3.2.4
### Aggregation of Proteins by SDS

Reports on the ability of SDS to aggregate proteins stretch back at last 30 years. Hepatitis B surface antigen polypeptides were reported in 1980 to form doublets and triplets on SDS-PAGE (polyacrylamide gel electrophoresis), although the sample buffer concentration (70 mM) was well above the cmc [33]. In 1981, SDS was reported to precipitate myelin basic protein under conditions where the net charge of the

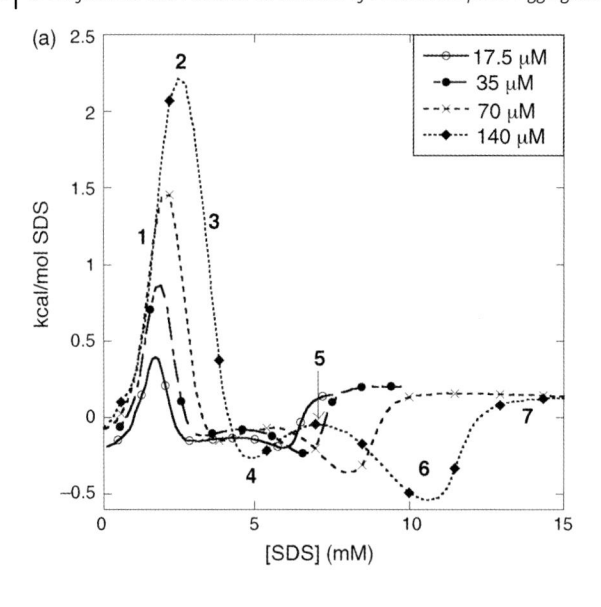

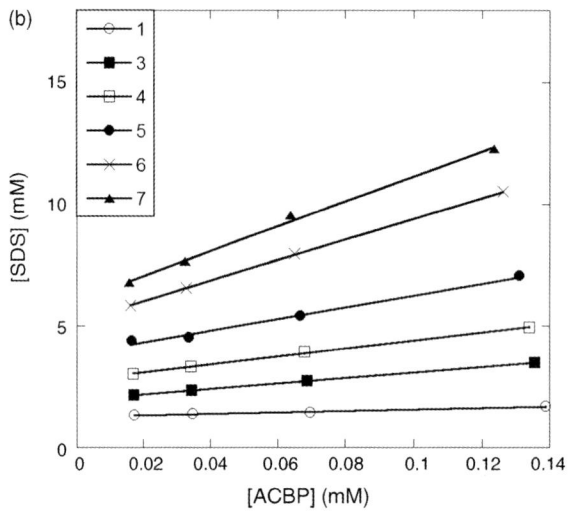

**Figure 3.2** (a) ITC enthalpograms for the titration of SDS (99 mM) into four different ACBP solutions (concentrations indicated in the legend). (b) Total concentrations of SDS, at the different transition points indicated in (a), plotted as a function of ACBP concentration. The slope of each plot gives the number of SDS molecules bound per ACBP molecule at that particular transition, and the intercept provides the free SDS concentration at that transition (cf. Eq. (3.1)). Adapted from [12].

complex changed from +18 to −9 [34]. For the same reason, zwitterionic surfactants had a much weaker effect [35]. Precipitation by SDS was presciently suggested to occur through noncovalent cross-linking through hydrophobic interactions between surfactant molecules electrostatically bound to different protein molecules, in effect shared micelles, although the structure of the complex was not reported [34]. In a more physiological context, in 1986 SDS was substituted for heparin to interact with collagen [36]. Collagen was known to fibrillate in the presence of anionic sulfate-containing heparin and other matrix components, which in this way could regulate the fibril width. The authors made an observation that was to form the template for many other SDS-induced protein aggregates, namely that aggregation of collagen peaked at 0.3–0.4 mM SDS and vanished above 0.5 mM SDS (very close to the cmc in a buffer of 30 mM phosphate and 200 mM NaCl), although the fibrils showed signs of decreasing order at 0.4 mM SDS. The nine-residue peptide bradykinin similarly aggregates with SDS in stoichiometries of one peptide to two SDS, consistent with one SDS binding to each of two Arg residues, one at either end [37].

Generally, a decrease in pH increases protein affinity for SDS, since protonation of the acidic side chains of Asp and Glu will reduce the electrostatic repulsion of SDS. Whereas micellar concentrations of S6 will still keep the proteins soluble at all pH values, a reduction in pH to below 5.5 combined with SDS titration leads to a precipitation phenomenon which peaks around 0.5–1.0 mM SDS and then slowly disappears around the cmc [13]. This is seen for a number of different proteins but is a very rapid process that does not involve the formation of fibrillar aggregates; rather, it probably represents a charge neutralization phenomenon where SDS binds strongly enough to let even low concentrations of SDS form clusters that can be shared very efficiently among several protein molecules, thus building up large networks that precipitate out. Interestingly, the phenomenon starts already around pH 5.5, which is well above the p$K_a$ of Asp and Glu in aqueous solution (3.5–4.0). This may be ascribed to the increase in p$K_a$ induced by the SDS environment, which will favor the protonated form due to the penalty of burying a charged group in a hydrophobic environment [38]. Provided that there is a sufficiently large number of cationic residues, aggregation can occur at neutral pH, as seen for the 17-residue antimicrobial peptide novispirin, which has eight cationic residues and no acidic side chains and precipitates massively in 0.5–2 mM SDS, but is solubilized in an α-helical form at higher SDS concentrations [39].

### 3.2.4.1 Aβ

We start with a detailed description of the SDS-induced aggregation of the Alzheimer peptide Aβ. Controllable aggregation of the Aβ peptide has been a major target since it was shown that diffusible Aβ oligomers (known as ADDLs) kill neurons at nanomolar concentrations [1] and inhibit hippocampal long-term potentiation [40]. Aβ naturally oligomerizes when expressed in CHO cells [41]; oligomers are also obtained in the soluble fraction when the peptide is incubated, for example, at 37 °C for 16 h [42], although yields are typically low. When titrated with SDS, Aβ(1–40) goes from random coil to β-sheet in a simple two-state transition with an isodichroic point, and then transforms into an α-helix above the cmc [28]. However, optimal conditions for oligomerization are tricky to find: nonionic, zwitterionic, and anionic detergents all induce Aβ fibrillation (rather than oligomerization) to essentially the same extent at concentrations corresponding to their cmc (although this is in the presence of 50 µM Aβ, so the cmc is in practice higher) [43]. At these concentrations, Aβ adopts a helical structure in SDS surrounded by solvent-exposed regions [44–50] and is mainly located at the water–micelle interface [49], although the C-terminus is closer to the interior [51].

Large-scale preparation of Aβ oligomers through SDS has been reported by two groups. Using size-exclusion chromatography (SEC) to obtain strictly monomeric Aβ as the starting species, Rosenberry's group found that Aβ(1–40) slowly forms soluble but fibrillar β-sheet aggregates over 2 weeks in 2 mM SDS (which at 100 µM Aβ probably contains only around 1 mM free SDS) [52], whereas only 0.5 mM SDS is required to induce the two residue longer peptide Aβ(1–42) to form insoluble fibrillar structures [53]. Unlike soluble protofibrils of Aβ prepared in the absence of SDS, the SDS-induced aggregates do not seed Aβ fibril growth despite a predominance of β-sheet structure, but are toxic to neuronal cell cultures [54]. Interestingly, Aβ(1–42) incubated in 2 mM SDS forms dimer–tetramer mixtures that remain soluble and do not form fibrils. These species represent a non-fibrillar aggregation pathway and show that varying SDS concentrations can terminate the aggregation process at an early stage. In this way, SDS resembles small-molecule inhibitors of fibrillation, many of which trap Aβ in an oligomeric state [55].

Barghorn's group obtained broadly similar results. Hexafluoroisopropanol (HFIP) was used to monomerize Aβ, which upon subsequent incubation with 7 mM SDS, followed by fourfold dilution, formed stable and SEC-certified homogeneous globulomers (globular oligomers) which were estimated to consist of ∼12 Aβ monomers [56]. The globulomers show biological activity, blocking long-term potentiation in rat hippocampal slices [56]. Solid-state NMR spectra indicate mixed parallel and antiparallel β-sheet structure, in contrast to the all-parallel mature fibrils [57]. Their stability, unreactivity towards thioflavin T (ThT) and Congo Red (which bind bona fide amyloids), inability to seed Aβ fibril growth, and resistance to aggregation inhibitors define them as the product of a separate aggregation pathway [58]. Protease digestion reveals a well-defined and well-protected C-terminal core (residues 24–42) whereas the N-terminus (1–19∼23) is more solvent exposed, similar to the structure of Aβ in SDS micelles [28]. The authors speculated that a conformational switch stabilized by SDS early in the aggregation process directs Aβ to soluble globulomers rather than fibrils (Figure 3.3). Low millimolar concentrations of lauric, oleic, and arachidonic acid also stimulated globulomer formation, indicating that numerous amphiphiles can stabilize these conformations. However, it should be noted that these concentrations are many orders of magnitude above the typical physiological concentrations of fatty acids (∼15 nM [59]). Hence the amphiphile approach here mainly provides a very convenient tool to control the formation of a biologically active species but does not in itself represent physiological conditions.

An interesting twist to the induction of Aβ fibrillar structure by surfactants has recently been provided by Lendel et al. [60]. The dye Congo Red has long been known to reduce the toxicity of various prefibrillar protein aggregates, including Aβ [61]. This

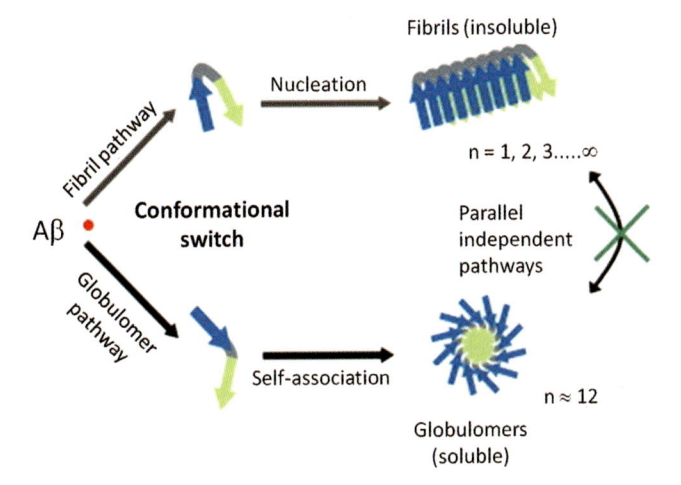

**Figure 3.3** Independent pathway of Aβ$_{1-42}$ oligomer formation in the presence of SDS. Conventional Aβ fibrils are formed in the absence of SDS when the monomer forms nucleation seeds which polymerize to protofibrils and mature Aβ fibrils. In the globulomer pathway, promoted by submicellar concentrations of SDS, Aβ multimerizes to structures with distinct Aβ numbers. This is hypothesized to form a well-defined core consisting of the hydrophobic C-terminal region, while the more hydrophilic N-terminal region is surface exposed. Adapted from [58].

appears to be linked to the molecule's ability to form micellar structures on its own [62]. Depending on the conditions, these micellar structures can either promote monomeric protein structures (by binding to parts of the protein that would otherwise be involved in intermolecular contacts) [62] or stimulate the formation of fibrillar structures [60] by the same type of mechanism as SDS. The latter appears to be the case for Congo Red *vis-à-vis* Aβ and it may just reflect serendipitous binding preferences that the aggregates formed in this way do not stabilize as cytotoxic oligomers but rather proceed to bona fide fibrils. In these types of experiments, the concentration of the aggregation-modulating compound is critical since that will influence its ability to form micellar structures or clusters on the protein surface. The cmc of Congo Red is around 5 µM in buffer, but binding to equimolar concentrations of Aβ will obviously affect this value.

Cationic surfactants generally interact more weakly with proteins than anionic surfactants and the opposite charge means that the pH profiles of two classes of surfactants with regard to protein interactions will be approximately inverted [38]. Nevertheless, the same overall type of interactions are observed, and this is illustrated nicely for Aβ, where alkyltrimethylammonium bromides stimulate aggregation below their cmc values but inhibit them above it [63]. This dual behavior also observed for the cationic gemini surfactants, which have twinned headgroups and hydrophobic tails linked by a spacer and consequently significantly lower cmc values than their monomeric counterparts [64]. High concentrations of gemini surfactants also induce the formation of amorphous aggregates, presumably by combining simple neutralization of Aβ's negative charge with solubilization to form α-helical structures which cannot associate to form cross-β structures [64]. Aβ is net negatively charged at neutral pH, and the combination of electrostatic complementarity and low cmc also allows gemini surfactant micelles, in contrast to single-chain cationic surfactants and anionic surfactants, to dissolve Aβ fibrils to nanoscopic aggregates [65].

### 3.2.4.2 β$_2$-Microglobulin and β$_2$-Glycoprotein I

The elegant and extensive work by Naiki, Goto, and co-workers on the aggregation of β$_2$-microglobulin (β2m) provides a more physiologically relevant example of the use of SDS and other amphiphiles to induce protein aggregation. β2m is the major component of amyloid deposits formed in the carpal tunnel of hemodialysis patients. Although the protein can form amyloid at low pH or in the presence of FOSs, β2m seeds are not elongated by monomeric β2m under physiological conditions [66] and slowly dissociate on their own [67]. However, over a narrow range of concentrations, SDS actually accelerates fibril seed extension and stabilizes these fibrils [68]. The effect peaks at 0.5 mM SDS, just below the cmc (0.67 mM), whereas higher concentrations disaggregate the fibrils to α-helical structures. The structural rearrangements needed for this to occur are unclear, but negative charge is crucial since other surfactant classes have no effect. The same group showed that negatively charged (but not zwitterionic) lysophospholipids induced not only the extension of β2m amyloid fibrils by monomers but also the formation of β2m amyloid fibrils from the β2m monomer at neutral pH. The mechanism involves partial unfolding of the compact structure of β2m to an amyloidogenic conformer in addition to stabilization of the extended fibrils [69]. This is physiologically relevant, since hemodialysis patients show increased levels of lysolipids [70]. In addition, the combination of hemodialysis with heparin treatment leads to a decrease in albumin [71] and a concomitant rise in free fatty acids from $\sim$15 nM to $\sim$1 μM [59]. Remarkably, these fatty acids induce the extension of β2m fibrils at concentrations above their cmc, also by partially unfolding β2m [72]. Albumin abolishes this extension, showing that free fatty acids are required. The structural similarity of free fatty acids (negative headgroup and alkyl chain) to that of SDS indicates a similar mechanism. As the authors did not measure the cmc in the presence of β2m, bulk micelles may in fact have been absent during the elongation step due to a high level of sequestration of free fatty acids by β2m. Nevertheless, this work is an excellent example of the physiological relevance of protein–amphiphile interactions in promoting unwanted aggregation.

Goto and co-workers continued in a similar vein with the protein β$_2$-glycoprotein I (β$_2$-GPI), which is an important co-factor in antiphospholipid syndrome (aLPS), helping the associated antibody bind phospholipid. β$_2$-GPI aggregates in 0.5 mM SDS (well below the cmc of 5.2 mM); similar effects are seen at 50 μM monooleoylphosphatidylserine and other anionic lipids [73], well within the physiological concentration range. The increased efficiency of the lysolipids compared with SDS may relate to their longer chain length, which will increase the binding affinity towards proteins and also the ability to form clusters on the protein surface. The aLPS antibody is bivalent, so lipids may help the antibody–$\beta_2$-GPI–lipid complex form large clusters on the membrane, enhancing the physiological response.

### 3.2.4.3 Tau Protein

In Alzheimer's disease, tau protein forms neurofibrillary tangles. Many agents can induce fibrillation of recombinant tau *in vitro*, including phosphotransferases, polyanionic compounds, fatty acids, and alkyl sulfate surfactants, which act as anionic condensing agents [74] to stabilize a fibrillation-competent intermediate state. Fibrillation requires an alkyl chain length of at least 12 carbons and a negative charge consisting of carboxylate, sulfonate, or sulfate moieties [75]. Lowering the amount of anionic amphiphiles in mixed systems to below ~50% abolished fibrillation, probably because of insufficient charge density of the bound clusters. The authors ascribed the lowering of surfactant cmc by tau to detergent aggregates with the same free energies of formation as free micelles, and therefore suggested that "tau fibrillization is proportional to the concentration of anionic detergent or fatty acid micelles." However, it must be noted that free micelles do *not* promote tau fibrillation, since fibrillation declines markedly above the apparent cmc. Rather, it is the surfactant clusters formed below the bulk cmc in the presence of tau (and which are stabilized by the same types of forces that stabilize neat micelles) that most likely drive association of tau. Thus different types of hydrophobic tails will show different clustering properties and therefore probably affect the details of tau aggregation. Consistent with this, the length of the hydrophobic tail affected filament morphology, arachidonic acid producing more and shorter filaments than alkyl sulfates [75].

### 3.2.4.4 Prion Protein

Prion proteins (PrPs) are the central components in the etiology of fatal neurodegenerative disorders such as sheep scrapie, human Creutzfeldt–Jakob disease (CJD), and bovine spongiform encephalopathy (BSE) [76]. Scrapie prions were originally purified as rod-shaped particles, which dissolved in micellar concentrations of SDS [77]. Although the cellular basis of their toxicity remains unknown, it may be linked to their membrane interactions. Whereas PrPs are membrane-anchored through a lipid group, recombinant PrPs without this anchor can still bind to membranes provided that acidic groups are present [78]. Prion diseases are thought to be propagated by the conversion of the benign cellular $PrP^C$ to the pathogenic $PrP^{Sc}$, catalyzed by existing $PrP^{Sc}$ or induced by other means [76]. It remains very difficult to produce infectious $PrP^{Sc}$ particles from $PrP^C$ *in vitro*. Although micromolar amounts of sarkosyl or SDS can convert $PrP^C$ to β-sheet-rich aggregates, these lack the scrapie-associated protease resistance, morphology, and infectivity [79]. However, these conformational changes give insight into the protein's conformational versatility. PrP(90–231) forms α-helical structures above 2 mM SDS (which is still well below the cmc), but the situation is more complex at lower concentrations [80]. Riesner and co-workers used analytical ultracentrifugation and methylene blue assays for stoichiometric measurements, showing that the complex containing 0.5 g of SDS per protein or 31 SDS molecules per prion (corresponding to complex A in the Tanford binding scheme [9]) is partially denatured, α-helix rich and contains monomeric PrP. Gradual removal of SDS leads to α-helical dimers at 21 SDS:PrP, β-structured oligomers at 19–20 SDS:PrP, and finally β-structured multimers with 5 SDS:PrP. The multimers aggregate irreversibly. Remarkably, the bacterial chaperone GroEL induces similar effects [80]. The hydrophobic surface area within GroEL's central cavity can be varied by structural rearrangements. Thus the limited number of hydrophobic contacts provided by low stoichiometries of SDS, which appears instrumental in triggering PrP conversion, may have direct parallels to the action of a biological macromolecule.

### 3.2.4.5 Acyl CoA Binding Protein (ACBP)

The 88-residue four helical bundle ACBP does not aggregate in SDS. However, we have recently obtained some intriguing insights into how SDS may induce association of proteins [12]. ACBP retains a characteristic helix-rich CD spectrum at all SDS concentrations, although the helices appear to dissociate. Hence there are no options for intermolecular β-sheet formation. The interaction with SDS has been characterized by a combination of Trp fluorescence and anisotropy (changes in tertiary structure and the polarity and immobilization of the aromatic residues), pyrene fluorescence (cluster formation), CD (secondary structure), ITC, and gel permeation chromatography (stoichiometry of SDS binding at different stages), and small-angle X-ray scattering (SAXS) to elucidate the low-resolution structures of the different species. This has led to a model with four distinct transitions (Figure 3.4). Initially, ACBP binds 1–3 SDS molecules while retaining its native structure, probably by simple electrostatic interactions promoted by the hydrophobic ligand-binding site. In

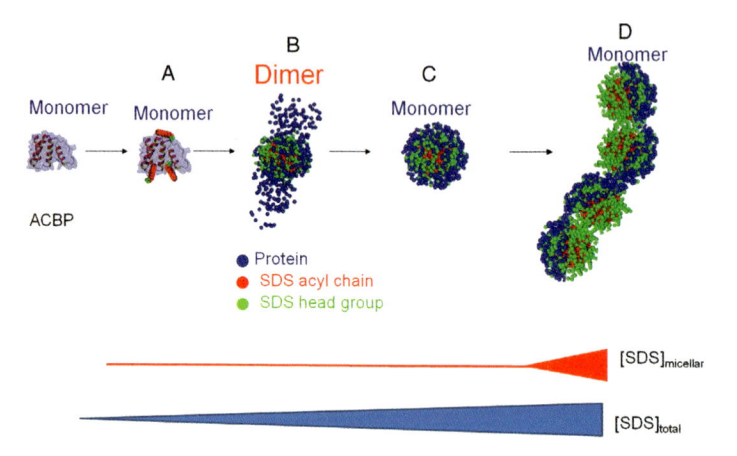

**Figure 3.4** Schematic representation of the different stages of ACBP denaturation. In stage A, ACBP binds between one and three SDS molecules without losing native structure. Stage B involves the formation of a decorated micelle of 37 SDS molecules that binds two ACBP molecules. Further binding of SDS to a total of 40 in stage C leads to monomeric ACBP with a shell-like structure of SDS. The structure presented in stage D is speculative, but it represents the "beads-on-a-string" model that has been proposed for protein interactions with SDS micelles above the cmc. Adapted from [12].

the second stage, 18–19 SDS molecules bind each ACBP molecule (0.4 g of SDS per gram of protein as in complex A [9]), accompanied by a strongly endothermic (denaturing) reaction that is caused by loss of tertiary structure, increase in Trp tumbling, formation of SDS clusters, and changes in secondary structure. The SAXS data indicate the formation of a decorated micelle of $\sim$37 SDS molecules. The 18–19:1 SDS:ACBP stoichiometry dictates that this micelle contains two ACBP molecules. Part of ACBP is bound to the decorated micelle and part protrudes into the solvent. A third stage leads to the uptake of enough SDS ($\sim$42 per ACBP) to allow a micelle to form on a single ACBP molecule, and in a fourth stage 56–60 SDS molecules are bound per ACBP molecule to form the characteristic micellar complex often modeled as beads-on-a-string.

The central observation is that a shared micelle leads to ACBP dimerization. Dimerization is driven by unsatisfied "hydrophobic bonds" that promote intermolecular contacts between protein molecules. At higher SDS concentrations, hydrophobic bonds can be satisfied internally on a single protein molecule and SDS reverts to its role as an agent of dispersion. The solvent-exposed disordered regions of ACBP outside the micellar region could in principle provide anchor points to other proteins, since conformational flexibility (in combination with a certain degree of structure) is a common prerequisite for protein aggregation [81, 82]. ACBP does not aggregate further, since the residues in question do not have a sufficiently strong propensity to form β-sheet structures. However, this is the case for αSN, the topic of the next section.

### 3.2.4.6 α-Synuclein (αSN)

αSN is the main constituent of cytoplasmic inclusions termed Lewy bodies that occur in the dopaminergic neurons of Parkinson's disease patients [83]. αSN is natively unfolded and fibrillates readily if somewhat unpredictably under physiological conditions [84]. The prefibrillar aggregates of the protein permeabilize synthetic membranes [85] and are toxic to neuronal cells [86]. αSN–amphiphile interactions are physiologically highly relevant, as the protein appears to be involved in vesicular transport [87]. This must involve the amphipathic N-terminal region, which binds to anionic lipid vesicles [88], inducing α-helical structure and inhibiting fibrillation [89]. Similarly, SDS micelles (50 mM SDS with 0.3 mM αSN) induce a helical structure where three distinct helices are separated by flexible stretches [90]. In these complexes, the C-terminal helix (residues 88–96), which is part of the NAC region with the highest amyloid propensity (residues 61–95), is rather deeply buried in the micelle. This screens it from aggregation. CD reveals at least three structural transitions in SDS [30, 91]. Above 0.5 mM SDS, there is a helical "overshoot" to a "most folded" state, which is completed when bulk micelles start to form (1.5–2 mM SDS). Around 4–5 mM, an intermediate folded state with slightly reduced helicity is formed, and this finally gives way to another folded state above 100 mM SDS. According to single-molecule fluorescence studies [92], the high-concentration state may correspond to an extended helix (as opposed to the broken helix state seen below the cmc). Such a state can also be induced in stacked bilayers and in phospholipid vesicles [93], indicating that different vesicle conditions may switch between broken and extended helices.

The aggregation properties of SDS have been ascribed to both induction of α-helical structure [91, 92, 94] and scaffolding of αSN nucleation by micelles [95]. Ahmad et al. disproved both theories, reporting a fibrillogenic ensemble formed in the very early stages of the first structural transition (0.5–0.75 mM SDS with 1 mg ml$^{-1}$ αSN), characterized by enhanced hydrophobic exposure and partially helical conformations. In contrast, the species formed at 2 mM SDS has a less accessible hydrophobic surface and a high helical content [96]. This makes it clear that an increase in α-helix content decreases the ability of αSN to fibrillate, so that aggregation of αSN only occurs below the cmc. The fastest aggregation rates occur around the midpoint of the first transition from disordered to α-helical secondary structure [97]. Under these conditions, pulsed-field gradient diffusion NMR spectra indicate that the proteins have same size as an SDS micelle [97], suggesting for simple stoichiometric reasons that multiple proteins are bound to each micelle under these conditions.

When we study αSN–SDS interactions with the same battery of techniques as for ACBP, we conclude the following. The fibrillogenic state (the accumulation of which peaks around 0.5 mM SDS) binds around 12 SDS molecules per αSN molecule [30] to form a distinct SDS cluster. In these clusters, αSN initially assumes a mixture of α-helices and random coil, but after a lag time of only about 30 min the complexes start to form ThT-binding aggregates that consist of a mixture of α-helices and β-sheets. SAXS studies revealed that the protein starts out as a tetrameric protein–SDS complex with ∼48 shared SDS molecules per micelle, similar to the ∼37 SDS molecules linking the slightly smaller ACBP molecules to dimers. The subsequent growth can be modeled as the formation of beads-on-a-string where the beads consist of a core of elongated micelles, stabilized by a coat of proteins on the micelle shell (Figure 3.5), probably in an α-helical conformation, and with individual beads linked by aggregated protein, presumably in a cross-β conformation. The number of proteins per micelle rises from the initial four to a stable level of around six.

The number of αSN molecules per molecule declines markedly in the four subsequent transitions documented by SDS. The next two species have 32 and 36 SDS molecules per αSN molecule, respectively, and consequently about two αSNs per micelle, with greatly reduced aggregation propensity. The final two species with 55 and 73 SDS per αSN both show a ∼1:1 αSN:micelle stoichiometry and do not form ThT-binding complexes. They highlight a number of important points:

- SDS only stimulates protein aggregation if more than two protein molecules are linked by shared micelles (the 2ACBP:1 micelle and 2αSN:1 micelle complexes are stable entities that do not grow over time).
- Aggregation by shared micelles promotes rapid and reproducible growth of fibrils by circumventing the classical nucleation step, whose stochastic appearance often bedevils the statistics of fibrillation. SDS-induced fibrillation is much better than other methods, including vigorous shaking agitation, at achieving reproducible fibrillation behavior [98]. This must be because SDS stabilizes a fibrillation-prone state through simple binding equilibria rather than complex and mechanically

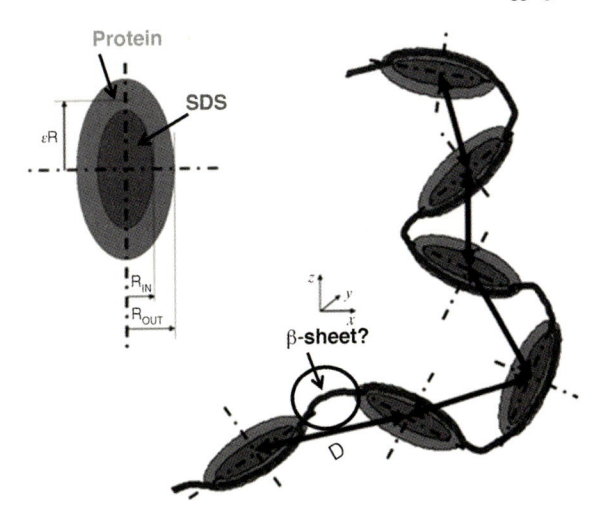

**Figure 3.5** Sketch of the beads-on-a-string model of α-synuclein aggregation in the presence of SDS. The aggregate structure can be modeled as ellipsoid micelles with a protein shell (presumably α-helical) and a surfactant interior linked together by flexible regions where the amyloid structures are probably formed. Adapted from [30].

dependent processes involving air–water interfaces, shear forces, surface binding, and so on [99].

- The short lag phase combines with a very long elongation phase, because the stepwise growth by extension through linker regions in effect means that each new elongation of the bead-on-a-string structure has to start anew without seeding from the existing fibril. The growing tip (which is presumably found at both ends of the string) does not form a simple complementary surface for incoming monomers but merely provides an anchoring surface for a new fibrillar module.
- The combination of micelle-bound and linking αSN molecules makes for a mixture of secondary structure rather than a uniform β-sheet. Most likely the α-helical structures form around the micelles while β-sheets are in the linker region.
- The morphology of the SDS-induced fibrils are much more worm-like and flexible than the straight fibrils prepared in its absence. This likely reflects the flexible structure of linker regions and the dynamics of the micellar states.

It seems unlikely that short peptides will fibrillate by the same mechanism, since the smaller peptides have a reduced surface area for SDS clustering and linkage with other peptides. For the 10-residue peptide hormone kisspeptin, the verdict is clear: while the polyanionic glucosaminoglycan heparin is absolutely required for heparin to form amyloid structures, submicellar SDS concentrations completely inhibit aggregation [100]. Heparin provides a contiguous and extensive anionic binding surface that presumably promotes the ordered assembly of higher order kisspeptin structures, whereas SDS monomers most likely bind to kisspeptin and out-compete heparin binding while not accumulating in sufficient amounts to form clusters that can coalesce to micelles, let alone provide linker regions for aggregate growth. In contrast, the 219-residue protein p25α is stimulated to aggregate just as much by SDS as by heparin, since its much larger size makes for more versatile interactions with both SDS and heparin [101]. For antimicrobial peptides such as the 17-residue novispirin, the massive surplus of cationic residues may lead to simple precipitation due to the "glue" of neutralizing electrostatic interactions [39], whereas in other cases, such as the equally sized SCR3 peptide, 3 mM SDS actually induces a β-sheet-rich fibrillar structure starting from a helical state [102]. Interestingly, the SDS-induced fibrils are shorter and thicker than the fibrils which form more slowly in the absence of SDS. There is no indication of increased fibril flexibility or helical structure in the final fibril state. Intermolecular β-sheet formation may be promoted by other means in this case, although it is likely to involve some degree of SDS clustering. This could be resolved by SAXS studies in combination with ITC.

## 3.3
### Palimpsests of Future Functions: Cytotoxic Protein–Lipid Complexes

With the possible exception of $β_2$m aggregation stimulated by elevated levels of free fatty acids, the amphiphile-induced aggregation stories presented in the previous section lack direct physiological parallels. However, an exciting recent phenomenon commands our attention, in which a naturally occurring fatty acid induces a biologically potent and self-contained protein aggregate structure. Work in this area was initiated by the serendipitous discovery that a fraction of human milk showed highly specific anti-tumor activity, and this activity could be assigned to a complex between the common milk protein α-lactalbumin and the most abundant fatty acid in human milk, oleic acid [103, 104]. The complex, discovered at Lund University little more than 50 km from Elsinore Castle as the crow flies, was aptly named HAMLET (human α-lactalbumin made lethal to tumor cells), while its bovine counterpart is called BAMLET. The details of the complex, although still unclear, are beginning to emerge [105]. To form HAMLET, α-lactalbumin must unfold partially by the loss of its $Ca^{2+}$ co-factor (e.g., by EDTA) and then forms a complex with oleic acid by passing the protein over an anion-exchange column (such as DEAE) conditioned with a suspension of oleic acid. It appears that oleic acid immobilized on the chromatographic surface provides a solid–liquid interface which facilitates both protein–oleic acid interactions and protein self-assembly [106]. This can be done at mildly alkaline pH. The ensuing complex is highly stable and the protein does not refold by the reintroduction of $Ca^{2+}$ at neutral pH, indicating that oleic acid has trapped the protein in a kinetically stable state. The stoichiometry of this complex is unlikely to be a simple 1:1 ligand–receptor model. Atomic force microscopy (AFM) images hint that the HAMLET complex may consist of protein-encapsulating lipid molecules, reminiscent of the shared micelles surrounded by a protein shell that we have reported for the αSN–SDS aggregates [30]. The inner core of oleic acid is probably key to the activity of HAMLET, which targets the cellular membrane and leads to an increase in membrane fluidity and eventual rupture [107].

Note that these studies are carried out well above the $pK_a$ of oleic acid (4.8), which means that the oleic acid in principle has access to a micellar structure. A different situation is observed at low pH. In the absence of other additives, α-lactalbumin will slowly fibrillate at pH values below 4.5 [108]. Between pH 4.5 and 3.0, α-lactalbumin forms a partially folded state which is more ordered than the classical molten globule state formed at pH 3.0 [109] and which acts as a fibrillation precursor. This process is inhibited by oleic acid, which forms an insoluble aggregate with α-lactalbumin. Below its $pK_a$, oleic acid coexists as protonated monomers and an oil phase and it is probably the monomers which render the protein insoluble. Although the biological efficacy of these precipitates has not been reported, they are unlikely to be cytotoxic in view of their physically inert state. Rather, they are testimony to the great variety of protein–amphiphile complexes formed under different conditions.

More information about the nature of these protein–oleic compounds has emerged from a study of the complexes between oleic acid and equine lysozyme, which is a homolog of α-lactalbumin [106]. Even in the absence of amphiphilic compounds, equine lysozyme forms a range of partially folded states under destabilizing conditions, similarly to α-lactalbumin [110]. These partially folded states, formed typically at low pH, are precursors to oligomers and amyloids [111, 112]. The oligomers are only populated to a few percent, converting rapidly to amyloid protofilaments. This makes it attractive for stabilizing oligomeric intermediates by oleic acid, leading to the complex ELOA (equine lysozyme with oleic acid) [106], which is formed by the same chromatographic approach as for HAMLET. Just like HAMLET, the resulting complex is stable over extended time periods, is much less structured than the native state, and possesses a spherical and ring shaped morphology. As might be expected for a fluctuating protein structure, it binds the hydrophobic dye ANS (8-anilino-1-naphthalene sulfonate). More surprisingly, it also shows affinity for the fibril-binding dye ThT (whether HAMLET does so is not known). This is a remarkable observation since most prefibrillar oligomers do not elicit such a response from ThT, and suggests that the oleic acid complexation induces a cross-β structure. The oligomeric nature of ELOA was confirmed by $^1H$ NMR spectroscopy, which showed that bound oleic acid exists in a different environment to free oleic acid and is bound within a large molecular complex at molar ratios of 11–48 oleic acid:lysozyme [106]. Cross-peaks between oleic acid and lysozyme reveal intermolecular binding, particularly via the aromatic residues. Diffusion gradient experiments indicate 4–30 molecules of lysozyme per complex, that is, complexes with up to 1000 oleic acid molecules can occur. These complexes show significant cytotoxic activity through an apoptotic mechanism. ELOA initially accumulates around the cell membrane, which is the primary target for ELOA toxic activity. These probably trigger an apoptotic cascade, leading to cell rupture and subsequent streaming of ELOA into the cell [106]. While the toxicity of oleic acid *per se* is significantly lower than that of ELOA [106], our more recent work has revealed that ELOA is able to transfer oleic acid to membranes and regain some of its original native structure, suggesting that the protein–lipid complex may simply act as a transporter of oleic acid and potentiate

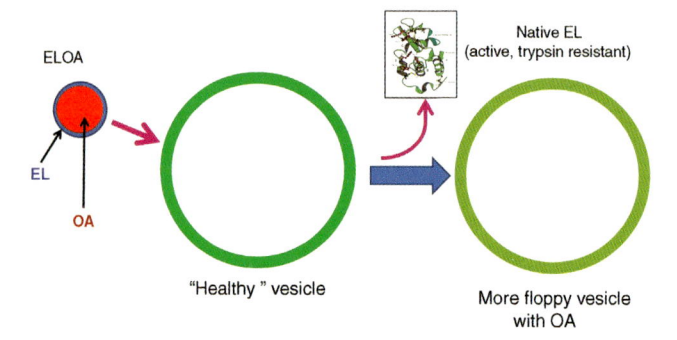

**Figure 3.6** Proposed mechanism for the interaction of the ELOA complex with phospholipid membranes. The ELOA complex is modeled as a lipid droplet stabilized by an outer shell of protein, which upon interacting with a phospholipid vesicle transfers the oleic acid contents to the membrane, increasing its fluidity and at the same time releasing (a fraction of) the partially unfolded equine lysozyme as folded protein.

its toxicity by targeted off-loading of oleic acid into the membrane [113]. This makes good sense from a structural point of view. Our own preliminary SAXS-based analyses of ELOA complexes outline a lipid droplet type of structure, in which the protein forms a stabilizing shell (C.L.P. Oliveira, D.E. Otzen, and J.S. Pedersen, Aarhus University, unpublished observations). Once this complex meets another lipid environment in the membrane, this will most likely destabilize the complex structure and release the oleic acid to the surrounding membrane (Figure 3.6).

Although HAMLET and ELOA are the most spectacular examples of novel properties arising from protein–oleic acid complexes, they are not the only ones. Oleic acid also induces aggregation of destabilized versions of superoxide dismutase (the wild-type apo form or destabilized mutants), leading to deposits of round or amorphous morphology with clustered tiny spherical aggregates [114]. Importantly, these aggregates are also cytotoxic. Aggregate formation probably occurs by the same mechanism as proposed for other anionic surfaces; although oxidation promotes aggregation of superoxide dismutase in general, this is unlike to play a role in oleic acid-induced aggregation due to the lack of effect of antioxidant radical scavengers. Other examples are likely to emerge in the future.

## 3.4
### Aggregation in Fluorinated Organic Solvents

Our discussion on fluorinated organic solvents will in practice concentrate on TFE. As an organic solvent, the dielectric constant of TFE is approximately one-third that of water, which means that hydrophobic interactions are weaker, but electrostatic interactions are stronger, in TFE. TFE also strengthens the intramolecular (polar) hydrogen bonds between residues close to each other [115], thereby stabilizing local secondary structures [116]. The reasons for this are controversial. The strong electron-withdrawing effect of the three fluorine atoms makes TFE a better proton or hydrogen bond donor, but a poorer acceptor, than water [115]. This will drive it to minimize amide solvent exposure and in this way strengthen intra-polypeptide hydrogen bonding of the amide group [117]. Alternatively, TFE may act as a cosmotrope which drives a reduction in solvent exposure [118] or selectively destabilizes planar resonance-stabilized amides involving water–peptide contacts, favoring compact states that contain internally hydrogen-bonded or solvent-sequestered amide functions [119]. The preferential accumulation of fluorinated alcohol molecules around peptide and protein surfaces, which is suggested by experiments and simulations, will displace water, thereby removing alternative hydrogen bonding partners and providing a low dielectric environment that favors the formation of intra-peptide hydrogen bonds, selectively destabilizing the unfolded structure [120–122]. The effect may also be seen as a stabilization of a structured (but not necessarily folded) state, in which TFE induces polypeptide structure formation by displacing water molecules from hydrogen-bondable sites around the polypeptide [123]. Remarkably, TFE also forms clusters in solution, peaking around 30% TFE, probably driven by the hydrophobic nature of the $CF_3$ group [124, 125]. These large clusters may provide proteins with high local alcohol concentrations that promote specific conformations.

Whatever the underlying basis for its action, the important point to note is that TFE affects proteins in the same two complementary ways as SDS: it weakens the protein's hydrophobic interactions and strengthens its hydrogen bonds. The relative strengths of these two effects will vary with TFE concentration, and this provides a fertile ground for perturbing protein structures in different ways, depending on the exact conditions [126, 127]. Put simply, TFE destabilizes the specific tertiary interactions of native proteins at low concentrations; at higher concentrations, it stabilizes secondary structures (α-helix) and induces non-native folded states of proteins, but the details will vary from protein to protein. TFE stabilizes helical structures in some peptides [127–129] and β-sheet structures in others [130, 131]. α-lactalbumin is first transformed into a molten state before it reaches the so-called TFE state, whereas RNase A directly forms the TFE state [132].

Amyloid induction by TFE will depend on how it affects the protein's balance between hydrophobic and polar interactions. However, as for SDS, the general trend is that intermediate (~20–30%) concentrations of TFE favor a β-sheet structure and subsequent amyloid formation, whereas concentrations above 50% stabilize α-helical structures and may dissolve existing fibrils [133, 134]. This is a general effect, also seen for membrane proteins despite their altered hydrophilic/hydrophobic balances [135]. It may be linked to the peaking in TFE clustering in this concentration range, which may provide "docking surfaces" for both β-sheet formation and aggregation. Whether TFE is able to increase the reproducibility of aggregation, as demonstrated for SDS, has not been demonstrated. Let us now turn to specific protein examples to illustrate TFE effects on aggregation.

## 3.4.1
### Protein Examples

#### 3.4.1.1 Acyl Phosphatase

Chiti, Dobson and co-workers first used this protein to systematically identify appropriate conditions to study amyloid formation *in vitro* [136]. Amyloid formation peaks around 20–30% TFE, but the range over which it occurs is dependent on the stability of the protein [82]. The more unstable the protein, the lower is the threshold concentration where amyloid formation starts. Combined with the upper limit of 30–35% TFE (above which α-helices predominate), it is concluded that aggregation occurs from an ensemble of denatured conformations under conditions in which non-covalent interactions are still favored. Put another way, it is necessary to destabilize the native state without destabilizing intermolecular contacts. Comparison of the disaggregation of acyl phosphatase aggregates in 5% TFE with the aggregation in 25% TFE reveals that a greater part of the protein residues is involved in the stabilization of the protein than in the actual kinetics of aggregation [137].

Hyp-N fibrillates optimally around 30% TFE [138]. However, even under conditions where a TFE-induced intermediate state is populated only to around 1% (based on kinetic analyses), aggregation will still occur over time [139]. Thus a small destabilization of the native state can be sufficient to drive aggregation. These reports illustrate the great use of different TFE concentrations to fine tune protein responses and finely shift conformational equilibria.

#### 3.4.1.2 $\beta_2$-Microglobulin

As part of their focus on the stimulation of β2m aggregation, Goto and co-workers have shown that TFE is the most efficient solvent to accelerate β2m fibril growth in the presence of heparin [140]. Dimethyl sulfoxide (DMSO) is much better at dissolving fibrils to the monomeric state [141, 142], whereas TFE and the related FOS HFIP change fibril conformation but do not dissolve the aggregates completely, although HFIP has often been used as an efficient solvent to monomerize stubbornly aggregative peptides such as Aβ and ADan [143]. This difference has been ascribed to DMSO's superior proton acceptor properties [144], which disrupt inter-/intra-protein hydrogen bonds better than the FOSs.

The K3 fragment, representing residues 20–41 of β2m, shows optimal aggregation around 20% TFE. The bell-shaped concentration curve is nicely explained by the decrease and increase in hydrophobic and polar interactions between groups in the peptide, respectively, with increasing TFE concentration [16] (Figure 3.7). This is supported by simulated aggregation within generalized Born/surface area (GBSA) force fields where the polar and hydrophobic forces are simple functions of the dielectric constant and surface tension [145]. Moderate amounts of alcohol weaken the native hydrophobic core and promote partial unfolding of the monomers. Too much alcohol makes polar interactions dominate, leading to increased intramolecular hydrogen bonds (α-helices, β-hairpins, or both). TFE led to thinner K3 fibrils than the fibrils formed in water, probably by weakening the hydrophobic interactions that promote protofibril association [146].

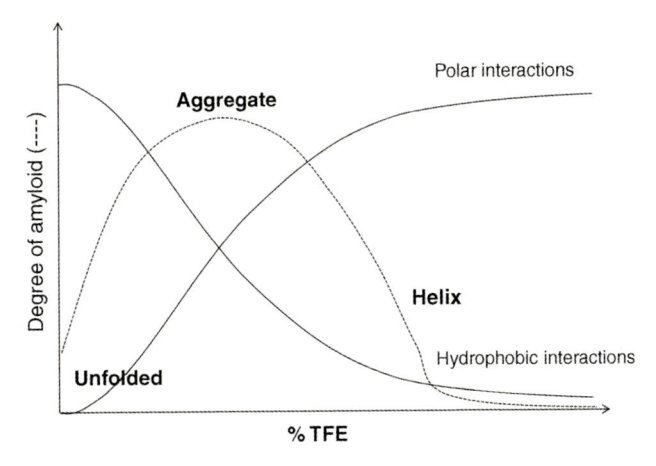

**Figure 3.7** Schematic representation of the TFE-induced formation of fibrils by the β₂m peptide K3 (residues 20–41). Increasing TFE concentrations lead to a decrease and increase in hydrophobic and polar interactions between peptide groups, respectively. At around 20% TFE this leads to optimal conditions for fibrillation, probably aided by clustering of TFE molecules. Adapted from [16].

### 3.4.1.3 α-Chymotrypsin

At neutral pH, this protein aggregates maximally around 35% TFE via a "sticky" non-native extended β-sheet conformation [147, 148], whereas higher concentrations induce a non-aggregation-prone helical state. Salts shift the equilibrium towards these states and away from the native state, so that less TFE is required to form amyloid at higher salt concentrations. At pH 2.5, only 12.5% TFE is needed to induce amyloid [149]. Thus TFE perturbations combine with charge neutralization and screening, providing a multitude of ways to modulate optimal amyloid conditions. Interestingly, the heat-denatured and TFE-induced states are structurally very different, with the TFE state showing significantly more helicity and formation of a helical oligomer whereas the heat-denatured state is much less structured. Nevertheless, both lead to the same amyloid, presumably because the regions concerned are flexible and do not involve strong intramolecular interactions [149].

### 3.4.1.4 Alteration of Fibril Structure by TFE

TFE can also alter the structure of amyloid fibrils, just as seen for αSN with SDS [30]. Similarly to K3 [146], TFE-denatured barstar forms single-layer β-sheets in a sigmoidal (cooperative) time course whereas heat-denatured barstar forms double-layered β-sheets without a lag phase [150]. The two aggregates also show differences in secondary structure and are formed from different precursors; heat-denatured barstar forms molten globule oligomers whereas the TFE state is helical and leads to structurally distinct protofibrils. Similarly, β-lactoglobulin forms worm-like aggregates in (a remarkably high concentration of) 50% TFE at pH 7, whereas the heat-denatured state leads to short, stiff rods [151]. These differences can be explained by the weakening of the hydrophobic bonds by TFE, which will modify and soften the quaternary structure of the amyloid ensemble.

#### 3.4.1.5 Other Proteins

TFE is now a widespread additive used to induce protein aggregation. In all cases, 20–30% TFE appears to be the "sweet range" for aggregation since this concentration range will stabilize a non-native but "sticky" conformation. Success stories include Aβ [152], αSN [153], TGFBIp [154], RNase Sa [155], protein L [156], domain 1 of the cell adhesion molecule [157], and a lysozyme-derived peptide Lys [158]. A few proteins such as acidic fibroblast growth factor (AFGF) [159], stefin [160], and insulin [161] aggregate optimally around 10% TFE and may even dissolve at slightly higher TFE concentrations [162]. However, this most likely reflects specific protein properties such as a high degree of existing β-sheet structure which pushes AFGF towards alternative β-sheet arrangements [159], the use of relatively low pH for stefin [160], and a complex quaternary structure in the case of insulin.

## 3.5
### From Mimetics to the Real Thing: Aggregation on Lipids

Let us now consider the extent to which SDS and TFE can, as is often claimed, be considered membrane mimetics. To do so, we need to consider the mechanisms thought to drive the aggregation of proteins on membranes.

### 3.5.1
#### Binding Surfaces and High Local Concentrations

Membranes differ from SDS and TFE in having a large contiguous membrane surface which provides many uniform binding sites. This can lead to a simple increase in local protein concentration upon binding at the membrane surface (disregarding any other conformational changes). Although the air–water interface can strongly stimulate aggregation {accelerating islet amyloid polypeptide (IAPP) aggregation 1400-fold [163]}, this interface is rarely present *in vivo*. However, the water–membrane interface constitutes a very attractive hydrophilic–hydrophobic interface. Kinnunen and co-workers showed that a large class of unrelated proteins could form fibrils in the presence of phospholipid vesicles containing anionic lipids such as phosphatidylserine (PS) and phosphatidylglycerol (PG) [164]. This was suggested to occur by recruiting proteins on to the membrane to the surface to obtain a high local concentration, in turn leading to conformational changes, self-aggregation, and seeding of subsequent fibrillation. A subsequent Fourier transform infrared (FTIR) microscopy study of fibrils of the heme protein cytochrome *c* fibrils formed with these lipids showed great variability in heme color even within one fiber, suggesting both ordered fibril spacing and striking variability in fibril structure [165].

Hence the expanses of the lipid surface may trap the proteins in different "frozen states," completely opposite to the leveling and averaging of surfactant effects which lead to highly reproducible and uniform fibrillar structures. An increase in local concentration is also suggested to drive the initiation of the aggregation of IAPP [98, 166].

Despite its large size, the membrane may still undergo major rearrangements when attacked by proteins. Lipid molecules are incorporated into fibrils [164, 165], and the original shape of the membrane can be entirely lost as a result, as seen for IAPP [167]. At lower protein:lipid ratios, the consequences need not be so extreme; the membrane surface can act catalytically by allowing the IAPP fibrils to form and then be released, leaving restituted membranes behind [166].

### 3.5.2
### Conformational Changes Associated with Binding

In addition to increasing the local peptide concentration, lipids can also align proteins on the surface to optimize binding and packing. This will depend on the lipid phase. In the gel phase, DPPG (dipalmitoylphosphatidylglycerol) induces the formation of crystalline β-sheet-rich protofibrillar assemblies of Aβ on the surface at peptide concentrations as low as 250 nM, whereas POPG (1-palmitoyl-2-oleoyl-sn-glycero-3-phosphoglycerol) (in the fluid phase) induces amyloid fibrils [168]. Given that a local increase in concentration of peptides is the start of the aggregation process, the lipid:peptide ratio will determine the local density of proteins/peptides on the vesicle surface and thus control which additional conformational changes (if any) can occur. Unsurprisingly, Aβ provides a well-investigated example. Aβ does not bind significantly to zwitterionic phosphocholine or ceramide lipids [169, 170], and the extent of its interaction with anionic lipids is not clear cut. There appear to be no interactions with PS [171] and PG lipids [172] at physiological salt concentrations. When the ionic strength is reduced (25 mM Tris, pH 7.5), several anionic lipids accelerate Aβ fibrillation and bind the peptide in a saturable manner, presumably through contacts between Lys28 and negative headgroups [173]. Similar ionic strength conditions (10 mM MOPS, pH 7.4) promote β-sheet structure in the presence of PG at low lipid:peptide (1:55) ratios [174] through insertion into the membrane, as seen by fluorescence peak shifts and a reduction in quenching efficiency [175]. This effect is equivalent to an eightfold increase in bulk Aβ concentration and corresponds to the effect of intermediate concentrations of TFE [176] and SDS. Aβ forms a more α-helical structure as the lipid:peptide ratio increases beyond 55, possibly by aligning five positive Aβ charges with the membrane interface rather than actual penetration [177], thus avoiding amyloid formation. This is equivalent to the structure in >50% TFE and micellar SDS.

In more recent work, Gröbner and co-workers suggested that acidic lipids can prevent the release of membrane-inserted Aβ by stabilizing its hydrophobic transmembrane C-terminal part (residues 29–40) in an α-helical conformation via an electrostatic anchor between its basic Lys28 residue and the negatively charged membrane interface [178]. Once Aβ is released as a soluble monomer, charged membranes may act as two-dimensional aggregation templates where an increasing amount of charged lipids (which may be pathological degradation products) leads to a dramatic accumulation of surface-associated Aβ followed by accelerated aggregation into toxic structures.

For Aβ, the choice lies between monomeric α-helix and amyloid β-sheet. α-helices can play a constructive role in the aggregation of other peptides. On anionic lipids, IAPP forms monomeric α-helices which assemble cooperatively into α-helical oligomers [179]. Remarkably, these oligomers permeabilize membranes and can form fibrils. Most likely they are arranged as parallel coiled-coil bundles that somehow make a cooperative transition to β-sheet amyloid. In the α-helical state, the N-terminal part of IAPP is presumably available to bind to and partially insert into the membrane, and the existence of parallel α-helices may be the best precursor to anti-parallel β-sheets [180].

αSN appears to have a genuine physiological function associated with membrane binding so that anionic lipids inhibit its aggregation, in particular ganglioside GM1 [181]. Although the lipids induce α-helical monomers and oligomers, this does not lead to fibrillation. Thus, the greater the degree of induced helicity, the more fibrillation is inhibited [89]. Completely abolishing α-helical propensity can have a very different effect, however. The hydrophobic hexapeptide $N$-Ac-Trp-Leu$_5$ is simply too short to complete a stable α-helical turn. Instead, it forms stable higher order β-sheet structures in POPC [182]. Although there is only a modest partitioning free energy (0.5 kcal mol$^{-1}$), this suffices together with multiple interactions to drive the formation of stable secondary structure. Whether submicellar concentrations of SDS or intermediate TFE concentrations will have the same effect has not been reported.

### 3.5.3
### Chemical Variability of the Lipid Environment

We should also note a fundamental distinction between membranes and AS/FOS: the rich and varied environment provided by the multitude of different lipid components. Domain structures formed by combinations of different lipids make possible novel interactions which are impossible to mimic in SDS or TFE. Cholesterol inhibits IAPP amyloid formation on membranes, possibly by stabilizing the random-coil conformation [183]. However, it increases the membrane affinity of transthyretin [184], probably by altering lipid packing, just as seen for Aβ [185]. The gangliosides (sialic acid-containing glycosphingolipids) constitute another important lipid class. In GM micelles, Aβ adopts conformations which contain 16–20% β-sheet structures and 6–10% α-helix structure [186]. Just like PG lipids, gangliosides induce β-sheet structure at low lipid:peptide ratios (<20) but α-helical structures at lipid: peptide ratios >40 [187]. Aβ bound to ganglioside clusters seeds fibrillation [188]. Based on this, it has been suggested that Aβ specifically binds to ganglioside clusters, which are stabilized by cholesterol [169]. Aβ undergoes a conformational transition from an α-helix-rich to a β-sheet-rich structure with increasing peptide density on the membrane (Figure 3.8). This β-sheet form serves as a seed for the formation of amyloid fibrils.

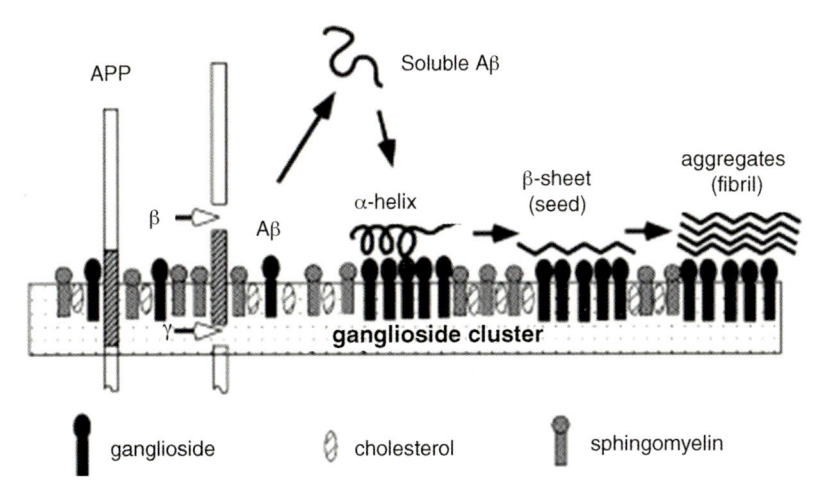

**Figure 3.8** Regulation of Aβ aggregation in a membrane environment. The peptide is generated, at least partly, in lipid rafts composed of sphingolipids and cholesterol by the enzymatic cleavage (β- and γ-secretases) of Amyloid Precursor Protein (APP). Aβ is soluble, and takes an unordered structure. Once ganglioside clusters are generated in a cholesterol-dependent manner, Aβ binds to the clusters, forming an α-helix-rich structure at lower peptide-to-ganglioside ratios, but a β-sheet at higher ratios due to intermolecular contacts. The β-sheet form is the basis for the formation of cytotoxic aggregates of Aβ. Adapted from [169].

## 3.6 Summary

SDS and TFE are unquestionably useful aggregation-inducing agents. This remarkable property is based on the modulation of opposing hydrophobic and hydrophilic interactions at different concentrations. An appropriate concentration range can usually be found where the loss of native interactions is balanced by the need to establish alternative hydrogen bonding patterns that will lead to amyloid formation. For SDS, this amyloid formation may be linked to the ability to form clusters on the protein and induce aggregation-prone structures, typically driven by the need to make shared micelles based on several protein molecules. In principle, the same effect could drive lipid-mediated aggregation, although here the limiting factor may be the lipids' lower solubility which opens up an alternative scenario: the formation of self-contained protein-coated lipid droplets where the protein effectively acts as a surfactant itself, shielding the lipid from the water phase and mobilizing it for targeted transport to the membrane. The situation is more complex at the lipid membrane itself. On the one hand, the membrane presents a contiguous rather than a "detachable" binding surface (which limits its effect on protein conformations), and on the other hand, it has access to a richness of different components which radically expand the landscape of possible lipid–protein interactions but at the same time provide a bewildering range of possibilities for experimental investigations. The most constructive approach is to increase our understanding of the physico-chemical properties of this landscape using membrane models of increasing complexity while linking it up to simple and convenient model systems such as surfactants and organic solvents. This will hopefully provide a firmer foundation for the underlying thermodynamic and structural features.

## References

1 Lambert, M.P., Barlow, A.K., Chromy, B.A., Edwards, C., Freed, R., Liosatos, M., Morgan, T.E., Rozovsky, I., Trommer, B., Viola, K.L., Wals, P., Zhang, C., FInch, C.E., Krafft, G.A., and Klein, W.L. (1998) Diffusible, nonfibrillar ligands derived from Abeta1-42 are potent central nervous system neurotoxins. *Proc. Natl. Acad. Sci. USA*, **95**, 6448–6453.

2 Arispe, N., Rojas, E., and Pollard, H.B. (1993) Alzheimer disease amyloid beta protein forms calcium channels in bilayer membranes: blockade by tromethamine and aluminum. *Proc. Natl. Acad. Sci. USA*, **90**, 567–571.

3 Tanford, C. (1980) *The Hydrophobic Effect. Formation of Micelles and Biological Membranes*, 2nd edn, John Wiley & Sons, Inc., New York.

4 Garavito, R.M. and Ferguson-Miller, S. (2001) Detergents as tools in membrane biochemistry. *J. Biol. Chem.*, **276**, 32403–32406.

5 Otzen, D.E., Sehgal, P., and Westh, P. (2009) α-Lactalbumin is unfolded by all classes of detergents but with different mechanisms. *J. Colloid Interface Sci.*, **329**, 273–283.

6 Putnam, F.W. and Neurath, H. (1945) Interaction between proteins and synthetic detergents. II. Electrophoretic analysis of serum albumin–sodium dodecyl sulfate mixtures. *J. Biol.Chem.*, **159**, 195–200.

7 Montserret, R., McLeish, M.J., Böckmann, A., Geourjon, C., and Penin, F. (2000) Involvement of electrostatic interactions in the mechanism of peptide folding induced by sodium dodecyl sulfate binding. *Biochemistry*, **39**, 8362–8374.

8 Yonath, A., Podjarny, A., Honig, B., Sielecki, A., and Traub, W. (1977) Crystallographic studies of protein denaturation and renaturation. 2. Sodium dodecyl sulfate induced structural changes in triclinic lysozyme. *Biochemistry*, **16**, 1418–1424.

9 Reynolds, J.A. and Tanford, C. (1970) Binding of dodecyl sulfate to proteins at high binding ratios. Possible implications for the state of proteins in biological membranes. *Proc. Natl. Acad. Sci. USA*, **66**, 1002–1005.

10 Kalyanasundaram, K. and Thomas, J.K. (1977) Environmental effects on vibronic band intensities in pyrene monomer fluorescence and their application in studies of micellar systems. *J. Am. Chem. Soc.*, **99**, 2039–2044.

11 Nielsen, M.M., Andersen, K.K., Westh, P., and Otzen, D.E. (2007) Unfolding of β-sheet proteins in SDS. *Biophys. J.*, **92**, 3674–3685.

12 Andersen, K.K., Oliveira, C.L.P., Larsen, K.L., Poulsen, F.M., Callisen, T.H., Westh, P., Pedersen, J.S., and Otzen, D.E. (2009) The role of decorated SDS micelles in sub-cmc protein denaturation and association. *J. Mol. Biol.*, **391**, 207–226.

13 Otzen, D.E., Nesgaard, L., Andersen, K.K., Hansen, J.H., Christiansen, G., Doe, H., and Sehgal, P. (2008) Aggregation of S6 in a quasi-native state by monomeric SDS. *Biochim. Biophys. Acta*, **1784**, 400–414.

14 Fersht, A.R. and Serrano, L. (1993) Principles of protein stability derived from protein engineering experiments. *Curr. Opin. Struct. Biol.*, **3**, 75–83.

15 White, S.H. and Wimley, W.C. (1999) Membrane protein folding and stability: physical principles. *Annu. Rev. Biophys. Biomol. Struct.*, **28**, 319–365.

16 Yamaguchi, K.-I., Naiki, H., and Goto, Y. (2006) Mechanism by which the amyloid-like fibrils of a beta2-microglobulin fragment are induced by fluorine-substituted alcohols. *J. Mol. Biol.*, **363**, 279–288.

17 Jirgensons, B. (1967) Effects of n-propyl alcohol and detergents on the optical rotatory dispersion of α-chymotrypsinogen, β-casein, histone fraction 1 and soybean trypsin inhibitor. *J. Biol. Chem.*, **242**, 912–918.

18 Gierasch, L.M. (1989) Signal sequences. *Biochemistry*, **28**, 923–930.

19 Jirgensons, B. (1961) Effect of detergents on the conformation of proteins. I. An abnormal increase of the optical rotatory dispersion constant. *Arch. Biochem. Biophys.*, **94**, 59–67.

20 Jirgensons, B. and Mori, E. (1981) Effect of long-chain alkyl sulfate binding on circular dichroism and conformation of soybean trypsin inhibitor. *Biochemistry*, **20**, 1630–1634.

21 Shirahama, K., Tsujii, K., and Takagi, T. (1974) Free-boundary electrophoresis of sodium dodecyl sulfate–protein polypeptide complexes with special reference to SDS–polyacrylamide gel electrophoresis. *J. Biochem.*, **75**, 309–319.

22 Reynolds, J.A. and Tanford, C. (1970) The gross conformation of protein–sodium dodecyl sulfate complexes. *J. Biol. Chem.*, **245**, 5161–5165.

23 Wright, A.K., Thompson, M.R., and Miller, R.L. (1975) A study of protein–sodium dodecyl sulfate complexes by transient electric birefringence. *Biochemistry*, **14**, 3224–3228.

24 Mattice, W.L., Riser, J.M., and Clark, D.S. (1976) Conformational properties of the complexes formed by proteins and sodium dodecyl sulfate. *Biochemistry*, **15**, 4264–4272.

25 Waterhous, D.V. and Johnson, W.C. (1994) Importance of environment in determining secondary structure in proteins. *Biochemistry*, **33**, 2121–2128.

26 Wu, C.S., Ikeda, K., and Yang, J.T. (1981) Ordered conformation of polypeptides and proteins in acidic dodecyl sulfate solution. *Biochemistry*, **20**, 566–570.

27 Wu, C.S. and Yang, J.T. (1981) Sequence-dependent conformations of short polypeptides in a hydrophobic environment. *Mol. Cell Biochem.*, **40**, 109–122.

28 Wahlström, A., Hugonin, L., Perálvarez-Marín, A., Jarvet, J., and Gräslund, A. (2008) Secondary structure conversions of Alzheimer's Abeta(1–40) peptide induced by membrane-mimicking detergents. *FEBS J.*, **275**, 5117–5128.

29 Andersen, K., Westh, P., and Otzen, D.E. (2008) A global study of myoglobin–surfactant interactions. *Langmuir*, **15**, 399–407.

30 Giehm, L., Oliveira, C.L.P., Christiansen, G., Pedersen, J.S., and Otzen, D.E. (2010) SDS-induced fibrillation of α-synuclein: an alternative fibrillation pathway. *J. Mol. Biol.*, **401**, 115–133.

31 Stigter, D. (1975) Electrostatic interactions in aqueous environments, in *Physical Chemistry: Enriching Topics from Colloid and Surface Science* (eds. H. van Olphen, K.J. Mysels, and International Union of Pure and Applied Chemistry. Commission I.6: Colloid and Surface Chemistry), Theorex, La Jolla, CA, 181–200.

32 Nielsen F A.D., Arleth, L., and Westh, P. (2005) Analysis of protein–surfactant interactions – a titration calorimetric and fluorescence spectroscopic investigation of interactions between *Humicola insolens* cutinase and an anionic surfactant. *Biochim. Biophys. Acta*, **1752**, 124–132.

33 Koistinen, V.U. (1980) Hepatitis B surface antigen polypeptides: artifactual bands in sodium dodecyl sulfate–polyacrylamide gel electrophoresis caused by aggregation. *J. Virol.*, **1980**, 20–23.

34 Burns, P.F., Campagnoni, C.W., Chaiken, I.M., and Campagnoni, A.T. (1981) Interactions of free and immobilized myelin basic protein with anionic detergents. *Biochemistry*, **20**, 2463–2469.

35 Smith, R. (1982) Self-association of myelin basic protein: enhancement by detergents and lipids. *Biochemistry*, **21**, 2697–2701.

36 Dombi, G.W. and Halsall, H.B. (1985) Collagen fibril formation in the presence of sodium dodecyl sulphate. *Biochem. J.*, **228**, 551–556.

37 Cann, J.R., Vatter, A., Vavrek, R.J., and Stewart, J.M. (1986) Interaction

of bradykinin with sodium dodecyl sulfate and certain acidic lipids. *Peptides*, **7**, 1121–1130.

38 Otzen, D.E. (2002) Protein unfolding in detergents: effect of micelle structure, ionic strength, pH, and temperature. *Biophys. J.*, **83**, 2219–2230.

39 Wimmer, R., Andersen, K., Vad, B., Davidsen, M., Mølgaard, S., Nesgaard, L.W., Kristensen, H.-H., and Otzen, D.E. (2006) Versatile interactions of the antimicrobial peptide Novispirin with detergents and lipids. *Biochemistry*, **45**, 481–497.

40 Walsh, D.M., Klyubin, I., Fadeeva, J.V., Cullen, W.K., Anwyl, M.S., Wolfe, M.S., Rowan, M.J., and Selkoe, D.J. (2002) Naturally secreted oligomers of amyloid beta protein potently inhibit hippocampal long-term potentiation *in vivo*. *Nature*, **416**, 535–539.

41 Podlisny, M.B., Ostaszewski, B.L., Squazzo, S.L., Koo, E.H., Rydell, R.E., Teplow, D.B., and Selkoe, D.J. (1995) Aggregation of secreted amyloid -protein into sodium dodecyl sulfate-stable oligomers in cell culture. *J. Biol. Chem.*, **270**, 9564–9570.

42 Hepler, R.W., Grimm, K.M., Nahas, D.D., Breese, R., Dodson, E.C., Acton, P., Keller, P.M., Yeager, M., Wang, H., Shughrue, P., kinney, G., and Joyce, J.G. (2006) Solution state characterization of amyloid beta-derived diffusible ligands. *Biochemistry*, **45**, 15157–15167.

43 Yamamoto, N., Hasegawa, K., Matsuzaki, K., Naiki, H., and Yanagisawa, K. (2004) Environment- and mutation-dependent aggregation behavior of Alzheimer amyloid β-protein. *J. Neurochem.*, **90**, 62–69.

44 Rodziewicz-Motowidło, S., Juszczyk, P., Kołodziejczyk, A.S., Sikorska, E., Skwierawska, A., Oleszczuk, M., and Grzonka, Z. (2007) Conformational solution studies of the SDS micelle-bound 11–28 fragment of two Alzheimer's beta-amyloid variants (E22K and A21G) using CD, NMR, and MD techniques. *Biopolymers*, **87**, 23–39.

45 Poulsen, S.A., Watson, A.A., Fairlie, D.P., and Craik, D.J. (2000) Solution structures in aqueous SDS micelles of two amyloid beta peptides of A beta(1–28) mutated at the alpha-secretase cleavage site (K16E, K16F). *J. Struct. Biol.*, **130**, 142–152.

46 Coles, M., Bicknell, W., Watson, A.A., Fairlie, D.P., and Craik, D.J. (1998) Solution structure of amyloid beta-peptide(1–40) in a water–micelle environment. Is the membrane-spanning domain where we think it is? *Biochemistry*, **37**, 11064–11077.

47 Kohno, T., Kobayashi, K., Maeda, T., Sato, K., and Takashima, A. (1996) Three-dimensional structures of the amyloid beta peptide (25–35) in membrane-mimicking environment. *Biochemistry*, **35**, 16094–16104.

48 Gaggelli, E., Janicka-Klos, A., Jankowska, E., Kozlowski, H., Migliorini, C., Molteni, E., Valensin, D., Valensin, G., and Wieczerzak, E. (2008) NMR studies of the $Zn^{2+}$ interactions with rat and human beta-amyloid (1–28) peptides in water–micelle environment. *J. Phys. Chem. B*, **112**, 100–109.

49 Shao, H., Jao, S., Ma, K., and Zagorski, M.G. (1999) Solution structures of micelle-bound amyloid beta-(1–40) and beta-(1–42) peptides of Alzheimer's disease. *J. Mol. Biol.*, **285**, 755–773.

50 Fletcher, T.G. and Keire, D.A. (1997) The interaction of beta-amyloid protein fragment (12–28) with lipid environments. *Protein Sci.*, **6**, 666–675.

51 Jarvet, J., Danielsson, J., Damberg, P., Oleszczuk, M., and Gräslund, A. (2007) Positioning of the Alzheimer Abeta (1–40) peptide in SDS micelles using NMR and paramagnetic probes. *J. Biomol. NMR*, **39**, 63–72.

52 Rangachari, V., Reed, D.K., Moore, B.D., and Rosenberry, T.L. (2006) Secondary structure and interfacial aggregation of amyloid-β(1–40) on sodium dodecyl sulfate micelles. *Biochemistry*, **45**, 8639–8648.

53 Rangachari, V., Moore, B.D., Reed, D.K., Sonoda, L.K., Bridges, A.W., Conboy, E., Hartigan, D., and Rosenberry, T.L. (2007) Amyloid-β(1–42) rapidly forms protofibrils and oligomers by distinct pathways in low concentrations of sodium dodecylsulfate. *Biochemistry*, **46**, 12451–12462.

54 Tew, D.J., Bottomley, S.P., Smith, D.P., Ciccotosto, G.D., Babon, J., Hinds, M.G., Masters, C.L., Cappai, R., and Barnham, K.J. (2008) Stabilization of neurotoxic soluble beta-sheet-rich conformations of the Alzheimer's disease amyloid-beta peptide. *Biophys. J.*, **94**, 2752–2766.

55 Necula, M., Kayed, R., Milton, S., and Glabe, C.G. (2007) Small-molecule inhibitors of aggregation indicate that amyloid beta oligomerization and fibrillization pathways are independent and distinct. *J. Biol. Chem.*, **282**, 10311–10324.

56 Barghorn, S., Nimmrich, V., Striebinger, A., Krantz, C., Keller, P., Janson, B., Bahr, M., Schmidt, M.A., Bitner, R.S., Harlan, J.E., Barlow, E., Ebert, U., and Hillen, H. (2005) Globular amyloid beta-peptide oligomer: a homogeneous and stable neuropathological protein in Alzheimer's disease. *J. Neurochem.*, **95**, 834–847.

57 Yu, L., Edalji, R., Harlan, J.E., Holzman, T.F., Lopez, A.P., Labkovsky, B., Hillen, H., Barghorn, S., Ebert, U., Richardson, P.L., Miesbauer, L., Solomon, L., Bartley, D., Walter, K., Johnson, R.W., Hajduk, P.J., and Olejniczak, E.T. (2009) Structural characterization of a soluble amyloid beta-peptide oligomer. *Biochemistry*, **48**, 1870–1877.

58 Gellermann, G.P., Byrnes, H., Striebinger, A., Ullrich, K., Mueller, R., Hillen, H., and Barghorn, S. (2008) Aβ-globulomers are formed independently of the fibril pathway. *Neurobiol. Dis.*, **30**, 212–220.

59 Davies, S.J. and Turney, J.H. (1987) Free fatty acids during extracorporeal circulation: the role of heparin. *Lancet*, **ii**, 1097–1098.

60 Lendel, C., Bolognesi, B., Wahlström, A., Dobson, C.M., and Gräslund, A. (2010) Detergent-like interaction of Congo Red with the amyloid beta peptide. *Biochemistry*, **49**, 1358–1360.

61 Frid, P., Anisimov, S.V., and Popovic, N. (2007) Congo Red and protein aggregation in neurodegenerative diseases. *Brain Res. Dev.*, **53**, 135–160.

62 Lendel, C., Bertoncini, C.W., Cremades, N., Waudby, C.A., Vendruscolo, M., Dobson, C.M., Schenk, D., Christodoulou, J., and Toth, G. (2009) On the mechanism of nonspecific inhibitors of protein aggregation: dissecting the interactions of alpha-synuclein with Congo Red and lacmoid. *Biochemistry*, **48**, 8322–8334.

63 Sabaté, R. and Estelrich, J. (2005) Stimulatory and inhibitory effects of alkyl bromide surfactants on β-amyloid fibrillogenesis. *Langmuir*, **21**, 6944–6949.

64 Cao, M., Han, Y., Wang, J., and Wang, Y. (2007) Modulation of fibrillogenesis of amyloid beta(1–40) peptide with cationic gemini surfactant. *J. Phys. Chem. B*, **111**, 13436–13443.

65 Han, Y., He, C., Cao, M., Huang, X., Wang, Y., and Li, Z. (2010) Facile disassembly of amyloid fibrils using gemini surfactant micelles. *Langmuir*, **26**, 1583–1587.

66 Naiki, H., Hashimoto, N., Suzuki, S., Kimura, H., Nakakuki, K., and Gejyo, F. (1997) Establishment of a kinetic model of dialysis-related amyloid fibril extension *in vitro*. *Amyloid*, **4**, 223–232.

67 Yamaguchi, H., Hasegawa, K., Takahashi, H., Gejyo, F., and Naiki, H. (2001) Apolipoprotein E inhibits the depolymerization of β$_2$-microglobulin-related amyloid fibrils at a neutral pH. *Biochemistry*, **40**, 8499–8509.

68 Yamamoto, S., Hasegawa, K., Yamaguchi, I., Tsutsumi, S., Kardos, J., Goto, Y., Gejyo, F., and Naiki, H. (2004) Low concentrations of sodium dodecyl sulfate induces the extension of beta-2-microglobulin-related amyloid fibrils at neutral pH. *Biochemistry*, **43**, 11075–11082.

69 Ookoshi, T., Hasegawa, K., Ohhashi, Y., Kumura, H., Takahashi, N., Yoshida, H., Miyazaki, R., Goto, Y., and Naiki, H. (2008) Lysophospholipids induce the nucleation and extension of beta2-microglobulin-related amyloid fibrils at a neutral pH. *Nephrol. Dial. Transplant.*, **23**, 3247–3255.

70 Sasagawa, T., Suzuki, K., Shiota, T., Kondo, T., and Okita, M. (1998) The significance of plasma lysophospholipids in patients with renal failure on hemodialysis. *J. Nutr. Sci. Vitaminol.*, **44**, 809–818.

71 Kaysen, G.A. (1998) Biological basis of hypoalbuminemia in ESRD. *J. Am. Soc. Nephrol.*, **9**, 2368–2376.

72 Hasegawa, K., Tsutsumi-Yasuhara, S., Ookoshi, T., Ohhashi, Y., Kimura, H., Takahashi, N., Yoshida, H., Miyazaki, R., Goto, Y., and Naiki, H. (2008) Growth of $\beta_2$-microglobulin-related amyloid fibrils by non-esterified fatty acids at a neutral pH. *Biochem. J.*, **416**, 307–315.

73 Hagihara, Y., Hong, D.P., Hoshino, M., Enjyoji, K., Kato, H., and Goto, Y. (2002) Aggregation of $\beta_2$-glycoprotein I induced by sodium lauryl sulfate and lysophospholipids. *Biochemistry*, **41**, 1020–1026.

74 Kuret, J., Chirita, C.N., Congdon, E.E., Kannanayakal, T., Li, G., Necula, M., Yin, H., and Zhong, Q. (2005) Pathways of tau fibrillization. *Biochim. Biophys. Acta*, **1739**, 167–178.

75 Chirita, C.N., Necula, M., and Kuret, J. (2003) Anionic micelles and vesicles induce tau fibrillization *in vitro*. *J. Biol. Chem.*, **278**, 25644–25650.

76 Prusiner, S.B. (1998) Prions. *Proc. Natl. Acad. Sci. USA*, **95**, 13363–13383.

77 Prusiner, S.B., McKinley, M.P., Bowman, K.A., Bolton, D.C., Bendheim, P.E., Groth, D.F., and Glenner, G.G. (1983) Scrapie prions aggregate to form amyloid-like birefringent rods. *Cell*, **35**, 349–358.

78 Morillas, M., Swietnicki, W., Gambetti, P., and Surewicz, W.K. (1999) Membrane environment alters the conformational structure of the recombinant human prion protein. *J. Biol. Chem.*, **274**, 36859–36865.

79 Xiong, L.-W., Raymond, L.D., Hayes, S.F., Raymod, G.J., and Caughey, B. (2001) Conformational change, aggregation and fibril formation induced by detergent treatments of cellular prion protein. *J. Neurochem.*, **79**, 669–678.

80 Leffers, K.W., Schell, J., Jansen, K., Lucassen, R., Kaimann, T., Nagel-Steger, L., Tatzelt, J., and Riesner, D. (2004) The structural transition of the prion protein into its pathogenic conformation is induced by unmasking hydrophobic sites. *J. Mol. Biol.*, **2004**, 839–853.

81 Pedersen, J.S., Christiansen, G., and Otzen, D.E. (2004) Modulation of S6 fibrillation by unfolding rates and gatekeeper residues. *J. Mol. Biol.*, **341**, 575–588.

82 Chiti, F., Taddei, N., Bucciantini, M., White, P., Ramponi, G., and Dobson, C.M. (2000) Mutational analysis of the propensity for amyloid formation by a globular protein. *EMBO J.*, **19**, 1441–1449.

83 Spillantini, M.G., Schmidt, M.L., Lee, V.M.-Y., Trojanowski, J.Q., Jakes, R., and Goedert, M. (1997) Alpha-synuclein in Lewy bodies. *Nature*, **388**, 839–840.

84 Fink, A.L. (2006) The aggregation and fibrillation of alpha-synuclein. *Acc. Chem. Res.*, **39**, 628–634.

85 Volles, M.J., Lee, S.J., Rochet, J.C., Shtilerman, M.D., Ding, T.T., Kessler, J.C., and Lansbury, P.T. (2001) Vesicle permeabilization by protofibrillar α-synuclein: implications for the pathogenesis and treatment of Parkinson's disease. *Biochemistry*, **40**, 7812–7819.

86 Liu, C.-W., Giasson, B.I., Lewis, K.A., Lee, V.M., DeMartino, G.N., and Thomas, P.J. (2005) A precipitating role for truncated alpha-synuclein and the proteasome in alpha-synuclein aggregation: implications for pathogenesis of Parkinson's disease. *J. Biol. Chem.*, **280**, 22670–22678.

87 Cooper, A.A., Gitler, A.D., Cashikar, A., Haynes, C.M., Hill, K.J., Bhullar, B., Liu, K., Xu, K., Stratearn, K.E., Liu, F., Cao, S., Caldwell, K.A., Caldwell, G.A., Marsischky, G., Kolodner, R.D., LaBaer, J., Rochet, J.C., Bonini, N.M., and Lindquist, S.L. (2006) Alpha-synuclein blocks ER-Golgi traffic and Rab1 rescues neuron loss in Parkinson's models. *Science*, **313** 234–238.

88 Davidson, W.S., Jonas, A., Clayton, D.F., and George, J.M. (1998) Stabilization of alpha-synuclein secondary structure upon binding to synthetic membranes. *J. Biol. Chem.*, **273**, 9443–9449.

89 Zhu, M. and Fink, A.L. (2003) Lipid binding inhibits alpha-synuclein fibril formation. *J. Biol. Chem.*, **278**, 16873–16877.

90 Bisaglia, M., Trolio, A., Bellanda, M., Bergantino, E., Bubacco, L., and Mammi, S. (2006) Structure and topology of the non-amyloid-beta component fragment of human alpha-synuclein bound to micelles: implications for the aggregation process. *Protein Sci.*, **15**, 1408–1416.

91 Ferreon, A.C.M. and Deniz, A.A. (2007) Alpha-synuclein multistate folding thermodynamics: implications for protein misfolding and aggregation. *Biochemistry*, **46**, 4499–4509.

92 Ferreon, A.C., Gambin, Y., Lemke, E.A., and Deniz, A.A. (2009) Interplay of alpha-synuclein binding and conformational switching probed by single-molecule fluorescence. *Proc. Natl. Acad. Sci. USA*, **106**, 5645–5650.

93 Georgieva, E.R., Ramlall, T.F., Borbat, P.P., Freed, J.H., and Eliezer, D. (2008) Membrane-bound alpha-synuclein forms an extended helix: Long-distance pulsed ESR measurements using vesicles, bicelles, and rodlike micelles. *J. Am. Chem. Soc.*, **130**, 12856–12857.

94 Veldhuis, G., Segers-Nolten, I., Ferlemann, E., and Subramaniam, V. (2009) Single-molecule FRET reveals structural heterogeneity of SDS-bound alpha-synuclein. *ChemBioChem*, **10**, 436–439.

95 Necula, M., Chirita, C.N., and Kuret, J. (2003) Rapid anionic micelle-mediated alpha-synuclein fibrillization *in vitro*. *J. Biol. Chem.*, **278**, 46674–46680.

96 Ahmad, M.F., Ramakrishna, T., Raman, B., and Rao Ch, M. (2006) Fibrillogenic and non-fibrillogenic ensembles of SDS-bound human alpha-synuclein. *J. Mol. Biol.*, **364**, 1061–1072.

97 Rivers, R.C., Kumita, J.R., Tartaglia, G.G., Dedmon, M.M., Pawar, A.P., Vendruscolo, M., Dobson, C.M., and Christodoulou, J. (2008) Molecular determinants of the aggregation behavior of alpha- and beta-synuclein. *Protein Sci.*, **17**, 887–898.

98 Giehm, L. and Otzen, D.E. (2010) Strategies to increase the reproducibility of α-synuclein fibrillation in plate reader assays. *Anal. Biochem.*, **400**, 270–281.

99 Frokjaer, S. and Otzen, D.E. (2005) Protein drug stability – a formulation challenge. *Nat. Rev. Drug Deliv.*, **4**, 298–306.

100 Nielsen, S.B., Franzmann, M., Basaiawmoit, R.V., Wimmer, R., Mikkelsen, J.D., and Otzen, D.E. (2010) β-Sheet aggregation of kisspeptin is stimulated by heparin but inhibited by amphiphiles. *Biopolymers*, **93**, 678–689.

101 Giehm, L., Sundbye, S., Jensen, P.H., and Otzen, D.E. (2010) Heparin induces the α-synuclein co-aggregator p25α to form non-amyloid aggregates. Submitted.

102 Pertinhez, T.A., Bouchard, M., Smith, R.A.G., Dobson, C.M., and Smith, L.J. (2002) Stimulation and inhibition of fibril formation by a peptide in the presence of different concentrations of SDS. *FEBS Lett.*, **529**, 193–197.

103 Håkansson, A., Zhivotovsky, B., Oreenius, S., Sabharwal, H., and Svanborg, C. (1995) Apoptosis induced by a human milk protein. *Proc. Natl. Acad. Sci. USA*, **92**, 8064–8068.

104 Svensson, M., Håkansson, A., Mossberg, A.-K., Linse, S., and Svanborg, C. (2000) Conversion of alpha-lactalbumin to a protein inducing apoptosis. *Proc. Natl. Acad. Sci. USA*, **97**, 4221–4226.

105 Mok, K.H., Pettersson, J., Orrenius, S., and Svanborg, C. (2007) HAMLET, protein folding, and tumor cell death. *Biochem. Biophys. Res. Commun.*, **354**, 1–7.

106 Wilhelm, K., Darinskas, A., Noppe, W., Duchardt, E., Mok, K.H., Vukojevic, V., Schleucher, J., and Morozova-Roche, L. (2009) Protein oligomerization induced by oleic acid at the solid–liquid interface – equine lysozyme cytotoxic complexes. *FEBS J.*, **276**, 3975–3980.

107 Mossberg, A.-K., Puchades, M., Halskau, Ø., Baumann, A., Lanekoff, I., Chao, Y., Martinez, A., Svanborg, C., and Karlsson, R. (2010) HAMLET interacts with lipid membranes and perturbs their structure and integrity. *PLoS ONE*, **23**, e9384.

108 Yang, F., Zhang, M., Zhou, B.-R., Chen, J., and Liang, Y. (2006) Oleic acid inhibits amyloid formation of the intermediate of alpha-lactalbumin at moderately acidic pH. *J. Mol. Biol.*, **362**, 821–834.

109 Kuwajima, K. (1989) The molten globule state as a clue for understanding the folding and cooperativity of globular-protein structure. *Proteins: Struct. Funct. Genet.*, **6**, 87–103.

110 Van Dael, H., Haezebrouck, P., Morozova, L., Aricomuendel, C., and Dobson, C.M. (1993) Partially folded states of equine lysozyme – structural characterization and significance for protein folding. *Biochemistry*, **32**, 11886–11894.

111 Malisauskas, M., Ostman, J., Darinskas, A., Zamotin, V., Liutkevicius, E., Lundgren, E., and Morozova-Roche, L.A. (2005) Does the cytotoxic effect of transient amyloid oligomers from common equine lysozyme *in vitro* imply innate amyloid toxicity? *J. Biol. Chem.*, **280**, 6269–6275.

112 Malisauskas, M., Zamotin, V., Jass, J., Noppe, W., Dobson, C.M., and Morozova-Roche, L. (2003) Amyloid protofilaments from the calcium-binding protein equine lysozyme: formation of ring and linear structures depends on pH and metal ion concentration. *J. Mol. Biol.*, **330**, 879–890.

113 Nielsen, S.B., Wilhelm, K., Vad, B.S., Schleucher, J., Morozova-Roche, L., and Otzen, D.E. (2010) The interaction of equine lysozyme: oleic acid complexes with lipid membranes suggests a cargo off-loading mechanism. *J. Mol. Biol.*, **398**, 351–361.

114 Kim, Y.J., Nakatomi, R., Akagi, T., Hashikawa, T., and Takahashi, R. (2005) Unsaturated fatty acids induce cytotoxic aggregate formation of amyotrophic lateral sclerosis-linked superoxide dismutase 1 mutants. *J. Biol. Chem.*, **280**, 21515–21521.

115 Buck, M. (1998) Trifluoroethanol and colleagues: cosolvents come of age. Recent studies with peptides and proteins. *Q. Rev. Biophys.*, **31**, 297–355.

116 Uversky, V.N., Narizhneva, N.V., Kirschstein, S.O., Winter, S., and Löber, G. (1997) Conformational transitions provoked by organic solvents in β-lactoglobulin: can a molten globule-like intermediate be induced by the decrease in dielectric constant? *Fold. Des.*, **2**, 163–172.

117 Rajan, R. and Balaram, P. (1996) A model for the interaction of trifluoroethanol with peptides and proteins. *Int. J. Pept. Protein Res.*, **48** 328–336.

118 Conio, G., Patrone, E., and Brighetti, S. (1970) The effect of aliphatic alcohols on the helix–coil transition of poly-L-ornithine and poly-L-glutamic acid. *J. Biol. Chem.*, **245**, 3335–3340.

119 Cammers-Goodwin, A., Allen, T.J., Oslick, S.L., McClure, K.F., Lee, J.H., and Kemp, D.S. (1996) Mechanism of stabilization of helical conformations of polypeptides by water containing trifluoroethanol. *J. Am. Chem. Soc.*, **118**, 3082–3090.

120 Kumar, S., Modig, K., and Halle, B. (2003) Trifluoroethanol-induced $\beta \rightarrow \alpha$ transition in β-lactoglobulin: hydration and cosolvent binding studied by $^2$H, $^{17}$O, and $^{19}$F magnetic relaxation dispersion. *Biochemistry*, **42**, 13708–13716.

121 Chatterjee, C. and Gerig, J.T. (2006) Interactions of hexafluoro-2-propanol with the Trp-Cage peptide. *Biochemistry*, **45**, 14665–14674.

122 Roccatano, D., Fioroni, M., Zacharias, M., and Colombo, G. (2005) Effect of hexafluoroisopropanol alcohol on the structure of melittin: a molecular dynamics simulation study. *Protein Sci.*, **14**, 2582–2589.

123 Jasanoff, A. and Fersht, A.R. (1994) Quantitative determination of helical propensities from trifluoroethanol titration curves. *Biochemistry*, **33**, 2129–2135.

124 Hong, D.-P., Hoshino, M., Kuboi, R., and Goto, Y. (1999) Clustering of fluorine-substituted alcohols as a factor responsible for their marked effects on proteins and peptides. *J. Am. Chem. Soc.*, **121**, 8427–8433.

125 Chitra, R., and Smith, P.E. (2001) Properties of 2,2,2-trifluoroethanol and water mixtures. *J. Chem. Phys.*, **114**, 426–435.

126 Thomas, P.D. and Dill, K.A. (1993) Local and nonlocal interactions in globular proteins and mechanisms of alcohol denaturation. *Protein Sci.*, **2**, 2050–2065.

127 Nelson, J.W. and Kallenbach, N.R. (1986) Stabilization of the ribonuclease

S-peptide alpha-helix by trifluoroethanol. *Proteins: Struct. Funct. Genet.*, **1**, 211–217.

128 Shiraki, K., Nishikawa, K., and Goto, Y. (1995) Trifluoroethanol-induced stabilisation of the alfa-helical structure of β-lactoglobulin: implication for non-hierarchical protein folding. *J. Mol. Biol.*, **245**, 180–194.

129 Luo, P. and Baldwin, R.L. (1997) Mechanism of helix induction by trifluoroethanol: a framework for extrapolating the helix-forming properties of peptides from trifluoroethanol/water mixtures back to water. *Biochemistry*, **36**, 8413–8421.

130 Ramírez-Alvarado, M., Blanco, F.J., and Serrano, L. (1996) De *novo* design and structural analysis of a model beta-hairpin peptide system. *Nat. Struct. Biol.*, **3**, 604–612.

131 Blanco, F.J., Jiménez, M.A., Pineda, A., Rico, M., Santoro, J., and Nieto, J.L. (1994) NMR solution structure of the isolated N-terminal fragment of protein-G B1 domain. Evidence of trifluoroethanol induced native-like beta-hairpin formation. *Biochemistry*, **33**, 6004–6014.

132 Gast, K., Zirwer, D., Muller-Frohne, M., and Damaschun, G. (1999) Trifluoroethanol-induced conformational transitions of proteins: insights gained from the differences between α-lactalbumin and ribonuclease A. *Protein Sci.*, **8**, 625–634.

133 MacPhee, C.E. and Dobson, C.M. (2000) Chemical dissection and reassembly of amyloid fibrils formed by a peptide fragment of transthyretin. *J. Mol. Biol.*, **297**, 1203–1215.

134 Chen, S. and Wetzel, R. (2001) Solubilization and disaggregation of polyglutamine peptides. *Protein Sci.*, **10**, 887–891.

135 Otzen, D.E., Sehgal, P., and Nesgaard, L. (2007) Alternative membrane protein conformations in alcohols. *Biochemistry*, **46**, 4348–4359.

136 Chiti, F., Webster, P., Taddei, N., Clark, A., Stefani, M., Ramponi, G., and Dobson, C.M. (1999) Designing conditions for *in vitro* formation of amyloid protofilaments and fibrils. *Proc. Natl. Acad. Sci. USA*, **96**, 3590–3594.

137 Calamai, M., Tartaglia, G.G., Vendruscolo, M., Chiti, F., and Dobson, C.M. (2009) Mutational analysis of the aggregation-prone and disaggregation-prone regions of acylphosphatase. *J. Mol. Biol.*, **387**, 965–974.

138 Bucciantini, M., Giannoni, E., Chiti, F., Baroni, F., Formigli, L., Zurdo, J., Taddei, N., Ramponi, G., Dobson, C.M., and Stefani, M. (2002) Inherent toxicity of aggregates implies a common mechanism for protein misfolding diseases. *Nature*, **416**, 507–511.

139 Marcon, G., Plakoutsi, G., Canale, C., Relini, A., Taddei, N., Dobson, C.M., Ramponi, G., and Chiti, F. (2005) Amyloid formation from HypF-N under conditions in which the protein is initially in its native state. *J. Mol. Biol.*, **347**, 323–335.

140 Yamamoto, S., Yamaguchi, I., Hasegawa, K., Tsutsumi, S., Goto, Y., Gejyo, F., and Naiki, H. (2004) Glycosaminoglycans enhance the trifluoroethanol-induced extension of β$_2$-microglobulin-related amyloid fibrils at a neutral pH. *J. Am. Soc. Nephrol.*, **15**, 126–133.

141 Hirota-Nakaoka, N., Hasegawa, K., Naiki, H., and Goto, Y. (2003) Dissolution of β$_2$-microglobulin amyloid fibrils by dimethylsulfoxide. *J. Biochem.*, **134**, 159–164.

142 Hoshino, M., Katou, H., Hagihara, Y., Hasegawa, K., Naiki, H., and Goto, Y. (2002) Mapping the core of the beta(2)-microglobulin amyloid fibril by H/D exchange. *Nat. Struct. Biol.*, **9**, 332–336.

143 Srinivasan, R., Jones, E.M., Liu, K., Ghiso, J., Marchant, R.E., and Zagorski, M.G. (2003) pH-dependent amyloid and protofibril formation by the ABri peptide of Familial British Dementia. *J. Mol. Biol.*, **333**, 1003–1023.

144 Kosower, E.M. (1958) The effect of solvent on spectra. I. A new empirical measure of solvent polarity: Z-values. *J. Am. Chem. Soc.*, **80**, 3253–3260.

145 Standley, D.M., Yonezawa, Y., Goto, Y., and Nakamura, H. (2006) Flexible docking of an amyloid-forming peptide from β$_2$-microglobulin. *FEBS Lett.*, **580**, 6199–6205.

146 Kanno, T., Yamaguchi, K., Naiki, H., Goto, Y., and Kawai, T. (2005) Association of thin filaments into thick filaments revealing the structural hierarchy of amyloid fibrils. *J. Struct. Biol.*, **149**, 213–218.

147 Pallares, I., Vendrell, J., Aviles, F.X., and Ventura, S. (2004) Amyloid fibril formation by a partially structured intermediate state of α-chymotrypsin. *J. Mol. Biol.*, **342**, 321–331.

148 Rezaei-Ghaleh, N., Ebrahim-Habibi, A., Moosavi-Movahedi, A.A., and Nemat-Gorgani, M. (2007) Role of electrostatic interactions in 2,2,2-trifluoroethanol-induced structural changes and aggregation of alpha-chymotrypsin. *Arch. Biochem. Biophys.*, **457**, 160–169.

149 Rezaei-Ghaleh, N., Zweckstetter, M., Morshedi, D., Ebrahim-Habibi, A., and Nemat-Gorgani, M. (2009) Amyloidogenic potential of alpha-chymotrypsin in different conformational states. *Biopolymers*, **91**, 28–36.

150 Kumar, S. and Udgaonkar, J.B. (2009) Structurally distinct amyloid protofibrils form on separate pathways of aggregation of a small protein. *Biochemistry*, **48**, 6441–6449.

151 Gosal, W.S., Clark, A.H., and Ross-Murphy, S.B. (2004) Fibrillar beta-lactoglobulin gels: Part 1. Fibril formation and structure. *Biomacromolecules*, **5**, 2408–2419.

152 Fezoui, Y. and Teplow, D.B. (2002) Kinetic studies of amyloid beta-protein fibril assembly. Differential effects of alpha-helix stabilization. *J. Biol. Chem.*, **277**, 36948–36954.

153 Li, H.T., Du, H.N., Tang, L., Hu, J., and Hu, H.Y. (2002) Structural transformation and aggregation of human alpha-synuclein in trifluoroethanol: non-amyloid component sequence is essential and beta-sheet formation is prerequisite to aggregation. *Biopolymers*, **64**, 221–226.

154 Grothe, H.L., Little, M.R., Cho, A.S., Huang, A.J., and Yuan, C. (2009) Denaturation and solvent effect on the conformation and fibril formation of TGFBI. *Mol. Vision*, **15**, 2617–2626.

155 Schmittschmitt, J.P. and Scholtz, J.M. (2003) The role of protein stability, solubility, and net charge in amyloid fibril formation. *Protein Sci.*, **12**, 2374–2378.

156 Cellmer, T., Douma, R., Huebner, A., Prausnitz, J., and Blanch, H. (2007) Kinetic studies of protein L aggregation and disaggregation. *Biophys. Chem.*, **125**, 350–359.

157 Carroll, A., Yang, W., Ye, Y., Simmons, R., and Yang, J.J. (2006) Amyloid fibril formation by a domain of rat cell adhesion molecule. *Cell Biochem. Biophys.*, **44**, 241–249.

158 Liu, W., Prausnitz, J., and Blanch, H.W. (2004) Amyloid fibril formation by peptide LYS (11–36) in aqueous trifluoroethanol. *Biomacromolecules*, **5**, 1818–1823.

159 Srisailam, S., Kumar, T.K., Rajalingam, D., Kathir, K.M., Sheu, H.S., Jan, F.J., Chao, P.C., and Yu, C. (2003) Amyloid-like fibril formation in an all beta-barrel protein. Partially structured intermediate state(s) is a precursor for fibril formation. *J. Biol. Chem.*, **278**, 17701–17709.

160 Zerovnik, E., Skarabot, M., Skerget, K., Giannini, S., Stoka, V., Jenko-Kokalj, S., and Staniforth, R.A. (2007) Amyloid fibril formation by human stefin B: influence of pH and TFE on fibril growth and morphology. *Amyloid*, **14**, 237–247.

161 Grudzielanek, S., Jansen, R., and Winter, R. (2005) Solvational tuning of the unfolding, aggregation and amyloidogenesis of insulin. *J. Mol. Biol.*, **351**, 879–894.

162 Murali, J. and Jayakumar, R. (2005) Spectroscopic studies on native and protofibrillar insulin. *J. Struct. Biol.*, **150**, 180–189.

163 Jean, L., Lee, C.F., Lee, C., Shaw, M., and Vaux, D.J. (2010) Competing discrete interfacial effects are critical for amyloidogenesis. *FASEB J.*, **24**, 309–317.

164 Zhao, H., Tuominen, E.K.J., and Kinnunen, P.K.J. (2004) Formation of amyloid fibers triggered by phosphatidylserine-containing membranes. *Biochemistry*, **43**, 10302–10307.

165 Alakoskela, J.-M.I., Jutila, A., Simonsen, A.C., Pirneskoski, J., Pyhäjoki, S.,

Turunen, R., Marttila, S., Mouritsen, O.G., Goormagntigh, E., and Kinnunen, P.K.J. (2006) Characteristics of fibers formed by cytochrome *c* and induced by anionic phospholipids. *Biochemistry*, **45**, 13447–13453.

166 Knight, J.D. and Miranker, A.D. (2004) Phospholipid catalysis of diabetic amyloid assembly. *J. Mol. Biol.*, **341**, 1175–1187.

167 Sparr, E., Engel, M.F.M., Sakharov, D.V., Sprong, M., Jacobs, J., de Kruijff, B., Höppner, J.W.M., and Killian, J.A. (2004) Islet amyloid polypeptide-induced membrane leakage involves uptake of lipids by forming amyloid fibers. *FEBS Lett.*, **577**, 117–120.

168 Chi, E.Y., Ege, C., Winans, A., Majewski, J., Wu, G., Kjaer, K., and Lee, K.Y. (2008) Lipid membrane templates the ordering and induces the fibrillogenesis of Alzheimer's disease amyloid-β peptide. *Proteins*, **72**, 1–24.

169 Matsuzaki, K. (2007) Physicochemical interactions of amyloid beta-peptide with lipid bilayers. *Biochim. Biophys. Acta*, **1768**, 1935–1942.

170 Matsuzaki, K. and Horikiri, C. (1999) Interactions of amyloid β-peptide (1–40) with ganglioside-containing membranes. *Biochemistry*, **38**, 4137–4142.

171 Curtain, C.C., ALi, F.E., Smith, D.G., Bush, A.I., Masters, C.L., and Barnham, K.J. (2003) Metal ions, pH and cholesterol regulate the interactions of Alzheimer's disease amyloid-beta peptide with membrane lipid. *J. Biol. Chem.*, **278**, 2977–2982.

172 Choo-Smith, L.P. and Surewicz, W.K. (1997) The interaction between Alzheimer amyloid β(1–40) peptide and ganglioside GM1-containing membranes. *FEBS Lett.*, **402**, 95–98.

173 Chauhan, A., Ray, I., and Chauhan, V.P. (2000) Interaction of amyloid beta-protein with anionic phospholipids: possible involvement of Lys28 and C-terminus aliphatic amino acids. *Neurochem. Res.*, **25**, 423–429.

174 Terzi, E., Hölzmann, G., and Seelig, J. (1995) Self-association of β-amyloid peptide (1–40) in solution and binding to lipid membranes. *J. Mol. Biol.*, **252**, 633–642.

175 Wong, P.T., Schauerte, J.A., Wisser, K.C., Ding, H., Lee, E.L., Steel, D.G., and Gafni, A. (2009) Amyloid-β membrane binding and permeabilization are distinct processes influenced separately by membrane charge and fluidity. *J. Mol. Biol.*, **386**, 81–96.

176 Barrow, C.J., and Zagorski, M.G. (1991) Solution structures of β peptide and its constituent fragments: relation to amyloid deposition. *Science*, **253**, 179–182.

177 Terzi, E., Hölzemann, G., and Seelig, J. (1997) Interaction of Alzheimer β-amyloid peptide(1–40) with lipid membrane. *Biochemistry*, **36**, 14845–14852.

178 Bokvist, M., Lindström, F., Watts, A., and Gröbner, G. (2004) Two types of Alzheimer's beta-amyloid (1–40) peptide membrane interactions: aggregation preventing transmembrane anchoring versus accelerated surface fibril formation. *J. Mol. Biol.*, **335**, 1039–1049.

179 Knight, J.D., Hebda, J.A., and Miranker, A.D. (2006) Conserved and cooperative assembly of membrane-bound alpha-helical states of islet amyloid polypeptide. *Biochemistry*, **45**, 9496–9508.

180 Lopes, D.H.J., Meister, A., Gohlke, A., Hauser, A., Blume, A., and Winter, R. (2007) Mechanism of IAPP fibrillation at lipid interfaces studied by infrared reflection absorption spectroscopy (IRRAS). *Biophys. J.*, **93**, 3132–3141.

181 Martinez, Z., Zhu, M., Han, S., and Fink, A.L. (2007) GM1 specifically interacts with alpha-synuclein and inhibits fibrillation. *Biochemistry*, **46**, 1868–1877.

182 Wimley, W.C., Hristova, K., Ladokhin, A.S., Silvestro, L., Axelsen, P.H., and White, S.H. (1998) Folding of β-sheet membrane proteins: a hydrophobic hexapeptide model. *J. Mol. Biol.*, **277**, 1091–1110.

183 Cho, W.-J., Trikha, S., and Jeremic, A.M. (2009) Cholesterol regulates assembly of human islet amyloid polypeptide on model membranes. *J. Mol. Biol.*, **393**, 765–775.

**184** Hou, X., Mechler, A., Martin, L.L., Aguilar, M.-I., and Small, D.H. (2008) Cholesterol and anionic phospholipids increase the binding of amyloidogenic transthyretin to lipid membranes. *Biochim. Biophys. Acta*, **1778**, 198–205.

**185** Subasinghe, S., Unabia, S., Barrow, C.J., Mok, S.S., Aguilar, M.-I., and Small, D.H. (2003) Cholesterol is necessary both for the toxic effect of Aβ peptides on vascular smooth muscle cells and for Aβ binding to vascular smooth muscle cell membranes. *J. Neurochem.*, **84**, 471–479.

**186** McLaurin, J., and Chakrabartty, A. (1996) Membrane disruption by Alzheimer β-amyloid peptides mediated through specific binding to either phospholipids or gangliosides. *J. Biol. Chem.*, **271**, 26482–26489.

**187** Kakio, A., Nishimoto, S., Yanagisawa, K., Kozutsumi, Y., and Matsuzaki, K. (2001) Cholesterol-dependent formation of GM1 ganglioside-bound amyloid β-protein, an endogenous seed for Alzheimer amyloid. *J. Biol. Chem.*, **276**, 24985–24990.

**188** Kakio, A., Nishimoto, S., Yanagisawa, K., Kozutsumi, Y., and Matsuzaki, K. (2002) Interactions of amyloid β-protein with various gangliosides in raft-like membranes: importance of GM1 ganglioside-bound form as an endogenous seed for Alzheimer amyloid. *Biochemistry*, **41**, 7385–7390.

# 4
# Interaction of hIAPP and Its Precursors with Model and Biological Membranes

*Katrin Weise, Rajesh Mishra, Suman Jha, Daniel Sellin, Diana Radovan, Andrea Gohlke, Christoph Jeworrek, Janine Seeliger, Simone Möbitz, and Roland Winter*

## 4.1
## Introduction

Type 2 diabetes mellitus (T2DM) is a complex degenerative disease that affects more than 150 million people worldwide. T2DM is characterized by a progressive deficit in β-cell function and mass with increased β-cell apoptosis [1]. In common with several neurodegenerative diseases, such as Parkinson's and Alzheimer's disease, the loss of β-cells is associated with the accumulation of amyloid deposits. Analysis of the extracellular amyloid plaques led to the identification of the 37-residue human islet amyloid polypeptide (hIAPP), also known as amylin, as the primary amyloidogenic and actual fibril-forming agent [2, 3]. hIAPP is a hormone that is synthesized in pancreatic islet β-cells and co-secreted with insulin under normal physiological conditions [4]. The exact physiological role of IAPP is still not clear, but it is known to be involved in glucose metabolism and to act as an insulin antagonist, that is, it is thought to play a paracrine inhibitory role in the regulation of insulin secretion [5, 6]. The sequence of islet amyloid polypeptide (IAPP) is closely homologous in humans and non-human primates, all of which develop T2DM. Conversely, rodent IAPP (mouse and rat) does not have the propensity to form amyloid fibrils due to proline substitutions in IAPP 20–29. Hence rodents do not spontaneously develop T2DM. Rat IAPP (rIAPP), which differs from the human variant only by six amino acid residues, is also often used as a control that does not form fibrils *in vitro* and is not toxic [7].

In the β-cells of the pancreas, hIAPP is synthesized as the prohormone precursor pro-islet amyloid polypeptide (proIAPP) along with pro-insulin (Scheme 4.1) [5, 6]. proIAPP, a 67 amino acid long polypeptide, is translated on the rough endoplasmic reticulum of pancreatic β-cells. After translation, it is transferred into the endoplasmic reticulum, where the disulfide bond between Cys13 and Cys18 is formed. The processing of proIAPP to IAPP occurs by the action of prohormone convertase enzymes, PC2 and PC1/3. Further post-translational events occur to remove the basic PC1/3 recognition motif from the N- and C-termini and Gly38 from the resulting

*Lipids and Cellular Membranes in Amyloid Diseases*, First Edition. Edited by Raz Jelinek.
© 2011 Wiley-VCH Verlag GmbH & Co. KGaA. Published 2011 by Wiley-VCH Verlag GmbH & Co. KGaA.

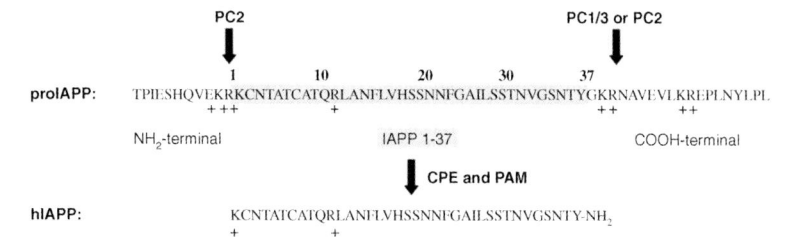

```
                    PC2                              PC1/3 or PC2
                     ↓                                    ↓
                  1       10        20        30    37
proIAPP:   TPIESHQVEKRKCNTATCATQRLANFLVHSSNNFGAILSSTNVGSNTYGKRNAVEVLKREPLNYLPL
                 +++     +                               ++         ++

             NH₂-terminal              IAPP 1-37                COOH-terminal
                                          ↓ CPE and PAM
hIAPP:                       KCNTATCATQRLANFLVHSSNNFGAILSSTNVGSNTY-NH₂
                                 +        +
```

**Scheme 4.1** Intracellular processing of pro-islet amyloid polypeptide (proIAPP) in islet β-cells is initiated by cleavage at its C-terminus by either prohormone convertase PC1/3 or PC2, although initiation by PC1/3 is favored. The two remaining basic residues are removed by carboxypeptidase E (CPE) and activation of the peptidyl amidating monooxygenase complex (PAM) leads to the removal of Gly38 and amidation of Tyr37. The 11 amino acid residues at the N-terminus are cleaved by PC2. Finally, a disulfide bridge is formed between Cys2 and Cys7, yielding mature hIAPP, which is stored in β-cell granules prior to its secretion.

C-terminus, and to amidate the free carboxyl end of Tyr37 that becomes exposed upon removal of Gly38 [8–10].

Although the initial stages of hIAPP amyloid formation are still unclear, autopsy studies of T2DM-affected human pancreas have indicated that the deposition of the islet amyloid is an extracellular event. However, studies on nude mice with transplanted human islet and on transgenic mice expressing hIAPP have shown that the early stages of islet amyloid formation may take place intracellularly [11, 12]. It has also been suggested that partially processed proIAPP may be important in early intracellular amyloid formation, and misprocessing of the peptide may trigger amyloid deposition [13–15]. As the amyloidogenic propensity decreases with increase in the number of charged residues directly flanking the hIAPP amyloid core, the partially processed and full-length proIAPP are less amyloidogenic, however [16].

Interestingly, increased levels of free fatty acids have also been observed in the prediabetic stage of T2DM [14]. Work from various groups [17–19] suggests that the β-cells exposed to hyperlipidemia remodulate their membranes by increasing the negatively charged lipid content. In the prediabetic stage, the anionic lipid content of the membrane seems to increase beyond the physiological range, that is, >10–30%. There is now clear evidence that hIAPP–lipid interactions might play an important role in the pathogenesis of islet amyloid disease by accelerating the formation of amyloid fibrils and toxic oligomers and triggering the permeabilization of lipid membranes – in fact, lipid membranes seem to constitute the preferred loci of aggregation and probably also of cytotoxicity [20–27]. For instance, it has been demonstrated that hIAPP aggregation is enhanced in the presence of membranes containing anionic lipids such as phosphatidylglycerol (PG) and phosphatidylserine (PS), and a mechanism of interaction has been proposed [22, 24, 27–29]. This phenomenon may be explained in terms of the interaction of membrane molecules with unfolded/partially folded peptide molecules, whereby the charged membrane acts as an anchor and screens the electrostatic repulsion between the charges of the monomeric conformations present in the proximity of the amyloidogenic sequence of the peptide. The sequence analysis of hIAPP shows that it has predominantly hydrophobic amino acid residues with only two positively charged amino acid residues at physiological pH. The precise molecular mechanism by which membranes catalyze the aggregation of hIAPP is not yet fully understood, however.

Although it is still under debate as to what species of hIAPP cause cellular toxicity, there are accumulating indications that an intermediate non-fibrillar oligomeric form of hIAPP rather than mature fibrils is responsible for membrane permeabilization which is proposed to be the main mechanism for hIAPP-induced cytotoxicity and death of insulin-producing β-cells [21, 30–34]. However, it has also been suggested that the process of hIAPP amyloid formation itself and not the presence of a particular species could be related to hIAPP cytotoxicity [22, 25, 26]. The disturbance of cellular homeostasis by hIAPP-induced cellular membrane disruption is thought to occur either through the formation of specific channels (pores) and/or through a nonspecific, detergent-like general disruption of the lipid membrane [25, 30, 31, 35–37].

Here, we discuss recent data on the amyloidogenic propensity and conformational properties of human proIAPP and hIAPP in the presence of neutral, negatively charged, and heterogeneous membrane interfaces as well as cellular membrane environments to reveal the underlying mechanisms of their aggregation/fibrillation reaction in the presence of various lipid membrane systems. Moreover, the effect of proIAPP on hIAPP fibrillation in the presence of lipid membranes is discussed. A combined spectroscopy and microscopy approach has been applied, using circular dichroism (CD), fluorescence spectroscopy and confocal microscopy, attenuated total reflection Fourier transform infrared (ATR-FTIR) and infrared reflection absorption (IRRA) spectroscopy, in addition to NMR spectroscopy and X-ray reflectivity (XRR), to gain a detailed understanding of peptide–membrane interactions. Atomic force microscopy (AFM) was used to characterize the size, growth kinetics, and morphology of hIAPP and proIAPP aggregates on a molecular level. Furthermore, to reveal the cytotoxic species of hIAPP aggregates, hIAPP species were isolated at distinct stages of the aggregation process and the thioflavin T (ThT) and water-soluble tetrazolium salt (WST-1) cell proliferation assay on INS-1E cells was used to characterize the kinetics of aggregation and the cytotoxicity of hIAPP aggregates, respectively.

## 4.2
## Results

### 4.2.1
### The Conformations of Native proIAPP and hIAPP in Bulk Solution

CD spectroscopy was applied to study the native conformations of hIAPP and proIAPP before and during aggregation. As an example, Figure 4.1 shows CD spectra of a 10 μM hIAPP solution (phosphate buffer, pH 7.4) at room temperature. The spectra indicate that the freshly prepared peptide adopts a predominantly random coil conformation along with some regular secondary structure

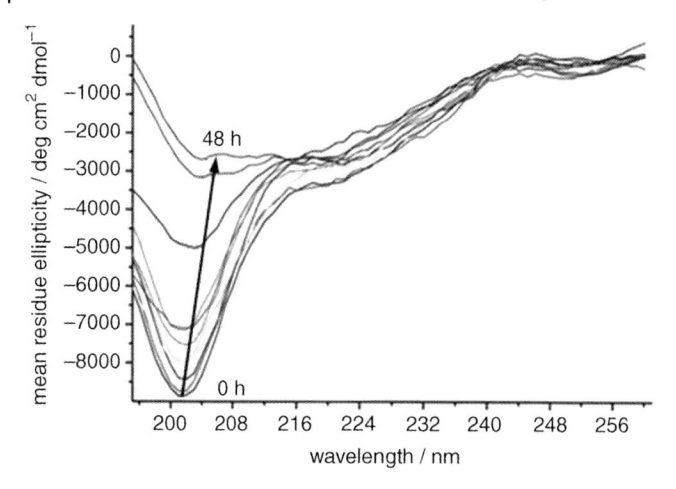

**Figure 4.1** Conformational transition monitored by CD spectroscopy for 10 μM hIAPP in 10 mM phosphate buffer, pH 7.4, at 21 °C. Freshly dissolved hIAPP displays a predominantly random coil conformation with a peak minimum at about 202 nm. This peak shifts upwards upon incubation for 48 h. The decrease in ellipticity with time indicates a decrease in the amount of soluble peptide molecules in the measured sample that is due to precipitation of fibrillar aggregates as revealed by AFM.

elements [38]. The CD analysis reveals a similar conformation for proIAPP, in agreement with FTIR data [38]. The monomeric structure of full-length hIAPP and the role of specific amino acid residues in the initiation of hIAPP fibril formation were recently elucidated by 2D-NMR spectroscopy (NOESY, TOCSY) [39]. The chemical shift dispersion of native hIAPP is characteristic of a largely disordered peptide. Time-lapse NMR data also strongly suggest that the N-terminal region of hIAPP (residues 1–17) is involved in the initial self-association of the peptide in bulk solution [39].

### 4.2.2
### Fibrillation Kinetics and Conformational Changes of hIAPP and proIAPP in the Presence of Anionic Lipid Bilayers

Amyloid fibril formation can be quantified by measuring the fluorescence intensity of the amyloid-specific, extrinsic fluorescence dye ThT at a wavelength of 482 nm after excitation at 440 nm [40]. The fluorescence emission of ThT is enhanced upon specific binding, most probably to the quaternary structure of amyloid fibrils that are rich in β-pleated sheets. Figure 4.2 shows the increase in the ThT emission maxima at 25 °C with time for 10 μM hIAPP and 40 μM proIAPP in the presence of 1,2-dioleoyl-sn-glycero-3-phosphocholine (DOPC) and 1,2-dioleoyl-sn-glycero-3-[phospho-rac-(1-glycerol)] (DOPG) at a ratio of 7:3 (w/w). For 10 μM proIAPP, no aggregation was observed over the time period shown. The detected fibrillation kinetics follow a typical nucleation-dependent aggregation mechanism, including a lag phase accompanied by nuclei formation, followed by an exponential growth phase and a

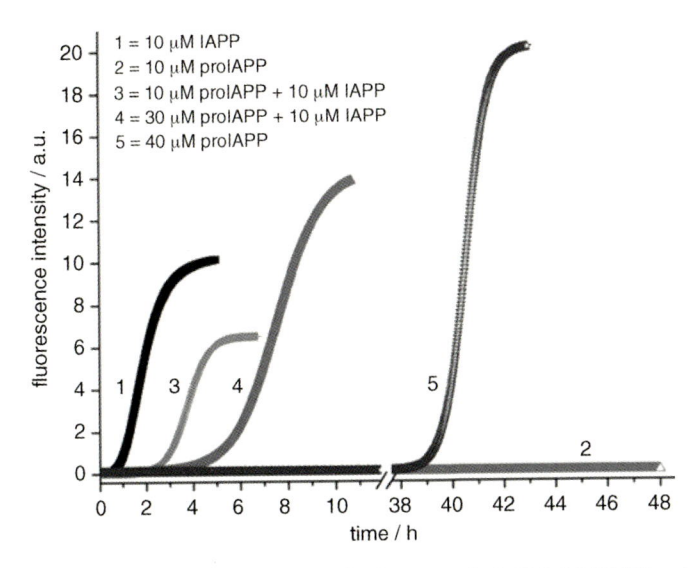

**Figure 4.2** ThT-monitored amyloid fibril formation kinetics of 10 μM hIAPP, 10 μM proIAPP, 10 μM proIAPP + 10 μM hIAPP, 30 μM proIAPP + 10 μM hIAPP, and 40 μM proIAPP in the presence of a DOPC–DOPG (7:3, w/w) lipid bilayer membrane. The experiments were carried out with 50 μM ThT in 10 mM phosphate buffer containing 100 mM NaCl at 25 °C, pH 7.4. Adapted from [38].

stationary phase, finally displaying a sigmoid-like curve. 10 μM hIAPP has a lag phase of 40 min under these experimental conditions, in comparison to 38 h for 40 μM proIAPP. Thus, hIAPP has a markedly higher amyloidogenic propensity than proIAPP. Still, the presence of the anionic DOPC–DOPG membrane is able to trigger the aggregation/fibrillation process of proIAPP. In the absence of the lipid bilayer, no aggregation was observed for 40 μM proIAPP at 25 °C for up to 7 days. Within this time period, aggregation of proIAPP could be determined neither at membranes with significantly less anionic lipid contents, nor at the pure zwitterionic DOPC lipid bilayer nor at heterogeneous model raft mixtures (see Section 4.2.4).

We also used XRR methodology to follow the temporal evolution of amyloid fibril formation at lipid interfaces on a molecular scale [41]. The XRR data for a DOPC–DOPG (7:3, w/w) lipid film at the water–air interface reveal marked hIAPP-induced changes in the presence of hIAPP (Figure 4.3a). The lipid film was spread on a solution of 1 μM hIAPP in 10 mM phosphate buffer, pH 7.0, in a Langmuir trough and immediately compressed to 30 mN m$^{-1}$. Inspection of the time evolution of the electron density profile perpendicular to the lipid interface, $\varrho(z)$, shows that immediately after the preparation of a 1 μM hIAPP solution, a film thickness of 22 Å is observed (Figure 4.3b), which is consistent with the thickness of the pure lipid film. Thereafter, $\varrho(z)$ of the headgroup and lipid chain region changes, indicating insertion of hIAPP into the lipid's upper chain region, where the aggregation process via oligomerization is initiated, which is followed by an increase in $\varrho(z)$ in the tail region. The size of the hIAPP oligomers is restricted to ~20 Å. The submicron-level

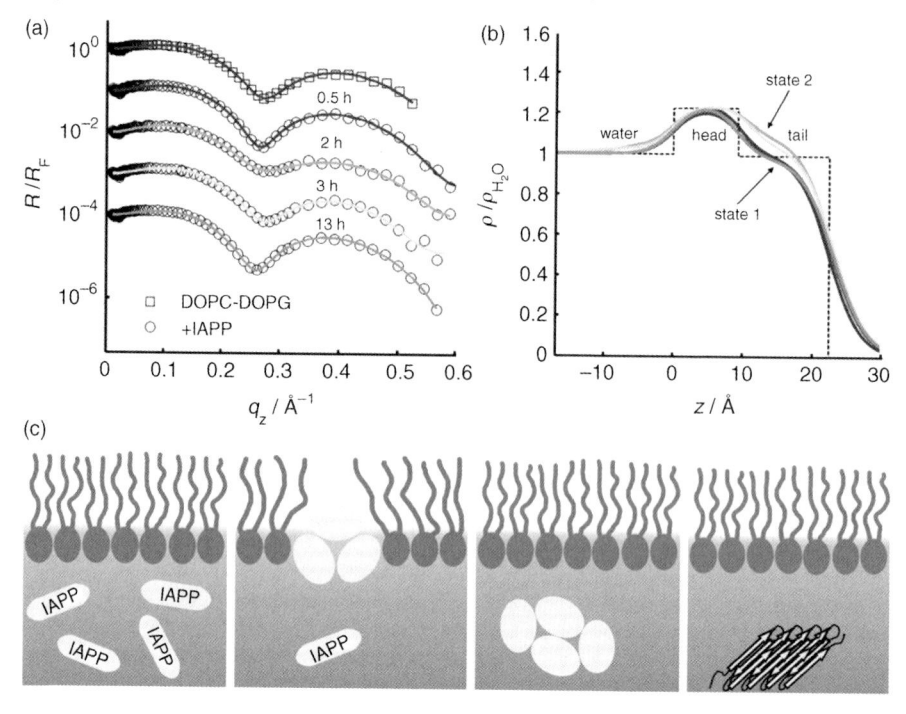

**Figure 4.3** Association of hIAPP with a negatively charged DOPC–DOPG (7:3, w/w) lipid film at the water–air interface. (a) Normalized XRR data at different time points (vertically shifted for better visibility by one order of magnitude). (b) Corresponding normalized electron density profiles. For a pure membrane and also for the membrane in the presence of hIAPP after 0.5 h no significant difference in the profile is observed (state 1). After 2 h, an enlarged headgroup region can be detected (state 2). Finally, the profile returns into its initial state 1 after 13 h. (c) Schematic illustration of the time evolution of the lipid-induced hIAPP aggregation process: hIAPP in bulk solvent, adsorption and oligomerization at the anionic lipid interface, detachment of larger aggregate structures, and fibril formation (arrows: cross β-sheet arrangement of fibrils) at hIAPP concentrations >5 µM. Adapted from [41].

morphology of hIAPP adsorbed at the anionic DOPC–DOPG lipid interface was also assessed by transferring the isolated hIAPP species onto mica and subsequent imaging with AFM (Figure 4.4a). The determined mean height ± standard deviation (s.d.) for the early hIAPP oligomers in the lipid monolayer is $5.2 \pm 3.0$ Å. Finally, after several hours, when the hIAPP aggregates increased in size, the electron density profile of the pure lipid film was restored, indicating that the larger hIAPP aggregates do not perturb the lipid film any longer, but rather detach from the lipid phase and enter the bulk solution (Figure 4.3c). An investigation of the bulk solution by AFM revealed that after 13.5 h the existing hIAPP oligomers had mean heights from $7.1 \pm 2.6$ to $26.7 \pm 10.2$ Å (Figure 4.4a). If the hIAPP concentration is high enough (>5 µM), fibrils are also formed (Figure 4.4b) [26].

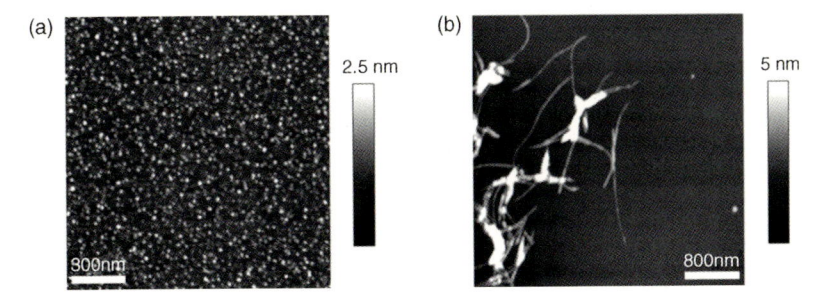

**Figure 4.4** (a) AFM image of isolated hIAPP species (1 μM hIAPP in 10 mM phosphate buffer, pH 7.0) aggregated for 13.5 h at an anionic DOPC–DOPG (7:3, w/w) lipid interface at room temperature. The determined mean height ± s.d. for the hIAPP oligomeric structures is 0.7 ± 0.3 nm. The vertical color scale from black to white corresponds to an overall height of 2.5 nm. (b) AFM image of fibrillar species grown from 10 μM hIAPP in 10 mM sodium phosphate, pH 7, at 25 °C for 96 h. hIAPP forms long unbranched fibrils showing heights of 3–5 nm, which is typical of hIAPP amyloid fibrils.

To evaluate the accompanying changes in secondary structure of hIAPP at the DOPC–DOPG membrane interface, ATR-FTIR spectra were collected over a time period of 20 h after injection of a solution of 10 μM hIAPP into an ATR cell containing the corresponding lipid bilayer system. In Figure 4.5a and b, spectra for selected time points are presented. During the aggregation process, the peak maximum of the amide-I′ band of hIAPP shifts from about 1646 cm$^{-1}$ towards 1624 cm$^{-1}$, indicating

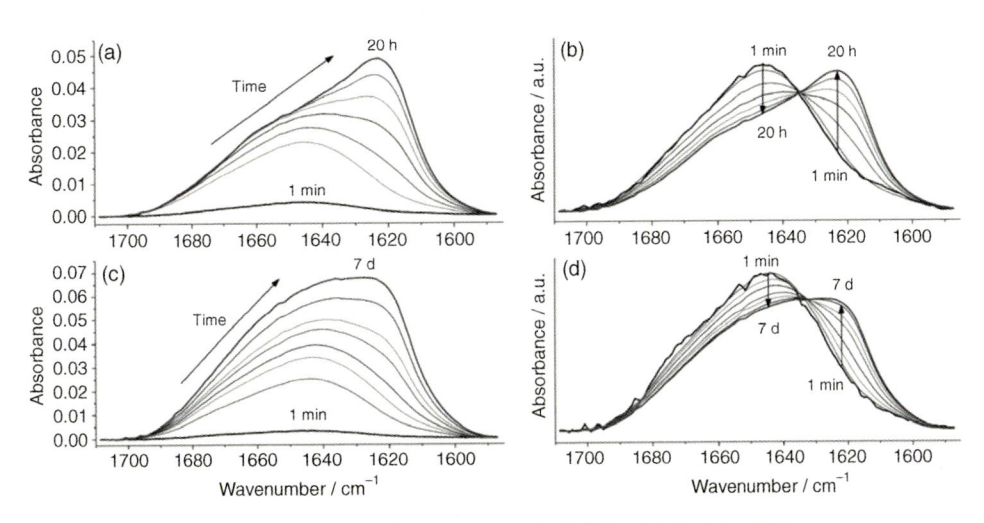

**Figure 4.5** Time evolution of the amide-I′ bands of hIAPP and proIAPP upon aggregation at 25 °C after injection into the ATR cell. (a, b) 10 μM hIAPP in the presence of phospholipid bilayers consisting of DOPC–DOPG (7:3, w/w). (c, d) 40 μM proIAPP in the presence of DOPC–DOPG (7:3, w/w). In (a) and (c), primary ATR-FTIR spectra after buffer and noise subtraction and baseline correction are shown. In (b) and (d), the normalized ATR-FTIR spectra are depicted. Adapted from [38].

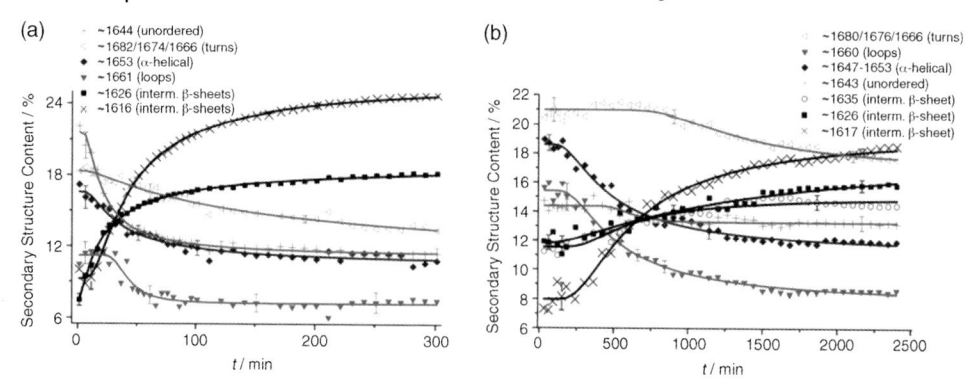

## 4 Interaction of hIAPP and Its Precursors with Model and Biological Membranes

**Figure 4.6** Time evolution of the secondary structural changes upon aggregation of 10 μM hIAPP (a) and 40 μM proIAPP (b) in the presence of a DOPC–DOPG (7:3, w/w) membrane derived from peak fitting of the normalized ATR-FTIR spectra. Adapted from [38].

a decrease in unordered conformations and a concomitant increase in β-structures. In Figure 4.5b, the corresponding normalized spectra are shown. Examination of the Fourier self-deconvolution (FSD) and second derivative of the normalized spectra revealed initially six distinct bands for the hIAPP conformation at about 1674, 1666, 1662, 1652, 1645, and 1638 cm$^{-1}$, with the most prominent band at 1645 cm$^{-1}$. During the aggregation process, new bands at ∼1684, 1626, and 1619 cm$^{-1}$ appeared. The strong bands at 1619 and 1626 cm$^{-1}$ indicate the development of intermolecular β-sheets. Peak fitting of the normalized spectra was carried out to allow quantitative analysis of the time-dependent changes in secondary structure, as shown in Figure 4.6a.

Recently, the association of hIAPP at various interfaces including neutral and charged phospholipids has been studied using infrared reflection absorption spectroscopy (IRRAS) [26]. The method uses polarization modulation of the IR light which allows the orientation of the protein at the lipid–water interface to be revealed. The results indicate that the interaction of hIAPP with the lipid interface is driven by the N-terminal part of the peptide and is largely driven by electrostatic interactions, as the protein is able to associate strongly only with negatively charged lipids. Furthermore, a two-step process is observed upon peptide binding, involving a conformational transition from a predominantly random coil structure via a largely α-helical to a β-sheet conformation, finally forming ordered fibrillar structures. The data thus support a multi-step process: membrane insertion through the N-terminus, which is followed by fibrillation mediated by the middle to C-terminal region via α-helical intermediate states. In fact, an initial helix-mediated association would lead to a high local concentration of an aggregation-prone sequence, which in turn would promote intermolecular β-sheet formation. As revealed by simulations of the IRRA spectra and complementary AFM studies, the fibrillar structures formed consist of parallel intermolecular β-sheets lying parallel to the lipid interface, but still contain a significant amount of turn structures.

The aggregation mechanism of proIAPP in contact with the anionic lipid membrane was analyzed in a similar manner. FTIR spectra were collected over a period of 7 days after the injection of a solution of 40 µM proIAPP into the ATR cell already containing the membrane. During the aggregation process, the peak maximum of the amide-I' band of proIAPP shifts from 1644 towards 1628 cm$^{-1}$. In Figure 4.5c and d, selected spectra of the time-dependent measurement are depicted. During the aggregation process, some of the bands shift to higher wavenumbers and two strong bands at ∼1628 and 1619 cm$^{-1}$ appear. The time evolution of the conformational changes is displayed in Figure 4.6b. Apart from the difference in time scale, the final aggregate structures of proIAPP and hIAPP at the lipid interface are similar. Compared with proIAPP, membrane-associated hIAPP exhibits a few percent less turn and more intermolecular β-sheet structures.

The amyloidogenic propensity of hIAPP has also been studied in the presence of its precursor protein proIAPP. The ThT binding assay for hIAPP–proIAPP peptide mixtures (1:1 and 1:3) after a mixing time of 4 min revealed aggregation, but with prolonged lag phases compared with pure hIAPP (Figure 4.2). The corresponding ATR-FTIR spectroscopic measurements also showed aggregation at the anionic lipid interface, the intermolecular β-sheet formation being slightly retarded compared with that of pure hIAPP. The addition of proIAPP to hIAPP delays the rate of hIAPP fibril formation, probably by some kind of complex formation. Hence it appears reasonable to speculate that the pro-region of proIAPP could serve to prolong the fibrillation kinetics of the highly amyloidogenic peptide hIAPP. In fact, it has been presumed that part of the function of hIAPP precursors, such as proIAPP, may be to prevent aggregation and amyloidogenesis in early stages of hIAPP biosynthesis and transport [16]. Interestingly, in particular in the case of excess proIAPP, increased aggregation levels have finally been observed, which could be attributed to a seed-induced co-fibrillation of proIAPP.

In addition to the negatively charged DOPC–DOPG membrane system, the interaction of hIAPP with a lipopolysaccharide (LPS) extracted from *Salmonella minnesota* strain R595 was investigated by synchrotron XRR. The deep rough mutant LPS Re that was used for this study is composed of a lipid moiety called lipid A, built of six hydrocarbon chains, two phosphorylated N-acetylglucosamines (GlcN), and two 2-keto-3-deoxyoctonoic (KDO) acid units connected by a 2–6 linkage to the lipid A. Each LPS molecule carries 3.5 negative net charges. For the XRR experiments, a monolayer of LPS Re was spread on a 1 µM solution of hIAPP in 10 mM phosphate buffer, pH 7.0. Subsequently, the lipid monolayer in the Langmuir trough was compressed to a film pressure of 30 mN m$^{-1}$, mimicking the physiological conditions of the outer membrane of Gram-negative bacteria. Inspection of the time evolution of the measured reflectivities revealed no significant changes of the LPS monolayer on the water–air interface within 10 h, indicating the absence of interaction of hIAPP with the negatively charged LPS film. This behavior could be explained by a compact configuration of the LPS molecules in the monolayer, where the negatively charged phosphate groups of the lipid A region are not exposed to the subphase and the partially negatively charged KDO subunits are not fully accessible to the bulk solution containing the hIAPP. Therefore, the repulsive interaction of the highly hydrophilic

KDO units and the hydrophobic hIAPP could outweigh the attractive electrostatic interaction of hIAPP and the partially accessible negative charges of the LPS.

### 4.2.3
### Effect of the Membrane-Mimicking Anionic Surfactant SDS on the Amyloidogenic Propensity of hIAPP and proIAPP

Micelle-forming amphiphilic molecules are often used as membrane mimetics or for encapsulating peptides, including amyloidogenic ones, in NMR spectroscopic studies. For example, sodium dodecyl sulfate (SDS) is an anionic detergent which has often been used as a membrane-mimicking agent and for studying whether anionic surfaces could be potential nucleating surfaces [42–44]. In phosphate buffer, pH 7.4, it exists either in a monomeric or in a micellar conformation, depending on its concentration [45]. SDS is known to interact with peptides and to induce order in the secondary structure of initially unordered peptides [46, 47]. Because of its mesophasic behavior, the different conformations adopted by SDS below and above the critical micelle concentration (cmc) ($\sim$8 mM in water [45]), can affect peptides differently. The unfolding of protein domains below the cmc value is largely driven by electrostatic interactions, followed by hydrophobic interactions. Above the cmc, the stabilization of unfolding transition states upon binding of cylindrical/spherical micelles to protein domains is the driving force to unfold the proteins. Cylindrical/spherical SDS micelles wrap themselves plastically around the protein and bind progressively tightly to the unfolding intermediates. In the same way, the effect of SDS concentration on amyloid fibril formation seems to vary. Several amyloidogenic peptides have been studied in the presence of SDS [48–50]. For α-synuclein, SDS concentration-dependent effects on the fibril formation process have been observed [48]. SDS at submicellar concentrations has been shown to induce spherical aggregates in the Aβ 1–42 peptide [49]. At low concentrations, SDS unfolds $\beta_2$-microglobulin into a helical-rich conformation, which serves as a precursor for amyloid fibril formation [50]. Recently, the interaction of SDS with hIAPP has been studied above the cmc to provide structural insights into micelle-bound hIAPP by NMR [51]. In the following, we discuss the effects of SDS on the conformational transitions and the amyloidogenic propensities of the physiologically active peptide hIAPP and its precursor peptide proIAPP in the presence of various concentrations of SDS.

The fibrillation kinetics followed by ThT binding to 10 μM hIAPP at room temperature start with an initial lag phase of $\sim$8 h that is followed by an $\sim$7 h exponential growth phase and is completed by a stationary phase. At equimolar concentrations of SDS and hIAPP (10 μM each), SDS is found to reduce the lag phase of the fibrillation kinetics of hIAPP to 4 h and leads to an increased slope of the exponential growth phase (Figure 4.7a). On raising the SDS concentration further to 100 μM, a reduction in the lag phase from 4 h to 30 min and also a further increase in the slope of the exponential growth phase are observed (Figure 4.7b). An additional increase in SDS concentration to 1 mM abolishes the lag phase and leads to a decrease in the slope of the exponential phase, thus indicating a slower elongation process in

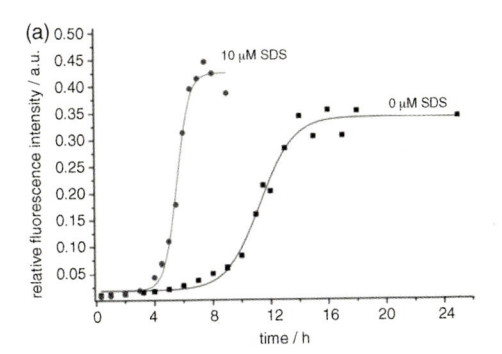

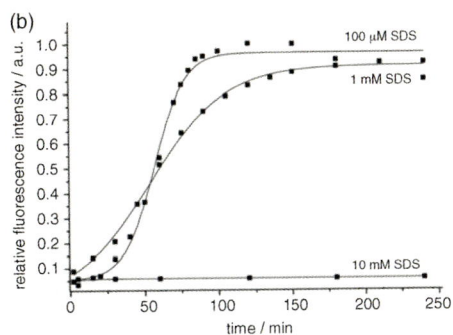

**Figure 4.7** Kinetics of 10 μM hIAPP amyloid fibril formation monitored by ThT fluorescence (10 μM ThT in 10 μM hIAPP, 10 mM phosphate buffer, pH 7.4, at 21 °C) in the presence of (a) 0 μM SDS (right) and 10 μM SDS (left), and (b) 100 μM SDS (top), 1 mM SDS (middle), and 10 mM SDS (bottom).

the presence of 1 mM versus 100 μM SDS. As revealed by AFM, all samples show the presence of amyloid fibrils with typical heights of ~3–6 nm, without any significant difference in fibrillar morphology. A representative image of fibrils grown in the presence of 100 μM SDS is shown in Figure 4.8a. Corresponding CD spectra reveal a distinct transition from a random coil to a β-sheet conformation in the presence of 100 μM SDS (Figure 4.8c). At concentrations from 10 μM to 1 mM, SDS is known to exist essentially as a monomer, which may facilitate hIAPP amyloid fibril formation by screening the electrostatic repulsion between monomeric conformations, thereby allowing a close proximity of the peptides.

At 10 mM SDS, the ThT intensity increase is only 6% of the maximum increase in intensity that was observed for the 100 μM SDS reaction mixture (Figure 4.7b), which indicates that this SDS concentration is able to inhibit the amyloid fibril formation. As revealed by CD analysis, already in the presence of 1 mM SDS, hIAPP starts to display α-helical conformation. This is because around the cmc, SDS micelles gradually appear that have the propensity to induce α-helical conformations in peptides. SDS above the cmc inhibits amyloid fibril formation by incorporating the peptides, thus preventing close proximity of neighboring peptide molecules, that is, amyloid formation [44, 51]. This binding of hIAPP to micelles induces an α-helical conformation in the peptide, as also demonstrated by NMR spectroscopy [51].

Different concentrations of SDS have differential effects on the fibrillation kinetics of hIAPP and proIAPP. hIAPP readily forms amyloid fibrils in bulk solution, whereas proIAPP does not. The presence of charged residues at the N-terminal site may be significant for the initiation of amyloid fibril formation. Another main difference between the primary sequence of hIAPP and proIAPP is the presence of two proline residues at positions 2 and 66 in the proIAPP sequence, which might inhibit the initiation of self-assembly of proIAPP molecules into amyloid fibrils as proline is known to be a β-sheet breaker [52]. rIAPP, which has three proline residues in its 20–29 amino acid sequence, also does not form amyloid fibrils, despite its 86% sequence homology with hIAPP [53].

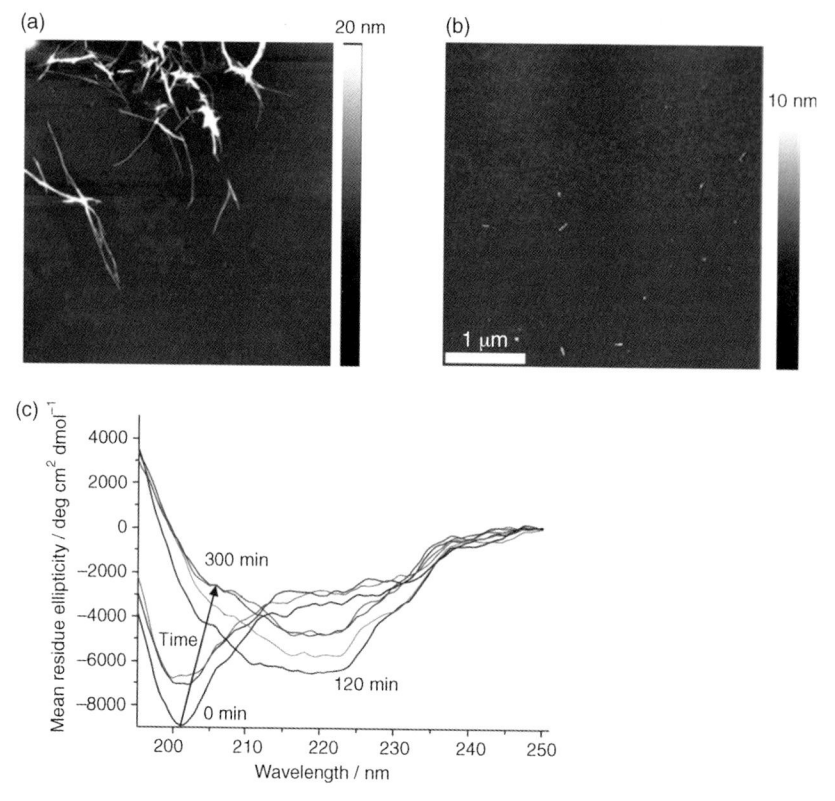

**Figure 4.8** AFM images of amyloid formation of (a) 10 μM hIAPP in the presence of 100 μM SDS and (b) of 10 μM proIAPP in the presence of 500 μM SDS. (c) Time-lapse CD spectra of 10 μM hIAPP in the presence of 100 μM SDS.

No fibrillation was observed for 10 μM proIAPP over 170 h as followed by ThT emission. Whereas SDS accelerates hIAPP fibril formation at a 1:1 molar ratio, it takes a 50-fold higher SDS concentration to initiate amyloid fibril formation in proIAPP. Since proIAPP has additional positive charges at pH 7.4 in comparison with hIAPP, it seems reasonable to assume that a higher concentration of SDS is needed to screen the electrostatic repulsion and thus foster the amyloidogenic propensity of proIAPP. In the presence of 500 μM SDS, 10 μM proIAPP shows an increase in the ThT emission intensity and the AFM images reveal short, twisted protofibrils with a height of 3–4 nm (Figure 4.8b). In the 1–10 mM SDS concentration range, retardation of proIAPP aggregation was observed. proIAPP attains a significant α-helical conformation (20%) at 500 μM SDS, whereby 100 μM SDS was still unable to induce α-helical structures in the initially unordered conformation. SDS at a concentration of 10 mM (slightly above the cmc) did not induce amyloidogenic propensity in proIAPP any longer within the time frame of this study.

To summarize these findings, SDS is able to accelerate the hIAPP amyloid fibril formation process only below its cmc. More importantly, SDS, below its cmc, can induce amyloidogenic propensity also in proIAPP, which does not self-assemble into amyloid fibrils in the absence of SDS, that is, in bulk solution. The presence of up to 1 mM SDS reduces the lag phase of hIAPP fibril formation. A further increase in SDS concentration to 10 mM causes the induction of α-helical conformations in the initially unordered conformation of hIAPP. In contrast to hIAPP, proIAPP itself does not self-assemble into amyloid fibrils, but in the range of 0.1–1.0 mM SDS it shows typical fibrillation kinetics and also fibrillar morphologies. SDS accelerates the amyloidogenic propensity of proIAPP only below its cmc. This indicates a putative role of negatively charged inner membranes or cellular moieties in the initiation of proIAPP amyloid formation, which in turn may act as a seed to trigger hIAPP amyloid formation upon secretion to the extracellular matrix. At higher surfactant concentrations, where the SDS exists predominantly as micelles, it prevents fibril formation and the peptides retain a stable α-helical conformation. The behavior of SDS described here is consistent with previously reported findings related to other amyloidogenic proteins and peptides [48–50]. What seems also to be clear is that protein fibrillation may occur by remarkably different mechanisms in the presence of SDS, testifying to the versatility of this process [54].

### 4.2.4
### hIAPP and proIAPP Aggregation and Fibrillation at Neutral Lipid Bilayers and Heterogeneous Model Raft Mixtures

For comparison, we also studied the interaction of hIAPP with a neutral, zwitterionic DOPC lipid monolayer. The XRR data revealed that the structure of the DOPC lipid layer is not significantly affected in the presence of hIAPP (data not shown [41]). This can be attributed to the absence of an electrostatically driven interaction between the positively charged hIAPP and an anionic lipid interface, in agreement with recent IRRA spectroscopic studies [26], and again demonstrates that the initial association of hIAPP with lipid membranes is mainly driven by electrostatic interactions [24, 26]. AFM was used to reveal the morphology of the peptides in the presence of the membranous interfaces. In the presence of the zwitterionic DOPC lipid monolayer, no significant aggregation was observed for hIAPP and a large amount of very small peptide particles could be detected by AFM (Figure 4.9). The mean particle height is $0.4 \pm 0.2$ nm, indicating the prevailing existence of hIAPP monomers [41]. In comparison, measurements under equivalent conditions revealed an aggregation process of hIAPP via oligomeric structures at the aggregation-fostering anionic DOPC–DOPG lipid interface (see Section 4.2.2). Corresponding ATR-FTIR spectroscopic measurements of 10 μM hIAPP in the presence of a neutral DOPC lipid bilayer confirmed these results in determining partial adsorption to the zwitterionic membrane only, but no significant aggregation at the DOPC interface [38]. The time evolution of the ATR-FTIR spectra of the amide-I′ band region shows a peak maximum at $\sim 1645$ cm$^{-1}$ and the appearance of a small shoulder in the intermolecular β-sheet region around 1615 cm$^{-1}$ after 20 h that does not evolve any further (Figure 4.10a). This indicates minor changes of the mainly unordered secondary structure of adsorbed hIAPP and the formation of smaller, less ordered (probably

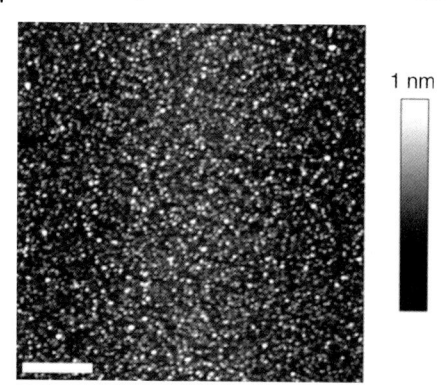

**Figure 4.9** AFM image of hIAPP species isolated after 4 h of incubation of 1 μM hIAPP in 10 mM phosphate buffer, pH 7.0, with a zwitterionic DOPC lipid monolayer. No aggregation of hIAPP is observed and hIAPP monomers can be detected that show a mean height ± s.d. of 0.4 ± 0.2 nm. In the AFM image, the vertical color scale from black to white corresponds to an overall height of 1 nm; the scale bar corresponds to 300 nm.

oligomeric) aggregate structures after long exposure times. Consequently, the initial association of hIAPP with lipid membranes that fosters hIAPP aggregation is mainly driven by electrostatic interactions, since the electrostatically driven interaction between the positively charged N-terminus of hIAPP and negatively charged lipids can only occur at anionic membrane interfaces. These results are in agreement with recent IRRA spectroscopic studies [26]. Finally, similarly to hIAPP, the ATR-FTIR spectra of the amide-I' band region of proIAPP reveal that 40 μM proIAPP does not aggregate in the presence of a pure DOPC bilayer (Figure 4.10b) [38].

In contrast to neutral one-component lipid bilayer systems, hIAPP is strongly adsorbed at heterogeneous lipid bilayer membranes. ATR-FTIR spectroscopic experiments were performed on both 10 μM hIAPP and 40 μM proIAPP in the

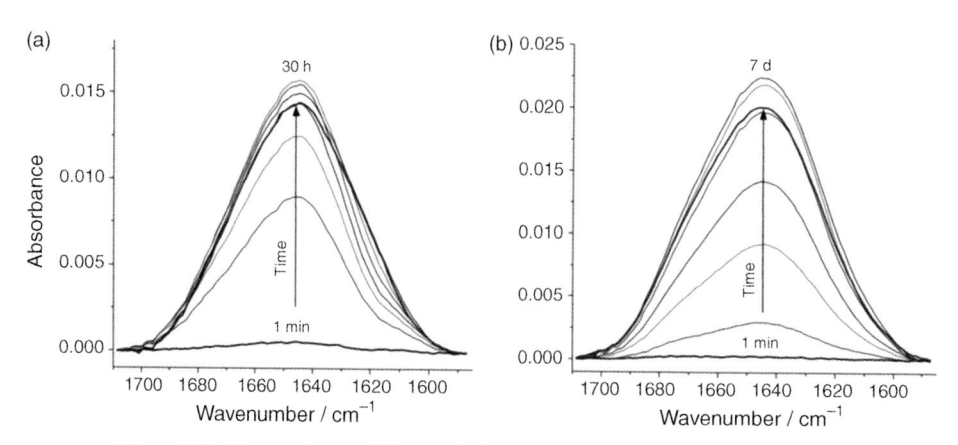

**Figure 4.10** Time evolution of the amide-I' bands of (a) 10 μM hIAPP and (b) 40 μM proIAPP in 10 mM phosphate buffer, pD 7.4 (total ionic strength 100 mM with NaCl) in the presence of a pure DOPC bilayer at 25 °C. Adapted from [38].

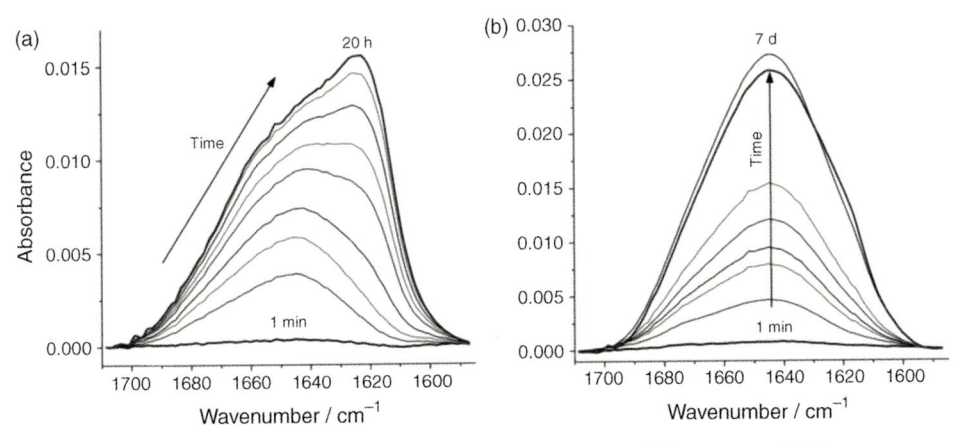

**Figure 4.11** Time evolution of the amide-I' bands of (a) 10 μM hIAPP and (b) 40 μM proIAPP in 10 mM phosphate buffer, pD 7.4 (total ionic strength 100 mM with NaCl) in the presence of the model lipid raft mixture consisting of DOPC–DPPC–Chol (1:2:1) at 25 °C. Adapted from [38].

presence of bilayers composed of the canonical lipid raft mixture DOPC–1,2-dipalmitoyl-sn-glycero-3-phosphocholine (DPPC)–cholesterol (Chol) (1:2:1), which segregates into liquid-ordered ($l_o$) and liquid-disordered ($l_d$) domains at room temperature [38]. Figure 4.11 shows the time evolution of the corresponding amide-I' bands. In the presence of the raft membrane, aggregation within 20 h can be observed for hIAPP, as indicated by the shift of the peak maximum to ~1624 cm$^{-1}$ (Figure 4.11a). The transition kinetics are only slightly slower than the aggregation reaction in the presence of the 30% anionic membrane, but the final ATR-intensity is lower (cf. Figure 4.5). For better visualization of the affinity of hIAPP to the different membrane systems, the adsorption kinetics are shown in Figure 4.12a as a plot of the amide-I' band intensity as a function of time. The data clearly demonstrate that the hIAPP–membrane interaction is more pronounced at anionic membranes. The corresponding time evolution of the intermolecular β-sheet content, which is characteristic of the aggregation propensity of hIAPP in the presence of the different membranes, is depicted in Figure 4.12b. In the case of the raft mixture, the content of hIAPP aggregate structures is about 60% lower than that in the presence of the anionic membrane [38].

The amide-I' band of proIAPP displays a small shoulder in the intermolecular β-sheet region after 3 days of exposure to the raft mixture and this shoulder is further evolved after 7 days (Figure 4.11b). Hence proIAPP seems to aggregate in the presence of the model raft membrane to only a minor extent, but the kinetics are very slow and the aggregate structure does not seem to be well ordered [38]. However, whereas both hIAPP and proIAPP do not aggregate significantly in the presence of the homogeneous zwitterionic DOPC membrane, there is evidence that aggregation takes place at heterogeneous zwitterionic membranes. One may speculate that an initial adsorption of hIAPP molecules at the rim of the lipid domains may serve as a nucleation site. The decrease in line tension at domain boundaries by peptide insertion might serve as a driving force of this process. In contrast, proIAPP seems to

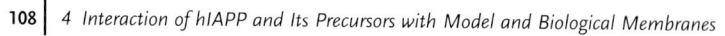

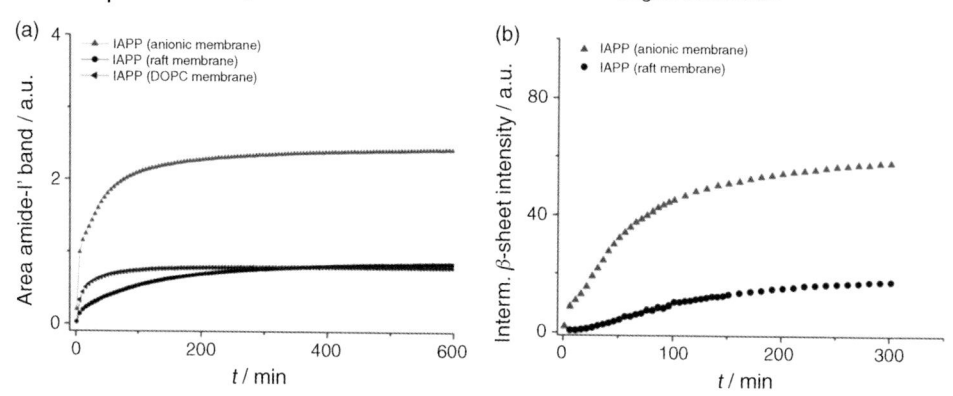

**Figure 4.12** (a) Adsorption kinetics of 10 µM hIAPP in 10 mM phosphate buffer, pD 7.4 (total ionic strength 100 mM with NaCl) in the presence of the anionic membrane DOPC–DOPG (7:3), the DOPC–DPPC–Chol (1:2:1) model raft membrane, and the pure DOPC lipid bilayer, followed by integration of the amide-I′ band between 1710 and 1585 cm$^{-1}$. (b) The corresponding intermolecular β-sheet content of hIAPP with respect to the whole amide-I′ area in the presence of the anionic and model raft membrane. Adapted from [38].

cluster into less-ordered structures and only to a minor extent in the presence of the heterogeneous raft membrane, with very slow kinetics.

In confocal fluorescence microscopy, phases can be identified as ordered or fluid-like disordered on the basis of the known partition behavior of appropriate fluorescence probes. While labeling the $l_d$ phase with the fluorescent marker N-Rh-DHPE (N-(lissamine rhodamine B sulfonyl)-1,2-dihexadecanoyl-sn-glycero-3-phosphoethanolamine triethylammonium salt; red channel), $l_o/l_d$ phase coexistence can be directly observed in giant unilamellar vesicles (GUVs) composed of the lipid mixture DOPC–DPPC–Chol (1:2:1) at room temperature [55]. For visualization of the hIAPP membrane interaction, 5 µM C-terminally, BODIPY-FL-labeled hIAPP (green channel) was added to the system. Partitioning of hIAPP into GUV membranes was already observed within 1–2 min after peptide addition and occurred preferentially into the $l_d$ phase, as indicated by the coinciding two channels (Figure 4.13a). Furthermore, the time-lapse confocal fluorescence microscopy study revealed drastic changes in the GUVs morphology due to the pronounced interaction with hIAPP, especially within the first hour of interaction [55, 56]. With time, hIAPP induces membrane permeabilization along with colocalization of the peptide and the fluid lipid domains (Figure 4.13b). After ∼72 h of incubation, intact GUVs are no longer detectable. However, superposition of the two fluorescent signals indicates the presence of amyloid–lipid colocalized complexes [55, 56]. Hence the fluorescence microscopy study on the model raft lipid system shows that hIAPP partitions preferentially into the $l_d$ phase of heterogeneous zwitterionic model membranes within minutes after addition of the peptide, which is followed by subsequent membrane permeabilization and pronounced association of lipids with the growing hIAPP fibrils.

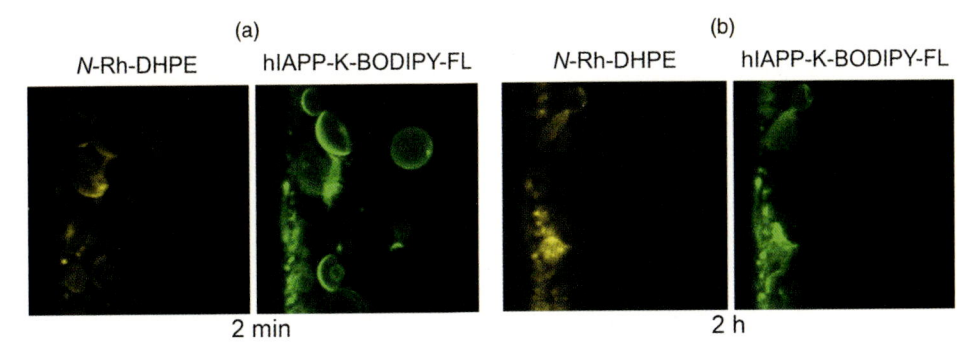

**Figure 4.13** Interaction of hIAPP with raft-like GUVs (DOPC–DPPC–Chol (1:2:1); ∼30 μm diameter) monitored at room temperature by confocal fluorescence microscopy. C-terminally labeled hIAPP (hIAPP-K-BODIPY-FL; right channel) inserts preferentially into the $l_d$ lipid phase (left channel with N-Rh-DHPE as marker for the $l_d$ phase; displayed in yellow for better visualization). (a) The insertion of hIAPP in the membrane is already detectable within 2 min after peptide addition (5 μM in water) and (b) the strongest effects are observable within the first hour of interaction. All images correspond to a scale of $40 \times 40$ μm. Adapted from [56].

To yield complementary structural data on a smaller, nanometer length scale, the interaction of hIAPP with zwitterionic lipid raft membranes was investigated by time-lapse tapping mode AFM. The results indicate a permeabilizing effect of hIAPP on DOPC–DPPC–Chol (1:2:1) lipid bilayers, which apparently occurs through a nonspecific, detergent-like mechanism that is presumably caused by hIAPP oligomers. Moreover, hIAPP fibril formation is not pronounced at zwitterionic raft membrane interfaces [56]. This is in agreement with the ATR-FTIR data, which showed slower and less pronounced adsorption and transition kinetics for hIAPP in the presence of zwitterionic raft membranes as compared with anionic membranes (see above). In addition, these results are in line with recent studies revealing an inhibitory effect of cholesterol on amylin aggregation [57]. Figure 4.14a shows a uniform membrane layer with isolated islands of fluid $l_d$ domains in a coherent pool of raft-like ($l_o$) protruding phase at the beginning of the AFM experiment ($t = 0$ min). The thickness of the lipid bilayer is ∼5.0 nm for the $l_o$ phase and ∼4.0 nm for the $l_d$ phase, that is, the coexisting phases can be clearly distinguished by a height difference of ∼1.0 nm, in good agreement with literature data [58]. After injection of a solution containing 1 μM hIAPP into the AFM fluid cell, disruption of the lipid raft membrane could be observed immediately (Figure 4.14a, $t \approx 29$ min). Since negatively charged surfaces are known to foster hIAPP fibrillation, the formation of small protofibrils on the exposed negatively charged mica surface can be detected within minutes after membrane disruption even at this low peptide concentration (cf. [56]). Unexpectedly, an hIAPP concentration-dependent rehealing of the membrane occurs subsequent to the initial membrane disruption, leading again to a coherent lipid rafts-containing membrane with small defects [56]. After long incubation times, mainly hIAPP protofibrils and oligomers can be detected that are preferentially localized in the fluid $l_d$ phase of the membrane or the rim of the demixed phases. Black areas in the AFM image represent membrane defects due to

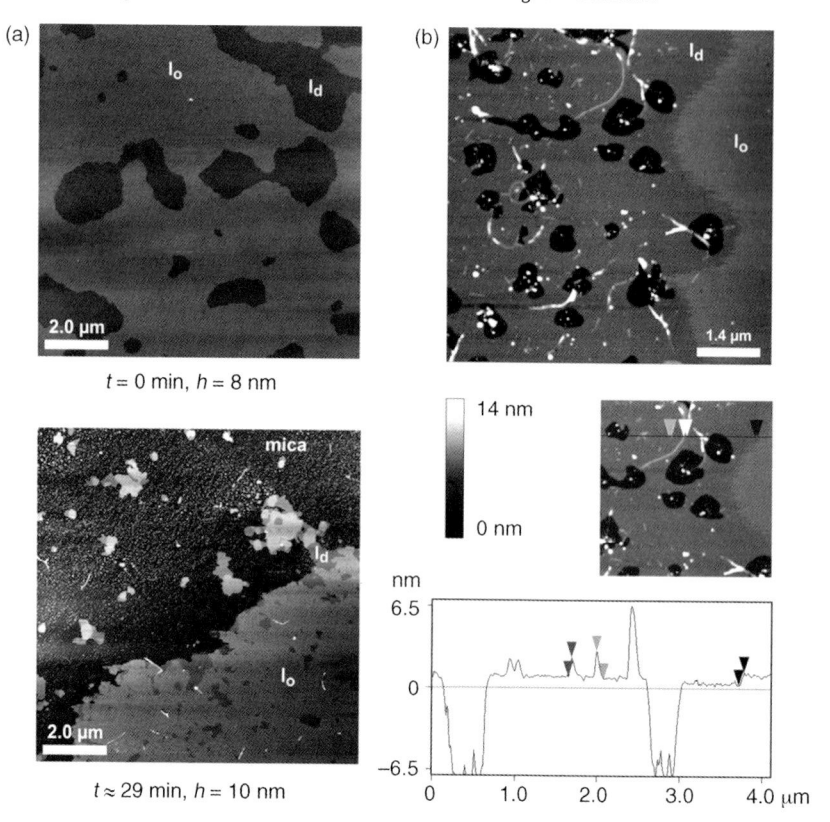

**Figure 4.14** (a) Tapping mode AFM images following the incubation of a DOPC–DPPC–Chol (1:2:1) membrane with 1 μM hIAPP in 10 mM phosphate buffer containing 100 mM NaCl, 5 mM $MgCl_2$, pH 7.4. Before injection of the peptide solution into the AFM fluid cell, a uniform lipid bilayer with coexisting domains in $l_o$ and $l_d$ phase can be detected ($t = 0$ min). The addition of hIAPP leads to an immediate disruption of the zwitterionic lipid raft bilayer ($t \approx 29$ min) and fibril formation is observed mainly at the negatively charged mica after membrane permeabilization. $h$ gives the overall height of the vertical color scale from black to white. (b) Localization of hIAPP in zwitterionic raft membranes. The AFM image indicates that even after long incubation times ($t \approx 30$ h) mostly short hIAPP protofibrils and oligomeric structures can be detected that reside mainly in the fluid $l_d$ phase of the DOPC–DPPC–Chol (1:2:1) membrane. The measurements were performed at room temperature with 1 μM hIAPP in 10 mM sodium phosphate, 100 mM NaCl, pH 7.4. In the upper part, the whole scan area is shown with a vertical color scale from black to white corresponding to an overall height of 14 nm. The concomitant section profile of the AFM image is given at the bottom. The horizontal black line in the figure is the localization of the section analysis shown at the bottom, indicating the vertical distances between pairs of arrows which are 1.51 nm (left), 2.18 nm (middle), and 1.01 nm (right) for the oligomer, protofibril and $l_o/l_d$ domain height difference, respectively. Adapted from [56].

hIAPP-induced membrane permeabilization (Figure 4.14b). The AFM measurements indicate a mean height of $3.3 \pm 2.1$ nm for hIAPP protofibrils with two main populations at around 1.9 and 5.3 nm, and a height of $3.6 \pm 2.3$ nm for hIAPP oligomers [56]. These values are in good agreement with data obtained from other AFM experiments [57–60].

The strong peptide–membrane interactions observed in these studies on the zwitterionic heterogeneous model membrane system, that involve rapid membrane damage by incorporation of hIAPP via oligomeric species, associated fibril growth, and lipid association with the growing fibrils, clearly demonstrate that the nature of the interaction is not necessarily electrostatic (a significant contribution in the case of anionic membranes), but rather that, once adsorbed and partially inserted (probably first with its N-terminus), hIAPP will interact with the membrane via hydrophobic interactions with its lipid chains. Several studies reported in the literature have previously indicated that the positively charged N-terminus is responsible for the partial insertion of hIAPP into negatively charged lipid membranes and that this process is mainly driven by electrostatic interactions [22, 24, 26, 28, 61]. Moreover, it has been suggested that interactions of hIAPP with lipid headgroups via charge interaction might lower the energy barrier for insertion of hIAPP between the lipid chains [61]. Hence the understanding of hIAPP–membrane interactions seems to be more complex, and electrostatic interactions are probably not the only type of interaction responsible for effective interaction of hIAPP with lipid membranes. In fact, hydrophobic interactions are also considered to be a major driving force in hIAPP fibril formation in the absence of membranes [62].

To diminish energetic costs, initial incorporation of the peptides at the rim of the domains of the heterogeneous membrane, where the volume and area fluctuations are most prominent, would be most likely, leading to a favorable decrease of the associated line energy [55]. The increase in membrane defects upon incorporation of the peptide will facilitate further peptide penetration, which leads to an increase in local peptide concentration in the membrane's core region, thus allowing condensation of oligomeric particles and fibril growth. As the biological membranes are composed of a plethora of more or less fluid-like and raft-like domains, the strong membrane-disrupting potency of hIAPP oligomers in heterogeneous membranes might be the reason for the hIAPP cytotoxicity. Furthermore, extraction of lipids may be a common feature of amyloid formation *in vivo*, since extracellular amyloid deposits from a number of diseases have been found to be associated with cellular lipids [63].

### 4.2.5
**Comparison with Insulin–Membrane Interaction Studies**

Compared with hIAPP–membrane interactions, a different scenario is observed for the interaction of the co-secreted hormone insulin with lipid membranes. The aggregation kinetics of insulin in the presence of various model biomembranes were investigated using the ThT fluorescence assay [27]. The lipid dynamics near the gel–fluid transition, the chain length of saturated lipids, and the presence of

1,2-dioleoyl-*sn*-glycero-3-phosphoethanolamine (DOPE) or 1,2-dioleoyl-*sn*-glycero-3-phospho-L-serine (DOPS) in DOPC vesicles modulate the aggregation kinetics of insulin in an indifferent, aggregation-accelerating or aggregation-inhibiting manner, subtly depending on the pH and the presence of salt. At variance with hIAPP, the rate of insulin aggregation in bulk solution dominates the overall aggregation process in most cases at low pH. The occurrence of dynamic line defects near the gel–fluid transition temperature facilitates partial membrane insertion of the protein, which in turn shields exposed hydrophobic protein patches from intermolecular association and hence inhibits aggregation. An exclusively aggregation-accelerating effect was observed in the presence of 0.1 M NaCl for all lipid additives investigated, which is likely due to enhanced surface accumulation of the protein. The release of curvature elastic stress in mixed DOPC–DOPE membranes and preferred interactions of insulin with carboxylic groups in DOPC–DOPS membranes favor increased surface accumulation. At neutral pH, partial insertion of insulin into the lipid bilayer is favored, which accounts for the aggregation-inhibiting effect of all lipid bilayer systems studied except those containing DOPS. Generally, the extent of inhibition increases with increase in the lipid chain length and the extent of curvature stress in mixed unsaturated lipid membranes and also when the gel–fluid transition temperature of pure phospholipids is approached. These results demonstrate that a delicate interplay between different physical and chemical properties of lipid membranes has to be taken into account in a detailed discussion of membrane-associated protein aggregation phenomena. The interactions of insulin with lipid bilayer membranes are rather weak, however.

### 4.2.6
### Cytotoxicity of hIAPP

To reveal the cytotoxic species of hIAPP aggregates, hIAPP species were isolated at distinct stages of the aggregation of 100 μM hIAPP in acetate buffer, pH 5.5, at 10 °C. The corresponding kinetics of the hIAPP aggregation process were determined by the ThT assay and the existing species were further characterized by AFM. The conditions were chosen based on previous NMR studies, indicating that at pH 5.5 His18 is protonated, which slows the fibrillation kinetics and thus allows for a more discrete identification and isolation of different on-pathway species [39]. Finally, a WST-1 cell proliferation assay on INS-1E cells revealed the cytotoxicity of the respective hIAPP aggregates.

In the ThT fluorimetric assay for 100 μM hIAPP in acetate buffer, pH 5.5, at 10 °C, slow kinetics of aggregation with a lag phase of ∼100 h were detected where the latter was followed by a slow exponential growth phase with the fibril formation process being almost complete after ∼400 h (Figure 4.15a). To further confirm the existing species at different stages of the aggregation process, hIAPP aliquots were taken on ice from the same stock solution at various time points and analyzed by tapping mode AFM. In the corresponding AFM images, small hIAPP oligomers appear nearly exclusively during the lag phase of the aggregation process, that is, between 0 and 100 h aggregation time [56]. The determined mean height $\pm$ s.d. is $0.7 \pm 0.2$ nm for

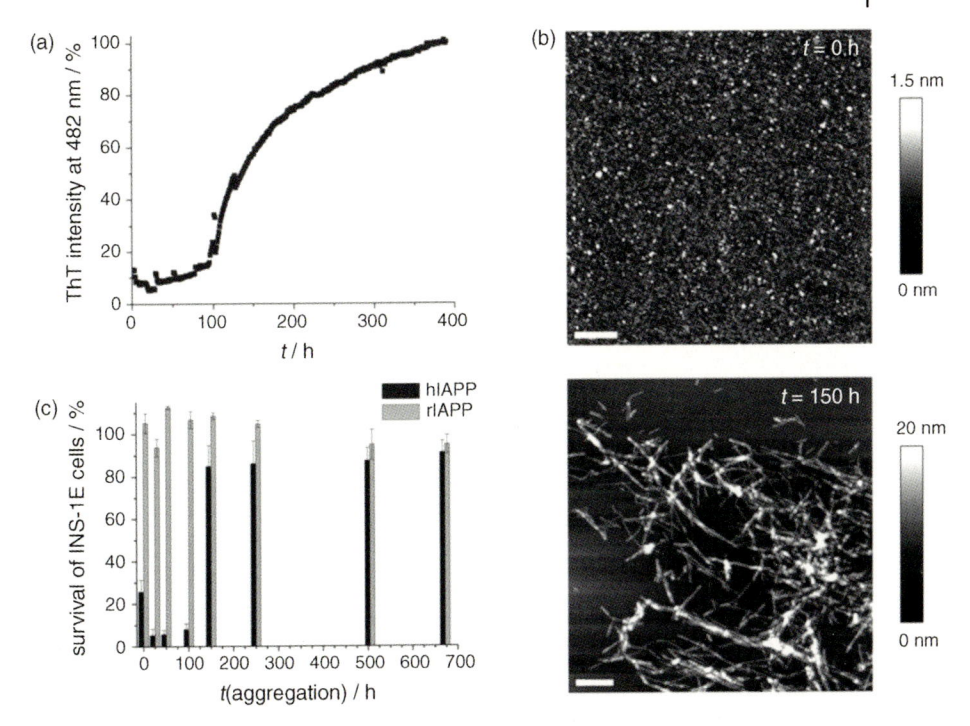

**Figure 4.15** (a) ThT assay monitoring the fibril formation of 100 µM hIAPP in 10 mM sodium acetate, pH 5.5, at 10 °C. hIAPP was incubated with a final ThT concentration of 50 µM and the aggregation process was followed for 400 h, after which the maximum ThT fluorescence intensity was almost reached, assuming the fibril formation process to be essentially complete. (b) Tapping mode AFM images of isolated hIAPP species at particular time points of the aggregation process (the scale bar corresponds to 250 nm). Aliquots were diluted to 5 µM hIAPP and 30 µl of this solution were deposited on mica and freeze-dried overnight. In line with the ThT assay results, mostly hIAPP oligomers can be detected at aggregation times up to 100 h, showing a mean height ± s.d. of $0.7 \pm 0.2$ nm at $t = 0$ h (top). In contrast, fibrils are the predominant species at aggregation times between 150 and 672 h (bottom). The determined mean height is $3.9 \pm 1.0$ nm at $t = 150$ h, indicative of protofibrils. (c) Cell viability of pancreatic β-cells (cell line INS-1E) after exposure to different isolated species of hIAPP and rIAPP at a final concentration of 10 µM in cell culture medium as estimated by the WST-1 cell proliferation assay. Cells were seeded into 96-well microtiter plates and allowed to grow for 24 h (5% $CO_2$, 37 °C) prior to a 24 h exposure to the isolated hIAPP and rIAPP species (aggregation time points: 0, 25, 50, 100, 150, 250, 504, and 672 h). The peptide was then removed and the cells were incubated for 24 h with the WST-1 reagent. Adapted from [56].

the oligomeric structures at $t = 0$ h (Figure 4.15b), which agrees well with the determined heights of isolated early oligomeric and larger oligomeric hIAPP species in a recent combined XRR and AFM study [41]. With time, fibrillation takes place that leads to an exponential growth phase in the ThT assay. As verified by AFM, hIAPP protofibrils are the causative species and a mean height of $3.9 \pm 1.0$ nm was determined for hIAPP protofibrils at $t = 150$ h (Figure 4.15b). Finally, higher order fibrillar structures are formed at long aggregation times up to 672 h, showing a mean height of 6.4 ± 1.8 nm [56]. The measured oligomer and fibril heights are in good agreement with previous AFM data for hIAPP [59, 60]. Overall, the findings of the ThT kinetics assay are in very good agreement with the verification of oligomeric and fibrillar hIAPP species at the particular stages of the aggregation process by AFM. Thereby, the slow kinetics of aggregation under such conditions allow the isolation of the different hIAPP aggregation states.

To reveal the cytotoxicity of the different hIAPP aggregates, hIAPP species from the same stock solution and at the same stages of the aggregation process were isolated to be tested for their cytotoxicity on INS-1E cells. In the WST-1 colorimetric assay, a color change to intense yellow can be quantified spectrophotometrically provided that viable cells with an intact mitochondrial succinate–tetrazolium–dehydrogenase system provoke an enzymatic reduction of the slightly red tetrazolium salt WST-1. Since the color change is directly proportional to the amount of mitochondrial dehydrogenase in the pancreatic β-cell culture, the net metabolic activity of the cells and thus the cell viability can be determined. The clonal β-cell line INS-1E, derived from the parental INS-1 cell line, represents a stable and reliable β-cell model that displays suitable insulin content and an adequate proliferation rate [64]. Figure 4.15c clearly shows that INS-1E cell viability depends on the aggregation time of hIAPP [56]. A strong correlation between the time frame of the lag phase (0–100 h) and the small cell survival rates between 3.5 and 25% detected for hIAPP species isolated in this time period implies an elevated cytotoxicity of the oligomeric hIAPP species. In contrast, the non-amyloidogenic rIAPP did not show any cytotoxicity in this time frame in the corresponding control experiments. As the nucleation and growth reaction proceeds, the ThT intensity increases, correlating with a significant decrease in cytotoxicity for samples tested within this time frame. Growing protofibrils (samples isolated after 150 and 250 h during the exponential growth phase) lead to 80–85% survival rates for 10 μM hIAPP-treated samples. As indicated by the AFM and ThT experiments, mostly mature hIAPP fibrils are present after an aggregation time of 3–4 weeks, which possess the lowest comparative cytotoxicity with cell viabilities reaching 90% [56].

Taken together, the findings support the recent hypothesis stating that membrane-permeant oligomers are the species responsible for hIAPP cytotoxicity since isolated oligomers showed the lowest cell survival rates in the WST-1 assay. Considering the proposed mechanism of interaction of hIAPP with lipid membranes (as discussed above), monomers and/or early oligomers probably accumulate at the membrane surface through electrostatic interactions between the peptide N-terminus and the zwitterionic lipid headgroups. For oligomeric species, hydrophobic interactions with the membrane acyl chain region might also play a significant role. Once initial binding is achieved, further peptides are recruited and elongation is promoted. The observed decrease in cytotoxicity with fibril growth would also imply that the toxic hIAPP oligomers are consumed during hIAPP fibril growth, suggesting that these cytotoxic oligomers are on-pathway and physiologically relevant [65]. In contrast, hIAPP fibrils were shown to be the least reactive species in the INS-1E cell assay, being already stabilized by intermolecular interactions between individual peptides. However, their increased size compared to individual monomeric peptides can still exert mechanical stress on the cell membranes, which might lead to perturbations in their barrier function and thus could account for the slight cytotoxicity observed. In addition, AFM experiments demonstrated that there are still a few oligomeric structures present, albeit outnumbered [56]. Amyloid fibrils also tend to accumulate extracellularly where they are relatively inert [21]. Furthermore, the results of these experiments are in line with a recent study showing that toxic IAPP oligomers form intracellularly within the secretory pathway, escaping and disrupting cellular membranes, including mitochondrial membranes [66]. Identifying and characterizing the major cytotoxic species is of great importance for amyloidogenic disease research, since understanding the mechanism of interaction and cytotoxicity on cellular membranes might provide a platform for successful strategies for the rational design of anti-amyloid (anti-fibrillogenesis, anti-cytotoxicity) inhibitors.

## 4.3 Conclusions

hIAPP, which is considered to be the primary culprit for β-cell loss in T2DM patients, is synthesized in the β-cells of the pancreas from its precursor, proIAPP, which may also be important in early intracellular amyloid formation. We compared the amyloidogenic propensities and conformational properties of hIAPP and proIAPP in the presence of various lipid membrane systems which have been discussed as primary loci for the initiation of the fibrillation reaction. CD and NMR studies verified the initial secondary structures of proIAPP and hIAPP to be predominantly unordered with small amounts of transient ordered secondary structure elements, and determined only minor differences between these two peptides. By using ATR-FTIR and ThT fluorescence spectroscopy as well as AFM, it has been shown that in the presence of negatively charged membranes, proIAPP exhibits a much higher amyloidogenic propensity than in the bulk solvent. Compared with hIAPP, it is far less amyloidogenic, however. Although the differences in the secondary structures of the aggregated species of hIAPP and proIAPP at the lipid interface are small, they are reflected in morphological changes. Unlike hIAPP, proIAPP essentially forms oligomeric-like structures at the lipid interface. The spectroscopic data also indicate that the interaction of hIAPP with the lipid interface is driven by the N-terminal part of the peptide and by electrostatic interactions, as the protein is able to associate most strongly with negatively charged lipids. A multi-step process is observed upon peptide binding, involving conformational transitions via α-helical structures. The formation of ordered secondary structure upon membrane binding is thought to result from the need to hydrogen bond the peptide backbone of the natively unstructured peptide when it partitions into the membrane [28].

In addition to the interaction with anionic membranes, the interaction with zwitterionic homogeneous (DOPC) and heterogeneous (DOPC–DPPC–Chol (1:2:1) model raft mixture) membranes has also been discussed. Both peptides did not aggregate significantly at DOPC bilayers. However, marked aggregation of hIAPP was observed in the presence of the heterogeneous model raft membrane. The content of hIAPP aggregate structures is about 60% lower than that in the presence of the anionic membrane. The aggregation may be initiated by adsorption of hIAPP at the boundaries of coexisting domains in the heterogeneous membrane, thus decreasing the unfavorable line tension between coexisting domains. The interface-induced self-assembly may thus serve as a heterogeneous nucleation-controlled aggregation mechanism. The concomitant increase in membrane defects upon incorporation of the peptide will facilitate further peptide incorporation and penetration leading to an increase in local peptide concentration in the membrane's core region, thus allowing condensation of oligomeric particles and finally fibril growth. Conversely, proIAPP clusters into less-ordered structures and to a minor extent at raft membranes. The addition of proIAPP to hIAPP retards the hIAPP fibrillation process also in the presence of negatively charged lipid bilayers. For excess proIAPP, increased aggregation levels are observed, which could be attributed to seed-induced co-fibrillation of proIAPP.

The rather strong hIAPP–membrane interaction observed in these studies on the zwitterionic heterogeneous model membrane system that involves rapid membrane damage by incorporation of hIAPP via oligomeric species, associated fibril growth, and lipid association with the growing fibrils clearly demonstrates that the nature of the interaction is not necessarily electrostatic (a significant contribution in the case of anionic membranes), but rather that, once adsorbed and partially inserted (probably first with its N-terminus), hIAPP will interact with the membrane via hydrophobic interactions with the lipid chains. As biological membranes are composed of a plethora of more or less fluid-like and raft-like domains (including negatively charged headgroups), the strong membrane-disrupting potency of hIAPP oligomers in heterogeneous membranes might be the reason for their cytotoxicity. Furthermore, extraction of lipids may be a common feature of amyloid formation *in vivo*. The recent hypothesis stating that membrane-permeant oligomers are the species responsible for hIAPP cytotoxicity is supported by the fact that isolated oligomers showed the lowest insulinoma cell survival rates in the WST-1 assay. The observed decrease in cytotoxicity with fibril growth also implies that the toxic hIAPP oligomers are consumed during hIAPP fibril growth, suggesting that these cytotoxic oligomers are on-pathway and physiologically relevant.

These conclusions are thus consistent with a recent study in which antibodies specific for toxic hIAPP oligomers and cryo-immunogold labeling in human IAPP transgenic mice, human insulinoma, and pancreas cells from humans with and without T2DM were studied. Gurlo *et al.* [66] concluded from these studies that IAPP toxic oligomers are formed intracellularly within the secretory pathway in T2DM. The hIAPP toxic oligomers appear to disrupt membranes of the secretory pathway and then, when adjacent to mitochondria, also disrupt mitochondrial membranes. Generally, the delivery of toxic oligomers to the cell membrane can, of course, also be expected to disrupt intra- and inter-cell communication. Furthermore, it is increasingly clear that the extracellular amyloid that originally drew attention to these conformational diseases plays a minor role in cytotoxicity, whereas small hydrophobic toxic oligomers play a more prominent role.

## Acknowledgments

Financial support from the Deutsche Forschungsgemeinschaft (DFG), the state of North-Rhine Westphalia and the EU (European Regional Development Fund) as well as the International Max Planck Research School of Chemical Biology (IMPRS-CB) is gratefully acknowledged. The INS-1E cell line was a gift from Dr. Pierre Maechler (University Hospital, Geneva, Switzerland) and the lipopolysaccharide (LPS) was provided by Prof. Dr. Klaus Brandenburg (Forschungszentrum Borstel, Germany).

## References

1 Kahn, S.E. (2001) The importance of β-cell failure in the development and progression of type 2 diabetes. *J. Clin. Endocrinol. Metab.*, **86** (9), 4047–4058.
2 Westermark, P., Wernstedt, C., Wilander, E., Hayden, D.W., O'Brien, T.D., and Johnson, K.H. (1987) Amyloid fibrils in human insulinoma and islets of Langerhans of the diabetic cat are derived from a neuropeptide-like protein also present in normal islet cells. *Proc. Natl. Acad. Sci. USA*, **84** (11), 3881–3885.
3 Cooper, G.J.S., Willis, A.C., Clark, A., Turner, R.C., Sim, R.B., and Reid, K.B.M. (1987) Purification and characterization of a peptide from amyloid-rich pancreases of type 2 diabetic patients. *Proc. Natl. Acad. Sci. USA*, **84** (23), 8628–8632.
4 Cooper, G.J.S., Willis, A.C., and Leighton, B. (1989) Amylin hormone. *Nature*, **340** (6231), 272.
5 Clark, A. and Nilsson, M.R. (2004) Islet amyloid: a complication of islet dysfunction or an aetiological factor in type 2 diabetes? *Diabetologia*, **47** (2), 157–169.
6 Kahn, S.E., D'Alessio, D.A., Schwartz, M.W., Fujimoto, W.Y., Ensinck, J.W., Taborsky, G.J. Jr., and Porte, D. Jr. (1990) Evidence of cosecretion of islet amyloid polypeptide and insulin by beta-cells. *Diabetes*, **39** (5), 634–638.
7 Moriarty, D.F. and Raleigh, D.P. (1999) Effects of sequential proline substitutions on amyloid formation by human amylin 20–29. *Biochemistry*, **38** (6), 1811–1818.
8 Sanke, T., Bell, G.I., Sample, C., Rubenstein, A.H., and Steiner, D.F. (1988) An islet amyloid peptide is derived from an 89-amino acid precursor by proteolytic processing. *J. Biol. Chem.*, **263** (33), 17243–17246.
9 Marcinkiewicz, M., Ramla, D., Seidah, N.G., and Chretien, M. (1994) Developmental expression of the prohormone convertases PC1 and PC2 in mouse pancreatic islets. *Endocrinology*, **135** (4), 1651–1660.
10 Marzban, L., Trigo-Gonzalez, G., Zhu, X., Rhodes, C.J., Halban, P.A., Steiner, D.F., and Verchere, C.B. (2004) Role of β-cell prohormone convertase (PC)1/3 in processing of pro-islet amyloid polypeptide. *Diabetes*, **53** (1), 141–148.
11 Westermark, G.T., Westermark, P., Eizirik, D.L., Hellerstrom, C., Fox, N., Steiner, D.F., and Anderson, A. (1999) Differences in amyloid deposition in islets of transgenic mice expressing human islet amyloid polypeptide versus human islets implanted into nude mice. *Metabolism*, **48** (4), 448–454.
12 Westermark, G.T., Steiner, D.F., Gebre-Medhin, S., Engstrom, U., and Westermark, P. (2000) Pro islet amyloid polypeptide (proIAPP) immunoreactivity in the islets of Langerhans. *Ups. J. Med. Sci.*, **105** (2), 97–106.
13 Paulsson, J.F. and Westermark, G.T. (2005) Aberrant processing of human proislet amyloid polypeptide results in increased amyloid formation. *Diabetes*, **54** (7), 2117–2125.
14 Paulsson, J.F., Andersson, A., Westermark, P., and Westermark, G.T. (2006) Intracellular amyloid-like deposits contain unprocessed pro-islet amyloid polypeptide (proIAPP) in β-cells of transgenic mice over expressing the gene

for human IAPP and tansplanted human islets. *Diabetologia*, **49** (6), 1237–1246.

15 Park, K. and Verchere, C.B. (2001) Identification of a heparin binding domain in the N-terminal cleavage site of pro-islet amyloid polypeptide. Implication for islet amyloid formation. *J. Biol. Chem.*, **276** (20), 16611–16616.

16 Yonemoto, I.T., Kroon, G.J., Dyson, H.J., Balch, W.E., and Kelly, J.W. (2008) Amylin proprotein processing generates progressively more amyloidogenic peptides that initially sample the helical state. *Biochemistry*, **47** (37), 9900–9910.

17 Clement, L., Kim-Sohn, K.A., Magnan, C., Kassis, N., Adnot, P., Kergoat, M., Assimacopoulos-Jeannet, F., Pénicaud, L., Hsu, F., Turk, J., and Ktorza, A. (2002) Pancreatic β-cell $\alpha_{2A}$ adrenoceptor and phospholipid changes in hyperlipidemic rats. *Lipids*, **37** (5), 501–506.

18 Wolf, B.A., Pasquals, S.M., and Turk, J. (1991) Free fatty acid accumulation in secretagogue-stimulated pancreatic islets and effects of arachidonate on depolarization-induced insulin secretion. *Biochemistry*, **30** (26), 6372–6379.

19 Turk, J., Wolf, B.A., Lefkowith, J.B., Stump, W.T., and McDaniel, M.L. (1986) Glucose-induced phospholipid hydrolysis in isolated pancreatic islets: quantitative effects on the phospholipid content of arachidonate and other fatty acids. *Biochim. Biophys. Acta*, **879** (3), 399–409.

20 Knight, J.D., Hebda, J.A., and Miranker, A.D. (2006) Conserved and cooperative assembly of membrane-bound alpha-helical states of islet amyloid polypeptide. *Biochemistry*, **45** (31), 9496–9508.

21 Janson, J., Ashley, R.H., Harrison, D., McIntyre, S., and Butler, P.C. (1999) The mechanism of islet amyloid polypeptide toxicity is membrane disruption by intermediate-sized toxic amyloid particles. *Diabetes*, **48** (3), 491–498.

22 Knight, J.D. and Miranker, A.D. (2004) Phospholipid catalysis of diabetic amyloid assembly. *J. Mol. Biol.*, **341** (5), 1175–1187.

23 Kurganov, B., Doh, M., and Arispe, N. (2004) Aggregation of liposomes induced by the toxic peptides Alzheimer's Aβs, human amylin and prion (106–126): facilitation by membrane-bound $G_{M1}$ ganglioside. *Peptides*, **25** (2), 217–232.

24 Engel, M.F., Yigittop, H., Elgersma, R.C., Rijkers, D.T., Liskamp, R.M., de Kruijff, B., Höppener, J.W., and Antoinette Killian, J. (2006) Islet amyloid polypeptide inserts into phospholipid monolayers as monomer. *J. Mol. Biol.*, **356** (3), 783–789.

25 Maarten, F.M., Engel, L.K., Kleijer, C.C., Meeldijk, H.J.D., Jacobs, J., Verkleij, A.J., de Kruijff, B., Killian, J.A., and Höppener, J.W.M. (2008) Membrane damage by human islet amyloid polypeptide through fibril growth at the membrane. *Proc. Natl. Acad. Sci. USA*, **105** (16), 6033–6038.

26 Lopes, D.H., Meister, A., Gohlke, A., Hauser, A., Blume, A., and Winter, R. (2007) Mechanism of islet amyloid polypeptide fibrillation at lipid interfaces studied by infrared reflection absorption spectroscopy. *Biophys. J.*, **93** (9), 3132–3141.

27 Grudzielanek, S., Smirnovas, V., and Winter, R. (2007) The effects of various membrane physical-chemical properties on the aggregation kinetics of insulin. *Chem. Phys. Lipids*, **149** (1–2), 28–39.

28 Jayasinghe, S.A. and Langen, R. (2007) Membrane interaction of islet amyloid polypeptide. *Biochim. Biophys. Acta*, **1768**, 2002–2009.

29 Apostolidou, M., Jayasinghe, S.A., and Langen, R. (2008) Structure of alpha-helical membrane-bound human islet amyloid polypeptide and its implications for membrane-mediated misfolding. *J. Biol. Chem.*, **283** (25), 17205–17210.

30 Anguiano, M., Nowak, R.J., and Lansbury, P.T. Jr. (2002) Protofibrillar islet amyloid polypeptide permeabilizes synthetic vesicles by a pore-like mechanism that may be relevant to type II diabetes. *Biochemistry*, **41** (38), 11338–11343.

31 Mirzabekov, T.A., Lin, M.-C., and Kagan, B.L. (1996) Pore formation by the cytotoxic islet amyloid peptide amylin. *J. Biol. Chem.*, **271** (4), 1988–1992.

32 Meng, X., Fink, A.L., and Uversky, V.N. (2008) The effect of membranes on the *in vitro* fibrillation of an amyloidogenic light-chain variable-domain SMA. *J. Mol. Biol.*, **381** (4), 989–999.

33 Porat, Y., Kolusheva, S., Jelinek, R., and Gazit, E. (2003) The human islet amyloid polypeptide forms transient membrane-active prefibrillar assemblies. *Biochemistry*, **42** (37), 10971–10977.

34 Kayed, R., Sokolov, Y., Edmonds, B., McIntire, T.M., Milton, S.C., Hall, J.E., and Glabe, C.G. (2004) Permeabilization of lipid bilayers is a common conformation-dependent activity of soluble amyloid oligomers in protein misfolding diseases. *J. Biol. Chem.*, **279** (45), 46363–46366.

35 Green, J.D., Kreplak, L., Goldsbury, C., Li Blatter, X., Stolz, M., Cooper, G.S., Seelig, A., Kistler, J., and Aebi, U. (2004) Atomic force microscopy reveals defects within mica supported lipid bilayers induced by the amyloidogenic human amylin peptide. *J. Mol. Biol.*, **342** (3), 877–887.

36 Quist, A., Doudevski, I., Lin, H., Azimova, R., Ng, D., Frangione, B., Kagan, B., Ghiso, J., and Lal, R. (2005) Amyloid ion channels: a common structural link for protein-misfolding disease. *Proc. Natl. Acad. Sci. USA*, **102** (30), 10427–10432.

37 Soong, R., Brender, J.R., Macdonald, P.M., and Ramamoorthy, A. (2009) Association of highly compact type II diabetes related islet amyloid polypeptide intermediate species at physiological temperature revealed by diffusion NMR spectroscopy. *J. Am. Chem. Soc.*, **131** (20), 7079–7085.

38 Jha, S., Sellin, D., Seidel, R., and Winter, R. (2009) Amyloidogenic propensities and conformational properties of ProIAPP and IAPP in the presence of lipid bilayer membranes. *J. Mol. Biol.*, **389** (5), 907–920.

39 Mishra, R., Geyer, M., and Winter, R. (2009) NMR spectroscopic investigation of early events in IAPP amyloid fibril formation. *ChemBioChem*, **10** (11), 1769–1772.

40 Naiki, H., Higuchi, K., Hosokawa, M., and Takeda, T. (1989) Fluorometric determination of amyloid fibrils *in vitro* using the fluorescent dye, thioflavin T[1]. *Anal. Biochem.*, **177** (2), 244–249.

41 Evers, F., Jeworrek, C., Tiemeyer, S., Weise, K., Sellin, D., Paulus, M., Struth, B., Tolan, M., and Winter, R. (2009) Elucidating the mechanism of lipid membrane-induced IAPP fibrillogenesis and its inhibition by the red wine compound resveratrol: a synchrotron X-ray reflectivity study. *J. Am. Chem. Soc.*, **131** (27), 9516–9521.

42 Pappalardo, G., Milardi, D., Magri, A., Attanasio, F., Impellizzeri, G., La Rosa, C., Grasso, D., and Rizzarelli, E. (2007) Environmental factors differently affect human and rat IAPP: conformational preferences and membrane interactions of IAPP17–29 peptide derivatives. *Chemistry*, **13** (36), 10204–10215.

43 Lindberg, M., Biverståhl, H., Gräslund, A., and Mäler, L. (2003) Structure and positioning comparison of two variants of penetratin in two different membrane mimicking systems by NMR. *Eur. J. Biochem.*, **270** (14), 3055–3063.

44 Mascioni, A., Porcelli, F., Ilangovan, U., Ramamoorthy, A., and Veglia, G. (2003) Conformational preferences of the amylin nucleation site in SDS micelles: an NMR study. *Biopolymers*, **69** (1), 29–41.

45 Henry, G.D. and Sykes, B.D. (1994) Methods to study membrane protein structure in solution. *Methods Enzymol.*, **239**, 515–535.

46 Reynolds, J.A. and Tanford, C. (1970) The gross conformation of protein–sodium dodecyl sulfate complexes. *J. Biol. Chem.*, **245** (19), 5161–5165.

47 Mattice, W.L., Riser, J.M., and Clark, D.S. (1976) Conformational properties of the complexes formed by proteins and sodium dodecyl sulfate. *Biochemistry*, **15** (19), 4264–4272.

48 Ahmad, M.F., Ramakrishna, T., Raman, B., and Rao, Ch.M. (2006) Fibrillogenic and non-fibrillogenic ensembles of SDS-bound human alpha-synuclein. *J. Mol. Biol.*, **364** (5), 1061–1072.

49 Rangachari, V., Moore, B.D., Reed, D.K., Sonoda, L.K., Bridges, A.W., Conboy, E., Hartigan, D., and Rosenberry, T.L. (2007) Amyloid-β(1–42) rapidly forms protofibrils and oligomers by distinct pathways in low concentrations of sodium dodecylsulfate. *Biochemistry*, **46** (43), 12451–12462.

50 Yamamoto, S., Hasegawa, K., Yamaguchi, I., Tsutsumi, S., Kardos, J., Goto, Y., Gejyo,

F., and Naiki, H. (2004) Low concentrations of sodium dodecyl sulfate induce the extension of $\beta_2$-microglobulin-related amyloid fibrils at a neutral pH. *Biochemistry*, **43** (34), 11075–11082.

51 Patil, S.M., Xu, S., Sheftic, S.R., and Alexandrescu, A.T. (2009) Dynamic α-helix structure of micelle-bound human amylin. *J. Biol. Chem.*, **284** (18), 11982–11991.

52 Wood, S.J., Wetzel, R., Martin, J.D., and Hurle, M.R. (1995) Prolines and amyloidogenicity in fragments of the Alzheimer's peptide β/A4. *Biochemistry*, **34** (3), 724–730.

53 Westermark, P., Engstrom, U., Johnson, K.H., Westermark, G.T., and Betsholtz, C. (1990) Islet amyloid polypeptide: pinpointing amino acid residues linked to amyloid fibril formation. *Proc. Natl. Acad. Sci. USA*, **87** (13), 5036–5040.

54 Giehm, L., Oliveira, C.L.P., Christiansen, G., Pedersen, J.S., and Otzen, D.E. (2010) SDS-induced fibrillation of α-synuclein: an alternative fibrillation pathway. *J. Mol. Biol.*, **401**, 115–133.

55 Radovan, D., Opitz, N., and Winter, R. (2009) Fluorescence microscopy studies on islet amyloid polypeptide fibrillation at heterogeneous and cellular membrane interfaces and its inhibition by resveratrol. *FEBS Lett.*, **583** (9), 1439–1445.

56 Weise, K., Radovan, D., Gohlke, A., Opitz, N., and Winter, R. (2010) Interaction of hIAPP with model raft membranes and pancreatic β-cells: cytotoxicity of hIAPP oligomers. *ChemBioChem*, **11** (9), 1280–1290.

57 Cho, W.-J., Jena, B.P., and Jeremic, A.M. (2008) Nano-scale imaging and dynamics of amylin–membrane interactions and its implication in type II diabetes mellitus. *Methods Cell. Biol.*, **90**, 67–286.

58 Rinia, H.A., Snel, M.M.E., van der Eerden, J.P.J.M., and de Kruijff, B. (2001) Visualizing detergent resistant domains in model membranes with atomic force microscopy. *FEBS Lett.*, **501** (1), 92–96.

59 Goldsbury, C., Kistler, J., Aebi, U., Arvinte, T., and Cooper, G.J.S. (1999) Watching amyloid fibrils grow by time-lapse atomic force microscopy. *J. Mol. Biol.*, **285** (1), 33–39.

60 Green, J.D., Goldsbury, C., Kistler, J., Cooper, G.J.S., and Aebi, U. (2004) Human amylin oligomer growth and fibril elongation define two distinct phases in amyloid formation. *J. Biol. Chem.*, **279** (13), 12206–12212.

61 Jeworrek, C., Hollmann, O., Steiz, R., Winter, R., and Czeslik, C. (2009) Interaction of IAPP and insulin with model interfaces studied using neutron reflectometry. *Biophys. J.*, **96** (3), 1115–1123.

62 Radovan, D., Smirnovas, V., and Winter, R. (2008) Effect of pressure on islet amyloid polypeptide aggregation: revealing the polymorphic nature of the fibrillation process. *Biochemistry*, **47** (24), 6352–6360.

63 Gellermann, G.P., Appel, T.R., Tannert, A., Radestock, A., Hortschansky, P., Schroeckh, V., Leisner, C., Lütkepohl, T., Shtrabsurg, S., Röcken, C., Pras, M., Linke, R.P., Diekmann, S., and Fändrich, M. (2005) Raft lipids as common components of human extracellular amyloid fibrils. *Proc. Natl. Acad. Sci. USA*, **102** (18), 6297–6302.

64 Merglen, A., Theander, S., Rubi, B., Chaffard, G., Wollheim, C.B., and Maechler, P. (2004) Glucose sensitivity and metabolism-secretion coupling studied during two-year continuous culture in INS-1E insulinoma cells. *Endocrinology*, **145** (2), 667–678.

65 Engel, M.F.M. (2009) Membrane permeabilization by islet amyloid polypeptide. *Chem. Phys. Lipids*, **160** (1), 1–10.

66 Gurlo, T., Ryazantsev, S., Huang, C.J., Yeh, M.W., Reber, H.A., Hines, O.J., O'Brien, T.D., Glabe, C.G., and Butler, P.C. (2010) Evidence for proteotoxicity in β cells in type 2 diabetes: toxic islet amyloid polypeptide oligomers form intracellularly in the secretory pathway. *Am. J. Pathol.*, **176** (2), 861–869.

# 5
# Amyloid Polymorphisms: Structural Basis and Significance in Biology and Molecular Medicine
*Massimo Stefani*

## 5.1
## Introduction

Amyloid diseases, the most clinically relevant protein deposition pathologies, include over 20 familial, sporadic, or transmissible conditions, in some cases with dramatic prevalence in the population of developed countries. A number of these pathologies affect the brain and the central nervous system (e.g., Alzheimer's, Parkinson's, Huntington, and Creutzfeldt–Jakob diseases, several ataxias, and other neurodegenerative conditions) or peripheral tissues and organs (type 2 diabetes mellitus, several systemic amyloidoses, dialysis-related amyloidoses, and others) (reviewed in [1, 2]). A shared feature of these conditions is the presence, in specific tissues and organs, of fibrillar deposits resulting from the polymerization of one out of a limited number of peptides or proteins characteristic of each specific disease or of a group of strictly similar pathological conditions (reviewed in [1–3]). In many systemic amyloidoses, the deposits are found in the extracellular environment as amyloid plaques of varying sizes and composition); however, in a subset of neurodegenerative diseases, the proteinaceous fibrillar deposits are found in the cytoplasm or in the nucleus as intracellular inclusions (inclusion bodies, aggresomes) (reviewed in [1–4]). Typically, amyloid fibrils are unbranched, micrometers in length, and 2–10 nm in width; they are typically composed of several interwined protofilaments, each displaying a shared regular, repetitive core structure that results from the propagation along the fibril of a double beta-sheet whose strands run perpendicular to the fibrils' main axis (the cross-beta structure).

Around 20 years ago, the amyloid hypothesis was proposed, stating the existence of a causative link between aggregate deposition in amyloid diseases and clinical symptoms due to the presence of fibrillar proteinaceous deposits of a specific protein/peptide in the tissues affected by any peculiar amyloid condition [5, 6]. Since then, the hypothesis has gained increasing support from many biochemical and genetic studies [7–9]; the latter specifically involved early-onset, familial forms of some of these diseases, where the aggregated polypeptides are variants resulting from genetic mutations of the same proteins found aggregated in the corresponding

*Lipids and Cellular Membranes in Amyloid Diseases*, First Edition. Edited by Raz Jelinek.
© 2011 Wiley-VCH Verlag GmbH & Co. KGaA. Published 2011 by Wiley-VCH Verlag GmbH & Co. KGaA.

sporadic forms. However, much remains still be learnt on the structural, environmental, functional, and biological factors contributing to cell/tissue functional impairment and cell demise. In particular, the structural and physical features of the pathogenic aggregated species (particularly those preceding the appearance of mature fibrils) and the molecular and cell biology basis of the functional and viability impairment of the exposed cells are under intense investigation [10–14]. The specific wild-type or mutant polypeptides found aggregated in the various amyloid diseases display significant or increased propensity, respectively, to misfold and to aggregate. Recent research has provided clues on the relation between amino acid sequence (and some resulting biophysical features), aggregation propensity of any unfolded polypeptide chain [15, 16], and, in case of polypeptides involved in amyloid diseases, the time of onset and severity of the clinical signs [17, 18].

A step forward in the understanding of the importance of protein misfolding and aggregation in biology and medicine was made 12 years ago [19], when it was first shown that the ability to misfold and to aggregate into amyloid assemblies could be considered a generic property shared by most, perhaps all, polypeptide chains under specific perturbing conditions (reviewed in [20]). These data prompted bioinformatic studies in the protein data bank to highlight the presence, in protein evolution and in the control of protein expression, folding and aggregation of specific structural features and constraints, and also of specific molecular machineries either limiting the tendency of natural proteins to aggregate into amyloid polymers (reviewed in [21]) or enabling the cell to clear any aggregation-prone species arising after protein synthesis (reviewed in [22]). More recently, this view has been modified by the discovery that amyloid fibrils with functional significance can be present not only in microorganisms and insects, but also in mammals under normal or altered physiological conditions [23]. Such a knowledge has led to the proposal that monomer conformation in the ordered amyloid fibrils (the "amyloid fold") can be considered one of the canonical folds available in the protein conformational space potentially exploitable for specific biological functions [2, 20]; however, when inappropriately taken, such a fold originates polymeric supramolecular structures energetically more stable, and hence thermodynamically favored, with respect to the monomeric compactly folded biologically active conformation of the parent polypeptide chain.

In spite of the basic tendency of any polypeptide chain to aggregate, protein aggregation in living organisms is a rare event, at least for most of the life span of an individual, and, accordingly, the number of different proteins and peptides found aggregated in specific human diseases is very limited. This apparent contradiction can be explained by several considerations; these include the highly cooperative nature of protein folding together with the previously mentioned structural and biological adaptations set up by protein evolution to disfavor (reviewed in [21]) or to keep under control protein aggregation by eliminating misfolded molecules and their early aggregates (reviewed in [22, 24]). Nevertheless, the build-up of the aggregation-prone conformations of a protein can be triggered by any alteration of the quality control of protein folding, of protein structure, or expression level following the presence of specific mutations or chemical modifications, increased synthesis, or reduced degradation. Moreover, not only major, but also minor, even subtle, changes of the environmental conditions can shift the folded–unfolded equilibria of specific polypeptide chains, favoring kinetically or thermodynamically aggregate nucleation and fibril growth. These changes include heat shock, oxidative stress, alterations of the intracellular macromolecular crowding, the presence of suitable surfaces, the absence of stabilizing ligands, and others (reviewed in [1, 2]).

A further step forward in the understanding of the molecular basis of amyloid toxicity was made in 2002, when it was first shown that amyloid cytotoxicity could not be considered a specific property of the peptides and proteins aggregated in the different amyloid diseases, possibly arising from their specific amino acid sequences; rather, similarly to the aggregation tendency, it appeared to be a generic property of the amyloid fold shared by aggregates grown from any peptide/protein [25]. Such a knowledge has established a new paradigm in the world of natural proteins featuring the latter not only as molecules of fundamental importance for life but also as potential toxins to living organisms [3]. It also provided further significance to the concept of life on the edge between folding and aggregation, and hence between function and dysfunction, and to the importance of the structural and functional adaptations discussed above set up by protein evolution to favor the functional side of such a border.

Mature amyloid fibrils, being the form of protein aggregates commonly found in extracellular plaques or intracellular inclusions associated with amyloid pathologies, have been considered the key factor responsible for cell impairment and tissue damage until a few years ago. In fact, on the basis of the amyloid hypothesis, it appeared that the pathogenic features of amyloid diseases resulted from some specific interaction with cell components of the deposits of the aggregated material. Such a view provided a theoretical frame to rationalize the molecular basis of amyloid diseases and to explore therapeutic approaches to amyloidoses mainly aimed at hindering fibril growth and deposition. However, the notion that the tissue deposits of mature fibrils are the main factor responsible for cell functional impairment and death is currently drawn into question by an impressive body of experimental data; according to the latter, the aggregation nuclei and prefibrillar assemblies transiently arising in the path of fibrillization of many peptides and proteins and preceding the growth of mature fibrils are the main toxic entities [10–14, 25–29]. The increasing importance of these prefibrillar aggregates stems from the fact that soluble oligomers comparable to those appearing at the onset of the fibrillization of several peptides and proteins *in vitro* have been repeatedly detected in, and purified from, cultured cells and animal tissues where the monomeric precursors are expressed [30–34]. It has been concluded that amyloid oligomers are really formed *in vivo* not only in cultured cells but also in diseased tissue, providing convincing proof that they can be directly and really associated, at least in part, with cell/tissue impairment [33]. Nevertheless, in some cases mature amyloid fibrils can directly affect cell viability [35, 36], whereas in other cases they can be a source of toxic prefibrillar species. However, the different assemblies appearing transiently or as final products in the fibrillization path from small oligomers to the highly polymeric mature fibrils differ remarkably not only in their cytotoxic potential but also in the cellular mechanisms and functions with which they interfere. For instance, Aβ42 oligomers found in the brains of people with Alzheimer's disease (AD) impair long-term potentiation [12, 37] and raise endoplasmic reticulum stress [38], whereas the fibrillar Aβ deposits appear to trigger a neuroinflammatory response [39, 40]. Moreover, Aβ aggregate cytotoxicity appears to depend on the aggregation state-specific uptake, only oligomer toxicity associated with oligomer internalization by endocytosis [41] or by passive diffusion.

The idea that mature fibrils are substantially inert, harmless deposits of the toxic precursors implies that their growth can be viewed as a cell defense mechanism and contributes to explaining the lack of direct correlation between amyloid load in the brains of AD patients and the severity of their clinical symptoms [42]. However, the growing information on the effects of amyloids on cell biology has not yet led to the proposal, for all forms of amyloidoses, of a unifying model describing the molecular basis of protein aggregation under physiological conditions, the identity of the species responsible for tissue damage, and the molecular mechanism(s) of the functional and viability impairment of tissue cells *in vivo*. In addition, the actual relevance to *in vivo* disease of amyloid toxicity data, mostly obtained using cultured cells, has not yet really been definitively established for any of the amyloidoses. Finally, in spite of their recognized importance as main cytotoxic entities, a severe lack of information on the structural features of the transient oligomeric assemblies still remains. Actually, currently it is only recognized that amyloid oligomers arising as in-path or off-path intermediates of fibril growth display increased flexibility and an exposed hydrophobic surface with respect to the parent natively folded monomers. However, in general, these species, when grown under different environmental conditions from the same peptide/protein, can exhibit remarkably variable conformational and biophysical features, eventually giving rise to fibrils with different structural, physical, and morphological properties. Such a notion is at the basis of the theme of oligomer/fibril polymorphisms and their effect on biological systems, an issue of increasing interest in protein aggregation science.

This chapter focuses on the conformational, biochemical, and immunological properties and the structural heterogeneity and polymorphisms of amyloid oligomers and fibrils, in an attempt to relate the variable amyloid cytotoxicity to those properties and to the biochemical and biophysical features of the exposed cells, and also to assess the relations among oligomer and cell membrane properties.

## 5.2
### Only Generic Data Are Currently Available on the Structural Features of Amyloid Oligomers

Even though, in general, prefibrillar amyloid aggregates display generic cytotoxicity, the work carried out in the last few years increasingly supports the idea that amyloid assemblies with variations of their basic supramolecular organization can display different cytotoxicity. Accordingly, the latter can no longer be considered directly associated with the shared basic structure of amyloids, as previously stated [27]; rather, it appears to result from the specific arrangements of the constituent misfolded monomers and, accordingly, from their different conformational and biophysical properties. Therefore, the amyloid structure and its relation to cytotoxicity are increasingly considered a key topic in the study of the molecular basis of amyloid diseases. Currently, considerable information is available on the structure of the ordered β-sheet-rich core of mature fibrils; the conformational features of the monomers providing it and the fibril supramolecular organization and hierarchical growth from structurally simpler precursors through a number of steps have also been extensively described [43–45]. However, a severe lack of knowledge still remains on the structural features at the molecular level of misfolded monomers, oligomers, and other fibril precursors, and also on their polymorphisms, highlighting the need for more information to describe better the molecular basis of amyloid toxicity.

The rate of the onset of protein aggregation and fibril growth appears limited by the kinetics of formation of oligomeric aggregation nuclei. However, little knowledge is currently available on the energetically favorable conformational states available to an aggregating polypeptide chain and its oligomeric assemblies; conversely, much more information has been acquired on the competing process of protein folding and on the structural features of folding intermediates and transition states. Actually, some of the energy minima in the energy landscape of protein aggregation are poorly defined, as suggested by the broad heterogeneity of the population of unstable, rapidly interconverting oligomeric states with comparable free energies, at variance with the much more structurally defined, increasingly stable, higher order species (protofibrils, protofilaments, and mature fibrils) [46]. Under the same conditions, mature amyloid fibrils and their structural variants are expected to display deeper energy minima and higher stabilities than the parent natively folded monomers; this can be explained by the reduced molecular dynamics of mature fibrils, the consistency of the ordered structure of their core, and the nucleation-dependent polymerization mechanism of fibril growth. Actually, a key difference between protein folding and protein fibrillization is marked by the physical basis of the latter, which approaches the ordered assembly occurring in crystal growth.

Many *in vitro* and *in silico* studies carried out in recent years have highlighted the morphological modifications occurring in amyloid assemblies made from different disease-related and disease-unrelated peptides and proteins in their path of fibrillization [47, 48], thus providing a theoretical frame to describe these modifications [49]. Several more-or-less defined steps are involved in amyloid fibril assembly. For example, in many cases electron microscopy (EM) and atomic force microscopy (AFM) studies have imaged the initial presence of transient, unstable, roundish, or tubular particles 2.5–5.0 nm in diameter, generally enriched in β-structure, often called "amorphous aggregates" [47, 48, 50, 51] and characterized as protein/peptide oligomers. These species frequently associate into bead-like chains, small annular rings ("donuts" or "pores"), or curvy protofibrils that appear as precursors of more organized amyloids such as large closed rings, ribbons [48, 50–53], or longer protofilaments, eventually generating mature fibrils (reviewed in [52]). As outlined above, the data emerging from these and other studies depict amyloid fibril growth as a hierarchical process in which both the initial natively folded structure of the monomers and the environmental parameters contribute to the general conformational properties of the misfolded/unfolded monomers involved in the aggregation pathway. These properties affect monomer arrangement into more or less compact oligomers with variable hydrophobic exposure and subsequently oligomer reorganization into increasingly more complex, compact, ordered, and stable β-sheet-rich assemblies. Similarly to crystal seeds, fragments of polymorphic mature fibrils or their intermediates grown from the same peptide/protein can speed up the growth from soluble monomers of oligomers and mature fibrils with the same conformational properties and morphological features as those of their seeding assemblies [46, 54, 55]. These data support the idea that amyloid self-propagation, first described in prion proteins [56], can be a general property of all amyloids.

As pointed out above, the heterogeneity, remarkable instability, and intrinsically disordered nature of amyloid prefibrillar aggregates, notably amyloid oligomers, make it very difficult to obtain reliable data on the conformational features of these species and hence to give convincing experimental support to any rationale for their structure–toxicity relation. Nevertheless, recently introduced techniques are of great value for obtaining information on the structural features and polymorphism of amyloid oligomers and other assemblies populated in the path of amyloid fibril growth [57]. Actually, biophysical techniques such as single-molecule spectroscopy (notably in fluorescence), small-angle X-ray scattering (SAXS) in solution, and solid-state nuclear magnetic resonance (ss-NMR) spectroscopy have provided valuable information on the conformational features of fibril structural nuclei (reviewed in [58]).

One of the most powerful tools recently introduced to overcome, at least in part, the lack of knowledge on the structure of prefibrillar amyloids is the use of conformation-specific polyclonal or monoclonal antibodies (Abs) raised against amyloid fibrils or their precursors. These Abs specifically recognize shared structural features present in comparable types of amyloid assemblies grown from different peptides and proteins; conversely, they allow one to discriminate among different types of aggregates grown from the same protein/peptide under the same or different conditions, thus allowing a sort of immunological classification of amyloid aggregates. The presence of shared structural features in amyloid fibrils grown from different peptides and proteins was first reported in 2002 in a study describing the generation of a conformational Ab recognizing a generic amyloid fibril epitope [59]. Other Abs raised subsequently were able to recognize specifically amyloid oligomers grown from differing peptides and proteins [27] or to cross-react either with amyloid pores [60] or with pores formed by pore-forming toxins [61]. Finally, a recent study described the generation of an Ab directed against Aβ amyloid fibrils that recognizes a shared, sequence-independent epitope present in fibrils grown from other peptides [62]. The same Ab cross-reacts with a subpopulation of Aβ oligomers (named soluble Aβ fibrillar oligomers) but not with other Aβ oligomers with apparently similar morphological features (named prefibrillar oligomers). These findings indicate the existence of structurally distinct families of amyloid oligomers of the same peptide/protein with potentially different cytotoxicities, suggesting that at least two alternative aggregation nuclei for amyloid fibrils can exist.

The use of conformational Abs has provided an invaluable tool that currently is widely used to discriminate between oligomers, prefibrillar aggregates, and mature fibrils. It has also allowed it to be definitively demonstrated (i) that the same protein/peptide can generate oligomeric and prefibrillar assemblies apparently indistinguishable yet with differences in their fine structural features [63] and (ii) that, conversely, structural similarities are present in comparable assemblies grown from different peptides and proteins [62–66]. Finally, it has provided a tool to classify amyloid oligomers [67] and to investigate the conformational modifications underlying the hierarchical growth of amyloid fibrils and also oligomer structural heterogeneity and polymorphisms (see Section 5.4).

The findings provided by the aforementioned new technical approaches indicate that the shared basic structural features of prefibrillar aggregates of different proteins/peptides vary in differing subtypes of oligomers or other prefibrillar assemblies of the same peptide/protein and are different from those displayed by the natively folded monomers or their fibrillar counterparts. These structural features can result from a number of consequences of protein unfolding, including increased chain flexibility and exposure to the aqueous environment of hydrophobic patches normally buried in the compact structure of the monomeric protein. Any change in the environmental conditions, including chemical modifications and alterations of the macromolecular crowding, can play a major role in promoting or hindering protein misfolding and aggregation; they can also increase, to different extents, the appearance of polymorphic amyloids by favoring the growth of aggregation nuclei composed of subpopulations of misfolded molecules with distinct conformational features and the organization of the latter into fibril precursors and mature fibrils with different structures, stabilities, and cytotoxicities [68–70]. In general, the structural and physical features of the transiently formed amyloid oligomers result from the partially unfolded condition of the associated monomers with progressive increase in β-structure. The consequently reduced compactness and increased hydrophobic exposure with respect to the natively folded monomers [71, 72] make the oligomers intrinsically unstable, strengthening their tendency to interact inappropriately with cellular components and, accordingly, their toxic potential.

## 5.3
### The Plasma Membrane Can Be a Primary Site of Amyloid Oligomer Generation and Interaction

The cell membranes can be the sites where the aggregating peptides can be generated by precursor membrane protein processing and the preformed toxic oligomers interact as a first step of the chain of events culminating with cell death; moreover, together with other biological surfaces, they can also be key sites where peptides and proteins misfold and reorganize into oligomeric aggregation nuclei. More generally, in addition to biological membranes and inorganic and biological surfaces, supramolecular protein structures and macromolecular assemblies such as collagen fibers, glycosaminoglycans (GAGs), and nucleic acids can be key modifiers of the conformational states of peptides and proteins and also sites where oligomeric assemblies of the latter are actively recruited. The strong gradient of hydrophobicity, polarity, or electrostatic potential from the outside to the inside of a synthetic or biological phospholipid bilayer or at the border of any amphipathic, hydrophobic, or charged surface can induce in an interacting protein/peptide local or more extensive unfolding to non-native conformations by weakening the intramolecular hydrophobic or electrostatic interactions [73]. Consequently, the unstable misfolded protein/peptide which exposes normally buried nonpolar groups undergoes structural reorganization with a reduction of the compactness of its hydrophobic core and an increase in flexibility [74, 75]. In addition, the active recruitment of the aggregation-prone conformers in the surface two-dimensional space results in a remarkable decrease of their mean distance. Both effects can remarkably accelerate aggregate nucleation, with a deeper interaction with the membrane of the monomers and the growing aggregates [76–78]. It is also possible that preformed amyloid oligomers interact directly with synthetic phospholipid bilayers, the cell plasma membrane, and the mitochondrial or the endoplasmic reticulum (ER) membranes (reviewed in [79]). In either case, the result is a loss of membrane integrity and selective permeability, with possible functional impairment of specific membrane proteins [80, 81].

The ability of biological surfaces, notably lipid membranes and macromolecules, to modify the fold of protein/peptide monomers and to nucleate and recruit amyloid oligomers suggests an important role of the membrane biophysical properties resulting from their lipid composition in modulating these effects. For example, it is known that anionic surfaces, such as those of GAGs and nucleic acids, and also phospholipid bilayers and cell membranes rich in anionic phospholipids, trigger protein/peptide fibrillization by acting as conformational catalysts inducing β-sheet structures [82–84]; anionic phospholipids have also been described as preferred interaction sites for amyloid aggregates, possibly by recognizing a shared fold [84]. In particular, the importance has recently emerged of the ganglioside GM1, whose sialic acid moiety provides the major negatively charged site in the outer leaflet of the cell membrane under normal conditions. The increased density of negative charge in GM1-rich membrane areas can reduce lipid compactness by charge repulsion, favoring interaction with, or insertion into, the plasma membrane of misfolded proteins or their oligomers, resulting in membrane disassembly [85]. The role of membrane cholesterol in modulating some physicochemical features of the cell membrane, such as compactness and rigidity, the proteolytic processing of membrane proteins whose products are involved in neurodegenerative conditions, protein/peptide aggregation at the membrane level, and aggregate interaction with the cell membrane, also appears remarkable. However, the effects of membrane cholesterol on amyloidogenic peptide generation and aggregation at the cell surface and also on the interaction of the latter with preformed oligomers can be variable and not easily predictable. Lipid rafts, ordered lipid domains in the plasma membrane that are increasingly recognized as preferred nucleation and interaction sites of amyloids, are particularly enriched in GM1 and cholesterol [85]. Accordingly, protein/peptide interaction with the cell surface, particularly with lipid rafts, is considered an important requirement for cytotoxicity (reviewed in [79, 86, 87]). In this regard, it has been reported that a loss of cholesterol in neuronal membranes enhances amyloid peptide generation and that the interaction of prefibrillar aggregates with the cell membrane, resulting in cytotoxicity, is impaired when the plasma membrane is enriched in cholesterol (reviewed in [79, 87]). Overall, the data currently available support the idea that membrane cholesterol and gangliosides can modulate in different ways the conformational changes underlying the aggregation of specific proteins/peptides and also the ability of the resulting aggregates to interact with, and to permeabilize, the cell membrane with impairment of cell viability (see Section 5.5) [88].

## 5.4
## Oligomer/Fibril Polymorphism Can Underlie Amyloid Cytotoxicity

Currently, the self-organization into oligomeric assemblies of misfolded monomeric peptides/proteins is described mechanistically by a number of theoretical models. The widely accepted "nucleated conformational conversion" model [54], supported by many experimental and theoretical observations [54, 89–91], describes the coalescence of a group of unstable misfolded monomers in solution into relatively disordered "molten" oligomers. The latter become increasingly structured upon extensive monomer reorganization, generating more compact, higher order assemblies eventually culminating in the appearance of mature fibrils [92]. However, in spite of the experimental support for this model, an in-depth description at the molecular level of the whole process, particularly of its earliest steps, is still lacking.

Experimental observations and theoretical considerations support, in most cases, a generic "two-step" mechanism that appears to depend closely on the hydrophobicity and the extent of exposed hydrophobic surface of the misfolded monomers. The latter favor monomer coalescence into conformationally disordered oligomers that subsequently reorganize into more ordered species with an increasingly compact hydrophobic core [49, 54, 89, 90, 92]. Monomers with low hydrophobicity or reduced hydrophobic exposure can skip the coalescence step, organizing directly into ordered oligomers where secondary interactions such as hydrogen bonds predominate [49]. This can be the case with protein aggregation from natively folded proteins (reviewed in [93]), where the aggregating monomers display a low exposure of hydrophobic surface. These considerations suggest that, together with protein concentration and temperature, the intermolecular hydrophobic interactions are major determinants of the rate of the hydrophobic collapse of the misfolded monomers into different types of oligomers. Accordingly, the appearance of disordered–ordered, or only ordered, oligomers from different proteins/peptides or from the same protein/peptide under different denaturing appears to depend on the balance between the rapidly formed intermolecular hydrophobic interactions and the slower exchange of the directional hydrogen bonds into the assembling oligomers [92].

The importance of the relative contributions of the hydrophobic forces and the hydrogen bonds rationalizes the findings that different aggregates (fibrillar or prefibrillar) with increased β-sheet content can be grown under different solution conditions favoring or disfavoring hydrogen bonding or tertiary interactions [94]. Taken together, these considerations provide a theoretical frame to investigate the polymorphisms of amyloid oligomers grown from the same peptide/protein under different conditions or in the presence of different destabilizing agents (high temperature, pressure and ionic strength, shear forces, urea, co-solvents, and others).

Oligomer (and mature fibril) heterogeneity and polymorphisms are key issues considering that these species can affect directly or indirectly the load of toxic amyloids in tissue. As mentioned above, there is increasing awareness that, under different conditions, the same peptide/protein can populate variously misfolded species whose initial aggregation generates oligomers with different conformational features, eventually resulting in polymorphic mature fibrils [63, 95, 96]. In addition to the environmental conditions, specific structural features of the natively folded monomers can determine the way in which they misfold and aggregate under suitable conditions and the toxicity of their aggregates; the aggregation under similar conditions of proteins/peptides carrying specific amino acid substitutions or chemical alterations modifying their physicochemical properties [68] is an example. However, in some cases these conformational peculiarities can be much more subtle. The recent report that, after release from the cell membrane upon amyloid precursor protein (APP) processing, A$\beta$42 can exist in two different conformations is an example. The normally prevalent, physiological, harmless conformation, with a turn at positions 25–26, does not aggregate by itself, but a second, less-populated, conformation, with a turn at positions 22–23, aggregates into toxic oligomers [97].

The increasingly recognized importance of oligomer polymorphisms explains several observations. These include the propagation of prion strain infectivity and other protein polymorphisms [46, 94, 98, 99], the structural heterogeneity [100–102] and the variable cytotoxicity [41, 103, 104] of amyloid fibrils and their precursors grown from the same peptide/protein under different conditions, and the appearance of in-path or off-path intermediates in fibril growth [51, 104]. The growth under different conditions of structurally different amyloid oligomers with different cytotoxicities appears to be of particular importance. As discussed above, in most cases amyloid cytotoxicity appears to be primarily associated with the ability of the unstable prefibrillar aggregates to interact with, and to permeabilize, the cell membranes, particularly the plasma membrane of the exposed cells (reviewed in [79]). However, such behavior can be remarkably different in amyloid aggregates, including harmless mature fibrils, grown under different conditions and displaying conformational polymorphisms.

Amyloid fibrils both *in vitro* and in tissue can be a potential source of misfolded monomers and/or toxic oligomers. As pointed out above, amyloid fibrils are currently considered far less cytotoxic than amyloid oligomers and other prefibrillar aggregates, even though, in some cases, direct [35, 36] or indirect [40] fibril cytotoxicity, or cytotoxicity associated with fibril assembly at, and growth on, lipid membranes [105], has been described. However, often the toxicity of the fibrillar material can result from fibril breakage, with the generation of shorter fibrils [106] or leakage of toxic species even *in vivo* [25, 33, 107]. Fibril breakage could result in an increase in the number of fibril fragments, and hence of free ends at which misfolded monomers can bind, thus competing with oligomerization into more toxic assemblies, but also in leakage of toxic species. Actually, recent data suggest that fibrils of the same protein/peptide grown under different conditions from structurally different oligomers, and hence with variable structures and stabilities, can display different cytotoxicities or abilities to leak toxic oligomers (see Section 5.5).

That amyloid fibrils are not as stable as previously considered is supported by the demonstration that monomers/oligomers can be recycled within fibrils both *in vitro* [108] and *in vivo* [33, 109], and that fibrils grown from the same monomer under different conditions display different stabilities [110]. The idea that amyloid fibrils can be disassembled under suitable conditions is supported by recent data on $\beta_2$-microglobulin fibril disassembly by agitation [106] and A$\beta$ fibril solubilization by lipid membranes reconstituted with lipids extracted from neuronal cells [107]. Taken together, the data on fibril disassembly by membranes suggest that surfaces, previously known to favor protein aggregation into oligomers and mature fibrils [79], can also result in the opposite, inducing fibril disassembly into toxic oligomers.

The idea that amyloid fibrils under a range of different conditions can be disassembled or, at least, can leak toxic oligomers is further reinforced by previous data on the effects on 4'-iodo-4'-deoxydoxorubicin on transthyretin fibrils [111] and by more recent data on transthyretin [112] and $\beta_2$-microglobulin (S. Giorgetti *et al.*, unpublished results) solubilization by a number of tetracyclines. Conversely, in some cases surfaces can promote the growth of highly stable fibrils. For example, a positive effect of sulfated polysaccharides on A$\beta$42 assembly into highly stable, thin, and harmless amyloid fibrils with very reduced hydrophobic exposure has been reported [70]. These data agree with the general idea that polysaccharides favor the growth of harmless amyloid oligomers and fibrils with increased rigidity due to the presence of stabilizing cross-links between glycated monomers or oligomers [113].

Overall, these studies highlight the importance, in amyloid cytotoxicity and synaptotoxicity, of the polymorphisms of amyloid oligomers and fibrils. The fact that amyloid fibrils can leak toxic oligomers implies that amyloid plaques in tissue can be not only protective recruiters of toxic species arising from peptide/protein misfolding but also, in some cases, potential sources of toxicity. Moreover, oligomer/fibril modification by surfaces strengthens the suggestion to treat with caution information on the structures and molecular masses of the amyloids, particularly oligomers, provided by chromatographic techniques, electrophoresis, and AFM. In fact, the small oligomers detected by these techniques could be absent in the original sample, merely resulting from some disassembly of pre-existing larger aggregates upon interaction with the surface of the AFM, chromatographic, or electrophoresis support, or with detergent micelles [114].

More information on a wider population of fibrils grown under different conditions from a variety of peptides and proteins is needed to explore the generality of these considerations. Nonetheless, an increasing number of coherent data indicate that amyloid deposits in tissue cannot be considered merely inert remnants of toxic species; more convincingly, together with misfolded monomers and toxic oligomers, they appear to be key components of complex equilibria and, as such, possible reservoirs of toxic species [108], providing additional clues useful in explaining the molecular basis of amyloid toxicity in tissue.

## 5.5
### Amyloid Oligomers Grown Under Different Conditions Can Display Variable Cytotoxicity by Interacting in Different Ways with the Cell Membranes

Several recently reported studies have investigated the structure–toxicity relation of amyloids indirectly, for example by establishing a link between their stability and the ability to impair cell viability. For example, a recent study has shown that apparently similar Aβ fibrils grown under normal conditions or under agitation displayed significantly variable stability to guanidinium hydrochloride, with different structural rearrangements and toxicities, the less stable being more cytotoxic [95]. Considering that many data support the idea that Aβ interaction with the cell membrane implies peptide conformational changes and is important for toxicity [73], it could be proposed that the toxicity of any Aβ species can, at least in part, be related to its structural flexibility and hence to its ability to change structure on, or within, the cell membrane. Similar conclusions were drawn in another study in which three different species of α-synuclein oligomers were prepared and assayed for the extent and mechanisms of cytotoxicity [69]. One of these, comprising spherical assemblies 2–6 nm in height and annular structures 45 nm in diameter similar to those found *in vivo*, was able to trigger membrane permeabilization, inducing elevation of intracellular $Ca^{2+}$ and impairing cell viability. The other two types of oligomers did not permeabilize the cell membrane, but entered into the cells directly without inducing caspase activation or cell loss [69]. A more recent investigation on these α-synuclein oligomers showed that the extracellular assemblies are dynamically related and can interconvert into each other, modifying their properties accordingly; however, only one type can specifically induce transmembrane seeding of intracellular aggregation of the protein in a dose- and time-dependent manner, suggesting a mechanism for tissue propagation of α-synuclein pathology [115]. The relevance of these findings is further stressed in the light of the recently reported data showing that α-synuclein levels are increased in the sera of Parkinson's disease patients [116] and that α-synuclein and its aggregates can, indeed, be secreted from the cells, thus rationalizing the possibility that they provide toxic insult to neighboring cells in tissue [117]. Taken together, these results indicate that various aggregated species of Aβ peptides and α-synuclein may cause neurotoxicity by distinct mechanisms contributing to different extents to neurodegeneration.

Our recent results on the structural features of amyloid oligomers grown from the N-terminal domain of the bacterial hydrogenase maturation factor F (HypF-N), a protein not associated with any amyloid disease, and their relation to cytotoxicity confirm the generality of these considerations. It was previously shown that HypF-N aggregates into prefibrillar and fibrillar assemblies indistinguishable from those associated with amyloid diseases and with the same cytotoxic effects [25, 28]. Recently, we have found that, under different destabilizing conditions affecting the strength of the electrostatic or the hydrophobic interactions, HypF-N misfolds generating oligomers that are morphologically similar, yet with different stabilities and structural features, in terms of exposure of hydrophobic surface, compactness, and flexibility; the less stable oligomers eventually grew into mature fibrils whereas the relatively more stable ones assembled into curvy protofibrils [118]. The two types of oligomers displayed different cytotoxicities and abilities to interact with, to permeabilize, and to cross the cell membrane, the less stable assemblies being remarkably more toxic. A closer analysis of the structural features of both oligomers carried out by fluorescence spectroscopy showed that in both cases the regions of the polypeptide chain involved in the intermolecular interactions substantially corresponded to those providing the hydrophobic core in the natively folded monomers; however, the two types of oligomers displayed remarkable conformational differences in terms of density of packing, flexibility, and extent of solvent-exposed hydrophobic surface [118]. These data establish a direct link between general structural features of prefibrillar oligomers generated under different conditions from differently misfolded monomers, their ability to grow into distinct, stable amyloid assemblies and to stick to, to disassemble, and to permeabilize the cell membrane, thus impairing cell viability. A recent implementation of this study has shown that modulating the lipid content and the biophysical features of the plasma membrane of cells exposed to either oligomer remarkably affects the ability of the latter to bind to and to permeabilize the plasma membrane and their relative toxicities (E. Evangelisti, unpublished results).

These results suggest that polymorphisms of amyloid assemblies arising from protein/peptide misfolding under different destabilizing conditions result in different cytotoxicities that can be traced back directly to the different biophysical and conformational features of the oligomers. Actually, it is not surprising that oligomers that are less compactly organized, less stable, more flexible, and with a more extended hydrophobic surface display a higher tendency to interact with, penetrate into, disassemble, and permeabilize the cell membrane, eventually impairing cell viability. On the other hand, more stable, compact oligomers with a reduced exposed hydrophobic surface are expected to display a modest tendency to interact with, and to penetrate inside, the bilayer. Conversely, the biophysical properties of the latter are expected to modulate the effects arising from the interaction with it of oligomers with different structural and biophysical properties. In conclusion, it can be proposed that the ways in which and the extent to which amyloid aggregates interact with the cell membrane are remarkably affected by the conformational and structural features of both, at least in terms of hydrophobicity and flexibility; accordingly, the latter can be considered major determinants of the resulting toxic effects in tissue.

The above considerations can be useful to explain the different toxicities associated with amyloid polymorphisms *in vivo*. As noted previously, amyloid aggregates of the same protein/peptide displaying different conformations, possibly as a consequence of their growth under different cell/tissue conditions, can display different mechanisms and variable severity of cell impairment. A recent study found that aggregates grown from Aβ42 carrying different mutations induced qualitatively different pathologies in a *Drosophila* model. The *Arctic* mutant caused greater neuronal loss than the wild-type form, whereas the peptide carrying an artificial mutation caused earlier onset memory defects and the strongest neurite degeneration. The cellular distribution of Thioflavin S-positive material agreed with these effects; in fact, the *Arctic* mutant formed large deposits in the cell body, whereas the artificial mutant accumulated preferentially in the neurites and the wild-type peptide in both sites [104]. It was shown previously that chronic exposure to high concentrations of fibrillar Aβ was needed to produce generalized dystrophic changes, but only modest death, in primary neuronal cultures; however, the latter, when exposed to nanomolar concentrations of Aβ oligomers or Aβ-derived diffusible ligands (ADDL), displayed chronic mitochondrial dysfunction but only minor changes of cell viability [103]. Other data have shown that oligomeric, but not fibrillar, Aβ also contributes to endoplasmic reticulum stress in neuroblastoma cells by inducing mild activation of the unfolded protein response [38] and that oligomeric and fibrillar Aβ induce neurodegeneration by two distinct apoptotic pathways proceeding through the activation of caspase-8 (fibrils) or of both caspase-8 and (mainly) caspase-9 (oligomers) [119].

More recent research has investigated the relation between different conformations of huntingtin exon-1 with an expanded polyglutamine stretch and their relative toxicities both to cultured cells and in animal tissue. At different temperatures (37 or 4 °C) the polypeptide aggregated *in vitro* into differing conformations, a more flexible and highly cytotoxic one, with exposed β-sheets and loop-turns, and a more rigid and substantially nontoxic conformation, with extended and buried β-sheet [107]. Extension of these findings to HD mice models by extracting amyloid fibrils from variously affected brain regions showed that huntingtin fibrils extracted from diseased brain areas displayed the same structural features as those of the *in vitro*-generated toxic conformation; conversely, fibrils extracted from unaffected brain regions were more rigid and substantially harmless, similarly to *in vitro*-grown fibrils with the less cytotoxic conformation [107]. These data suggest that the same protein in different brain areas experiences differing conditions that modulate its stability and aggregation pathway, eventually resulting in distinct mature fibrils with different physical and biological properties. Similar considerations can apply also to prefibrillar aggregates. For example, in multiple system atrophy, it has been reported that soluble 30–50 nm-sized annular α-synuclein oligomers are released from glial cytoplasmic inclusions purified from brain tissue by mild detergent treatment. In contrast to pathological α-synuclein, recombinant α-synuclein yielded only spherical oligomers after detergent treatment, indicating some structural difference in the fibrils grown from the pathological protein underlying their greater propensity to leak stable annular oligomers [120].

More information on a wider range of differently grown oligomers and of exposed biological systems is needed to elucidate more general considerations on the relation between the structural and physicochemical features of polymorphic amyloid oligomers or fibrils and the extent to which, and the way in which, they impair cell physiology and viability. However, the data currently available support the idea that the biological effects of these assemblies are ultimately related to the way in which they interact with the cell membrane and this, in turn, appears to depend on aggregate and membrane physical and structural features. These considerations further stress the importance of the early interaction with the cell membrane(s) in amyloid toxicity and are expected to stimulate increasing efforts to provide more knowledge on the structural features and the structure–toxicity relations of the amyloid oligomer-cell membrane system.

## 5.6 Conclusions

Only very recently has information started to appear on the structural features of amyloid oligomers and the relation of the latter to oligomer toxicity and ability to grow into mature fibrils with different stabilities. Recent data have also led to reconsider the widely accepted idea that amyloid fibrils are substantially harmless reservoirs of toxic oligomers; rather, under appropriate conditions, and to different extents, fibrils grown from the same peptide/protein under various conditions display different stabilities and tendencies to disassemble with leakage of toxic oligomers. This makes the theme of oligomer and fibril polymorphisms a key issue in the attempt to establish the type of toxic aggregated species and also the conditions affecting their origins and toxicity both *in vitro* and in tissue. However, more research is needed to increase the knowledge required to be able to propose experimentally supported hypotheses and models of the oligomer dynamics and structure and their relation to toxicity. In this respect, it can be of interest to provide reliable data on the relative contributions of the biophysical features of the oligomer and the phospholipid bilayer of the cell membrane(s) in determining the way in which they interact with each other and the outcome of such interaction in terms of cell functional and viability impairment. No information is currently available on the possible interplay between oligomer and membrane biochemical and biophysical features as a key factor determining the final outcome in terms of extent and intensity of oligomer–membrane interactions and the resulting effects on membrane function. Research trying to address these and similar questions is now starting to appear in the literature, and it can be expected that increased information on this theme will open the way in the near future to put forward hypotheses based on reliable data to describe the structural features of the most highly toxic species grown under specific tissue conditions and their relation to tissue function and viability impairment. Such a knowledge is expected to strengthen new approaches providing the rationale for designing new molecules aimed at modifying the aggregation path in tissue of peptides and proteins so as to avoid the appearance of the subset of misfolded conformers giving rise to the most toxic amyloid variants.

## Acknowledgments

The author's research was funded by grants from the Italian MIUR (PRIN 2007XY59ZJ_001), the Ente Cassa di Risparmio di Firenze, and the Fondazione Monte dei Paschi di Siena.

## References

1 Stefani, M. and Dobson, C.M. (2003) Protein aggregation and aggregate toxicity: new insights into protein folding, misfolding diseases and biological evolution. *J. Mol. Med.*, **81**, 678–699.

2 Chiti, F. and Dobson, C.M. (2006) Protein misfolding, functional amyloid, and human disease. *Annu. Rev. Biochem.*, **75**, 333–366.

3 Stefani, M. (2004) Protein misfolding and aggregation: new examples in medicine and biology of the dark side of the protein world. *Biochim. Biophys. Acta*, **1739**, 5–25.

4 Selkoe, D.J. (2003) Folding proteins in fatal ways. *Nature*, **426**, 900–904.

5 Selkoe, D.J. (1991) The molecular pathology of Alzheimer's disease. *Neuron*, **6**, 487–498.

6 Hardy, J.A. and Higgins, C.A. (1992) Alzheimer's disease: the amyloid cascade hypothesis. *Science*, **256**, 184–185.

7 Reilly, M.M. (1998) Genetically determined neuropathies. *J. Neurol.*, **245**, 6–13.

8 Kelly, J. (1998) Alternative conformation of amyloidogenic proteins and their multi-step assembly pathways. *Curr. Opin. Struct. Biol.*, **8**, 101–106.

9 Dobson, C.M. (2001) The structural basis of protein folding and its links with human disease. *Philos. Trans. R. Soc. Lond.*, **B356**, 133–145.

10 Lambert, M.P., Barlow, A.K., Chromy, B.A., Edwards, C., Freed, R., Liosatos, M., Morgan, T.E., Rozovsky, I., Trommer, B., Viola, K.L., Wals, P., Zhang, C., Finch, C.E., Krafft, G.A., and Klein, W.L. (1998) Diffusible nonfibrillar ligands derived from Aβ-42 are potent central nervous system neurotoxins. *Proc. Natl. Acad. Sci. USA*, **95**, 6448–6453.

11 Walsh, D.M., Hartley, D.M., Kusumoto, Y., Fezoui, Y., Condron, M.M., Lomakin, A., Benedek, G.B., Selkoe, D.J., and Teplow, D.B. (1999) Amyloid β-protein fibrillogenesis. Structure and biological activity of protofibrillar intermediates. *J. Biol. Chem.*, **274**, 25945–25952.

12 Walsh, D.M., Klyubin, I., Fadeeva, J.V., Cullen, W.K., Anwyl, R., Wolfe, M.S., Rowan, M.J., and Selkoe, D.J. (2002) Naturally secreted oligomers of amyloid β protein potently inhibit hippocampal long-term potentiation *in vivo*. *Nature*, **416**, 535–539.

13 Conway, K.A., Lee, S.-J., Rochet, J.C., Ding, T.T., Williamson, R.E., and Lansbury, P.T. (2000) Acceleration of oligomerization not fibrillization is a shared property of both alpha-synuclein mutations linked to early-onset Parkinson's disease. Implication for pathogenesis and therapy. *Proc. Natl. Acad. Sci. USA*, **97**, 571–576.

14 Reixach, N., Deechingkit, S., Jiang, X., Kelly, J.W., and Buxbaum, J.N. (2004) Tissue damage in the amyloidoses: transthyretin monomers and nonnative oligomers are the major cytotoxic species in tissue culture. *Proc. Natl. Acad. Sci. USA*, **101**, 2817–2822.

15 Chiti, F., Stefani, M., Taddei, N., Ramponi, G., and Dobson, C.M. (2003) Rationalization of the effects of mutations on peptide and protein aggregation. *Nature*, **424**, 805–808.

16 Pawar, A.P., Dubay, K.F., Zurdo, J., Chiti, F., Vendruscolo, M., and Dobson, C.M. (2005) Prediction of "aggregation-prone" and "aggregation-susceptible" regions in proteins associated with neurodegenerative diseases. *J. Mol. Biol.*, **350**, 379–392.

17 Clarke, G., Collins, R.A., Leavitt, B.R., Andrews, D.F., Hayden, M.R., Lumsden, C.J., and McInnes, R.R. (2000) A one-hit model of cell death in inherited neuronal degeneration. *Nature*, **406**, 195–199.

18 Perutz, M.F. and Windle, A.H. (2001) Cause of neuronal death in neurodegenerative diseases attributable to expansion of glutamine repeats. *Nature*, **412**, 143–144.

19 Gujiarro, J.I., Sunde, M., Jones, J.A., Campbell, I.D., and Dobson, C.M. (1998) Amyloid fibril formation by an SH3 domain. *Proc. Natl. Acad. Sci. USA*, **95**, 4224–4228.

20. Dobson, C.M. (2003) Protein folding and misfolding. *Nature*, **426**, 884–890.
21. Monsellier, E. and Chiti, F. (2007) Prevention of amyloid-like aggregation as a driving force of protein evolution. *EMBO Rep.*, **8**, 737–742.
22. Sherman, M.Y. and Goldberg, A.L. (2001) Cellular defences against unfolded proteins: a cell biologist thinks about neurodegenerative diseases. *Neuron*, **29**, 15–32.
23. Fowler, D.M., Koulov, A.V., Alory-Jost, C., Marks, M.S., Balch, W.E., and Kelly, J.W. (2006) Functional amyloid formation within mammalian tissue. *PLoS Biol.*, **4**, 1–8.
24. Dobson, C.M. (1999) Protein misfolding, evolution and disease. *Trends Biochem. Sci.*, **24**, 329–332.
25. Bucciantini, M., Giannoni, E., Chiti, F., Baroni, F., Formigli, L., Zurdo, J., Taddei, N., Ramponi, G., Dobson, C.M., and Stefani, M. (2002) Inherent toxicity of aggregates implies a common mechanism for protein misfolding diseases. *Nature*, **416**, 507–511.
26. Hirakura, Y. and Kagan, B.L. (2001) Pore formation by beta-2-microglobulin: a mechanism for the pathogenesis of dialysis-associated amyloidosis. *Amyloid*, **8**, 94–100.
27. Kayed, R., Head, E., Thompson, J.L., McIntire, T.M., Milton, S.C., Cotman, C.W., and Glabe, C.G. (2003) Common structure of soluble amyloid oligomers implies common mechanisms of pathogenesis. *Science*, **300**, 486–489.
28. Bucciantini, M., Calloni, G., Chiti, F., Formigli, L., Nosi, D., Dobson, C.M., and Stefani, M. (2004) Pre-fibrillar amyloid protein aggregates share common features of cytotoxicity. *J. Biol. Chem.*, **279**, 31374–31382.
29. Walsh, D.M. and Selkoe, D.J. (2004) Oligomers on the brain: the emerging role of soluble protein aggregates in neurodegeneration. *Protein Pept. Lett.*, **11**, 1–16.
30. Cleary, J.P., Walsh, D.M., Hofmeister, J.J., Shankar, G.M., Kuskowski, M., Selkoe, D.J., and Ashe, K.H. (2005) Natural oligomers of the amyloid-β specifically disrupt cognitive function. *Nat. Neurosci.*, **8**, 79–84.
31. Billings, L.M., Oddo, S., Green, K.N., McGaugh, J.L., and LaFerla, F.M. (2005) Intraneuronal Aβ causes the onset of early Alzheimer's disease-related cognitive deficits in transgenic mice. *Neuron*, **45**, 675–688.
32. Lesné, S., Koh, M.T., Kotilinek, L., Kayed, R., Glabe, C.G., Yang, A., Gallagher, M., and Ashe, K.H. (2006) A specific amyloid β protein assembly in the brain impairs memory. *Nature*, **440**, 352–357.
33. Koffie, R.M., Meyer-Luehmann, M., Hashimoto, T., Adams, K.W., Mielke, M.L., Garcia-Alloza, M., Micheva, K.D., Smith, S.J., Kim, M.L., Lee, V.M., Hyman, B.T., and Spires-Jones, T.L. (2009) Oligomeric amyloid beta associates with postsynaptic densities and correlates with excitatory synapse loss near senile plaques. *Proc. Natl. Acad. Sci. USA*, **106**, 4012–4017.
34. Nekooki-Machida, Y., Kurosawa, M., Nukina, N., Ito, K., Oda, T., and Tanaka, M. (2009) Distinct conformations of *in vitro* and *in vivo* amyloids of huntingtin-exon1 show different cytotoxicity. *Proc. Natl. Acad. Sci. USA*, **106**, 9679–9684.
35. Gharibyan, A.L., Zamotin, V., Yanamandra, K., Moskaleva, O.S., Margulis, B.A., Kostanyan, I.A. and Morozova-Roche, L.A. (2007) Lysozyme amyloid oligomers and fibrils induce cellular death via different apoptotic/necrotic pathways. *J. Mol. Biol.*, **365**, 1337–1348.
36. Novitskaya, V., Bocharova, O.V., Bronstein, I., and Baskakov, I.V. (2006) Amyloid fibrils of mammalian prion protein are highly toxic to cultured cells and primary neurons. *J. Biol. Chem.*, **281**, 13828–13836.
37. Klyubin, I., Walsh, D.M., Lemere, C.A., Cullen, V.K., Shankar, G.M., Betts, V., Spooner, E.T., Jiang, L., Amwyl, R., Selkoe, D.J., and Rowan, M.J. (2005) Amyloid beta protein immunotherapy neutralizes Abeta oligomers that disrupt

synaptic plasticity *in vivo*. *Nat. Med.*, **11**, 556–561.

38 Chafekar, F.M., Hoozemans, J.J., Zwart, R., Baas, F., and Scheper, W. (2007) Abeta 1–42 induces mild endoplasmic reticulum stress in an aggregation state-dependent manner. *Antiox. Red. Signal.*, **9**, 2245–2254.

39 Eikelemboom, R., Bate, C., Van Gool, W.A., Hoozemans, J.J., Rozemuller, J.M., Veerhuis, R., and Williams, A. (2002) Neuroinflammation in Alzheimer's disease and prion disease. *Glia*, **40**, 232–239.

40 Fuhrmann, M., Bittner, T., Jung, C.K.E., Burgold, S., Page, R.M., Mitteregger, G., Haass, C., LaFerla, F.M., Kretzschmar, H., and Herms, J. (2010) Microglial Cx3cr1 knockout in a mouse model of Alzheimer's disease. *Nat. Neurosci.*, **13**, 411–413.

41 Chafekar, S.M., Baas, F., and Scheper, W. (2008) Oligomer-specific Aβ toxicity in cell models is mediated by selective uptake. *Biochim. Biophys. Acta*, **1782**, 523–531.

42 Dickson, D.W. (1995) Correlation of synaptic and pathological markers with cognition of the elderly. *Neurobiol. Aging*, **16**, 285–298.

43 Serpell, L.C., Sunde, M., Benson, M.D., Tennent, G.A., Pepys, M.B., and Fraser, P.E. (2000) The protofilament substructure of amyloid fibrils. *J. Mol. Biol.*, **300**, 1033–1039.

44 Jiménez, J.L., Nettleton, E.J., Bouchard, M., Robinson, C.V., Dobson, C.M., and Saibil, H.R. (2002) The protofilament structure of insulin amyloid fibrils. *Proc. Natl. Acad. Sci. USA*, **99**, 9196–9201.

45 Lührs, T., Ritter, C., Adrian, M., Riek-Loher, D., Bohrmann, B., Döbeli, H., Schubert, D., and Riek, R. (2005) 3D structure of Alzheimer's amyloid-β(1–42) fibrils. *Proc. Natl. Acad. Sci. USA*, **102**, 17342–17347.

46 Petkova, A.T., Leapman, R.D., Guo, Z., Yau, W.-M., Mattson, M.P., and Tycko, R. (2005) Self-propagating, molecular-level polymorphism in Alzheimer's β-amyloid fibrils. *Science*, **307**, 262–265.

47 Quintas, A., Vaz, D.C., Cardoso, I., Saraiva, M.J.M., and Brito, R.M.M. (2001) Tetramer dissociation and monomer partial unfolding precedes protofibril formation in amyloidogenic transthyretin variants. *J. Biol. Chem.*, **276**, 27207–27213.

48 Relini, A., Torrassa, S., Rolandi, R., Gliozzi, A., Rosano, C., Canale, C., Bolognesi, M., Plakoutsi, G., Bucciantini, M., Chiti, F., and Stefani, M. (2004) Monitoring the process of HypF fibrillization and liposome permeabilization by protofibrils. *J. Mol. Biol.*, **338**, 943–957.

49 Cheon, M., Chang, I., Mohanty, S., Luheshi, L.M., Dobson, C.M., Vendruscolo, M., and Favin, G. (2007) Structural reorganisation and potential toxicity of oligomeric species formed during the assembly of amyloid fibrils. *PLoS Comp. Biol.*, **3**, 1727–1738.

50 Lashuel, H.A., Petre, B.M., Wall, J., Simon, M., Nowak, R.J., Walz, T., and Lansbury, P.T. (2002) α-Synuclein, especially the Parkinson's disease-associated mutants, forms pore-like annular and tubular protofibrils. *J. Mol. Biol.*, **322**, 1089–1102.

51 Poirier, M.A., Li, H., Macosko, J., Cail, S., Amzel, M., and Ross, C.A. (2002) Huntingtin spheroids and protofibrils as precursors in polyglutamine fibrillization. *J. Biol. Chem.*, **277**, 41032–41037.

52 Caughey, B. and Lansbury, P.T. (2003) Protofibrils, pores, fibrils and neurodegeneration: separating the responsible protein aggregates from the innocent bystanders. *Annu. Rev. Neurosci.*, **26**, 267–298.

53 Lin, H., Bhatia, R., and Lal, R. (2001) Amyloid β protein forms ion channels: implications for Alzheimer's disease pathophysiology. *FASEB J.*, **15**, 2433–2444.

54 Serio, T.R., Cashikar, A.G., Kowal, A.S., Sawicki, G.J., Moslehi, J.J., Serpell, L., Arnsdorf, M.F., and Lindquist, S. (2000) Nucleated conformation conversion and the replication of conformational information by a prion determinant. *Science*, **289**, 1317–1321.

55 Bitan, G., Kirkitadze, M.D., Lomakin, A., Vollers, S.S., Benedek, G.B., and Teplow, D.B. (2003) Amyloid beta-protein (Abeta) assembly: Abeta40 and Abeta42 oligomerize through distinct pathways. *Proc. Natl. Acad. Sci. USA*, **100**, 330–335.

56 Chang, H.Y., Lin, J.Y., Lee, H.C., Wang, H.L., and King, C.Y. (2008) Strain-specific sequences required for yeast [PSI+] prion propagation. *Proc. Natl. Acad. Sci. USA*, **105**, 13345–13350.

57 Orte, A., Birkett, N.R., Clarke, R.W., Devlin, G.L., Dobson, C.M., and Klenerman, D. (2008) Direct characterization of amyloidogenic oligomers by single-molecule fluorescence. *Proc. Natl. Acad. Sci. USA*, **105**, 14424–14429.

58 Langkilde, A.E. and Vestergaard, B. (2009) Methods for structural characterization of prefibrillar intermediates and amyloid fibrils. *FEBS Lett.*, **583**, 2600–2609.

59 O'Nuallain, B. and Wetzel, R. (2002) Conformational Abs recognizing a generic amyloid fibril epitope. *Proc. Natl. Acad. Sci. USA*, **99**, 1485–1490.

60 Demuro, A., Mina, E., Kayed, R., Milton, S.C., Parker, I., and Glabe, C.G. (2005) Calcium dysregulation and membrane disruption as a ubiquitous neurotoxic mechanism of soluble amyloid oligomers. *J. Biol. Chem.*, **280**, 17294–17300.

61 Yoshiike, Y., Kayed, R., Milton, S.C., Takashima, A., and Glabe, C.G. (2007) Pore-forming proteins share structural and functional homology with amyloid oligomers. *Neuromol. Med.*, **9**, 270–275.

62 Kayed, R., Head, E., Sarsoza, F., Saing, T., Cotman, C.W., Necula, M., Margol, L., Wu, J., Breydo, L., Thompson, J.L., Rasool, S., Gurlo, T., Butler, P., and Glabe, C.G. (2007) Fibril specific, conformation dependent antibodies recognize a generic epitope common to amyloid fibrils and fibrillar oligomers that is absent in prefibrillar oligomers. *Mol. Neurodegen.*, **2**, 1–18.

63 Kumar, S. and Udgaonkar, J.B. (2009) Structurally distinct amyloid protofibrils form on separate pathways of aggregation of a small protein. *Biochemistry*, **48**, 6441–6449.

64 Glabe, C.G. (2004) Conformation-dependent antibodies target diseases of protein misfolding. *Trends Biochem. Sci.*, **29**, 542–547.

65 Mamikonyan, G., Necula, M., Mkrtichyan, M., Ghochikyan, A., Petrushina, I., Movsesyan, N., Mina, E., Kiyatkin, A., Glabe, C.G., Cribbs, D.H., and Agadjanyan, M.G. (2007) Anti-A beta 1–11 antibody binds to different beta-amyloid species, inhibits fibril formation, and disaggregates preformed fibrils but not the most toxic oligomers. *J. Biol. Chem.*, **282**, 22376–22386.

66 Kayed, R. and Glabe, C.G. (2006) Conformation-dependent anti-amyloid oligomer antibodies. *Methods Enzymol.*, **413**, 326–344.

67 Glabe, C.G. (2008) Structural classification of toxic amyloid oligomers. *J. Biol. Chem.*, **283**, 29639–29643.

68 Chen, Y. and Kokholyan, N. (2005) A single disulfide bond differentiates aggregation pathways of β2-microglobulin. *J. Mol. Biol.*, **354**, 473–482.

69 Danzer, K.M., Haasen, D., Karow, A.R., Moussaud, S., Habeck, M., Giese, A., Kretzschmar, H., Hengerer, B., and Kostka, M. (2007) Different species of α-synuclein oligomers induce calcium influx and seeding. *J. Neurosci.*, **271**, 9220–9232.

70 Bravo, R., Arimon, M., Valle-Delgado, J.J., Garcia, R., Durany, N., Castel, S., Cruz, M., Ventura, S., and Fernandez-Busquets, X. (2008) Sulfated polysaccharides promote the assembly of amyloid $β_{1-42}$ peptide into stable fibrils of reduced cytotoxicity. *J. Biol. Chem.*, **283**, 32471–32483.

71 Kaylor, J., Bodner, N., Edridge, S., Yamin, G., Hong, D.P., and Fink, A.L. (2005) Characterization of oligomeric intermediates in alpha-synuclein fibrillation: FRET studies of Y125W/Y133F/Y136F alpha-synuclein. *J. Mol. Biol.*, **353**, 357–372.

72 Dusa, A., Kaylor, J., Edridge, S., Bodner, N., Hong, D.P., and Fink, A.L. (2006) Characterization of oligomers

during alpha-synuclein aggregation using intrinsic tryptophan fluorescence. *Biochemistry*, **2006**, *45*, 2752–2760.
73 Bokvist, M., Lindström, F., Watts, A., and Gröbner, G. (2004) Two types of Alzheimer's β-amyloid (1–40) peptide membrane interactions: aggregation preventing transmembrane anchoring versus accelerated surface fibril formation. *J. Mol. Biol.*, **335**, 1039–1049.
74 Zhu, M., Souillac, P.O., Ionesco-Zanetti, C., Carter, S.A., and Fink, A.L. (2002) Surface-catalyzed amyloid fibril formation. *J. Biol. Chem.*, **277**, 50914–50922.
75 Sethuraman, A. and Belfort, G. (2005) Protein structural perturbation and aggregation on homogeneous surfaces. *Biophys. J.*, **88**, 1322–1333.
76 Yip, C.M., Elton, E.A., Darabie, A.A., Morrison, M.R., and McLaurin, J. (2001) Cholesterol, a modulator of membrane-associated Aβ-fibrillogenesis and neurotoxicity. *J. Mol. Biol.*, **311**, 723–734.
77 Porat, Y., Kolusheva, S., Jelinek, R., and Gazit, E. (2003) The human islet amyloid polypeptide forms transient membrane-active prefibrillar assemblies. *Biochemistry*, **42**, 10971–10977.
78 Knight, J.D. and Miranker, A.D. (2004) Phospholipid catalysis of diabetic amyloid assembly. *J. Mol. Biol.*, **341**, 1175–1187.
79 Stefani, M. (2007) Generic cell dysfunction in neurodegenerative disorders: role of surfaces in early protein misfolding, aggregation, and aggregate cytotoxicity. *Neuroscientist*, **13**, 519–531.
80 Kourie, J.I. and Shorthouse, A.A. (2000) Properties of cytotoxic peptide-induced ion channels. *Am. J. Physiol. Cell Physiol.*, **278**, C1063–C1087.
81 Zhu, Y.J., Lin, H., and Lal, R. (2000) Fresh and nonfibrillar amyloid β protein (1–40) induces rapid cellular degeneration in aged human fibroblasts: evidence for AβP-channel-mediated cellular toxicity. *FASEB J.*, **14**, 1244–1254.
82 Necula, M., Chirita, C., and Kuret, J. (2002) Rapid anionic micelle-mediated α-synuclein fibrillization *in vitro*. *J. Biol. Chem.*, **278**, 46674–46680.
83 Jayakumar, R., Jayaraman, M., Koteeswarl, D., and Gomath, K. (2004) Cytotoxic and membrane perturbation effects of a novel amyloid forming model, peptide poly(leucine-glutamic acid). *J. Biochem. (Tokyo)*, **136**, 457–462.
84 Zhao, H., Tuominen, E.K.J., and Kinnunen, P.K.J. (2004) Formation of amyloid fibers triggered by phosphatidylserine-containing membranes. *Biochemistry*, **43**, 10302–10307.
85 Pike, L.J. (2003) Lipid rafts: bringing order to chaos. *J. Lipid Res.*, **44**, 655–667.
86 Stefani, M. and Liguri, G. (2009) Cholesterol in Alzheimer's disease: unresolved questions. *Curr. Alz. Res.*, **6**, 15–29.
87 Harris, J.R. and Milton, N.G. (2010) Cholesterol in Alzheimer's disease and other amyloidogenic disorders. *Subcell. Biochem.*, **51**, 47–75.
88 Wakabayashi, M. and Matsuzaki, K. (2009) Ganglioside-induced amyloid formation by human islet amyloid polypeptide in lipid rafts. *FEBS Lett.*, **583**, 2854–2858.
89 Plakoutsi, G., Bemporad, F., Calamai, M., Taddei, N., Dobson, C.M., and Chiti, F. (2005) Evidence for a mechanism for amyloid formation involving reorganization within native-like precursor aggregates. *J. Mol. Biol.*, **351**, 910–922.
90 Petty, S.A. and Decatur, S.M. (2005) Experimental evidence for the reorganization of beta strands within aggregates of the Abeta (16–22) peptide. *J. Am. Chem. Soc.*, **127**, 13488–13489.
91 Cerdà-Costa, N., Esteras-Chopo, A., Avilés, F.X., Serrano, L., and Villegas, V. (2007) Early kinetics of amyloid fibril formation reveals conformational reorganization of initial aggregates. *J. Mol. Biol.*, **366**, 1351–1363.
92 Carulla, N., Zhou, M., Arimon, M., Gairi, M., Giralt, E., Robinson, C.V., and Dobson, C.M. (2009) Experimental characterization of disordered and ordered aggregates populated during the process of amyloid fibril formation. *Proc. Natl. Acad. Sci. USA*, **106**, 7828–7833;Bader, R., Bamford, R.,

Zurdo, J., Luisi, B.F., and Dobson, C.M. (2006) Probing the mechanism of amyloidogenesis through a tandem repeat of the pi3–sh3 domain suggests a generic model for protein aggregation and fibril formation. *J. Mol. Biol.*, **356**, 189–208.

93 Chiti, F. and Dobson, C.M. (2009) Amyloid formation by globular proteins under native conditions. *Nat. Chem. Biol.*, **5**, 15–22.

94 Calamai, M., Chiti, F., and Dobson, C.M. (2005) Amyloid fibril formation can proceed from different conformations of a partially unfolded protein. *Biophys. J.*, **89**, 4201–4210.

95 Lee, S., Fernandez, E.J., and Good, T.A. (2007) Role of aggregation conditions in the structure, stability, and toxicity of intermediates in the Aβ fibril formation pathway. *Protein Sci.*, **16**, 723–732.

96 Meinhardt, J., Sachse, C., Hortschansky, P., Grigorieff, N., and Fandrich, M. (2009) Abeta(1–40) fibril polymorphism implies diverse interaction patterns in amyloid fibrils. *J. Mol. Biol.*, **386**, 869–877.

97 Masuda, Y., Uemura, S., Ohashi, R., Nakanishi, A., Takegoshi, K., Shimizu, T., Shirasawa, T., and Irie, K. (2009) Identification of physiological and toxic conformations in Aβ42 aggregates. *ChemBioLChem*, **10**, 287–295.

98 Chien, P., Weissman, J.S., and DePace, A.H. (2004) Emerging principles of conformation-based prion inheritance. *Annu. Rev. Biochem.*, **73**, 617–656.

99 Jones, E.M. and Surewicz, W.K. (2005) Fibril conformation as the basis of species- and strain-dependent seeding specificity of mammalian prion amyloids. *Cell*, **121**, 63–72.

100 Meyer-Luheman, M., Coomaraswamy, J., Bolmont, T., Kaeser, S., Schaefer, C., Kilger, E., Neuenschwander, A., Abramowski, D., Frey, P., Jaton, A.J., Vigouret, J.-M., Paganetti, P., Walsh, D.M., Mathews, P.M., Ghiso, J., Staufenbiel, M., Walker, L.C., and Jucker, M. (2006) Exogeneous induction of cerebral β-amyloidogenesis is governed by agent and host. *Science*, **313**, 1781–1784.

101 Gosal, W.S., Morten, J.J., Hewitt, E.W., Smith, D.A., Thomson, N.H., and Radford, S.E. (2005) Competing pathways determine fibril morphology in the self-assembly of $\beta_2$-microglobulin into amyloid. *J. Mol. Biol.*, **351**, 850–864.

102 Goldsbury, C., Frey, P., Olivieri, V., Aebi, U., and Müller, S.A. (2005) Multiple assembly pathways underlie amyloid-β fibril polymorphism. *J. Mol. Biol.*, **352**, 282–298.

103 Deshpande, A., Mina, E., Glabe, C., and Busciglio, J. (2006) Different conformations of amyloid β induce neurotoxicity by distinct mechanisms in human cortical neurons. *J. Neurosci.*, **26**, 6011–6018.

104 Iijima, K., Chiang, H.-C., Hearn, S.A., Hakker, I., Gatt, A., Shenton, C., Granger, L., Leung, A., Iijima-Ando, K., and Zhong, Y. (2008) Aβ42 mutants with different aggregation profiles induce distinct pathologies in *Drosophila*. *PLoS ONE*, **3**, 1–8.

105 Engel, M.F., Khemtémourian, L., Kleijer, C.C., Meeldijk, H.J., Jacobs, J., Verkeij, A.I., de Kruijff, B., Killan, J.A., and Höppener, J.K. (2008) Membrane damage by human islet amyloid polypeptide through fibril growth at the membrane. *Proc. Natl. Acad. Sci. USA*, **105**, 6033–6038.

106 Xue, W.-F., Hellewell, A.L., Gosal, W.S., Homans, S.W., Hewitt, E.W., and Radford, S.E. (2009) Fibril fragmentation enhances amyloid cytotoxicity. *J. Biol. Chem.*, **284**, 34272–34282.

107 Martins, I.C., Kuperstein, I., Wilkinson, H., Maes, E., Vambrabant, M., Jonckheere, W., Van Gelder, P., Hartmann, D., D'Hooge, R., De Strooper, B., Schymkowitz, J., and Rousseau, F. (2008) Lipids revert inert Abeta amyloid fibrils to neurotoxic protofibrils that affect learning in mice. *EMBO J.*, **27**, 224–233.

108 Carulla, N., Caddy, G.L., Hall, D.R., Zurdo, J., Gairi, M., Feliz, M., Giralt, E., Robinson, C.V., and Dobson, C.M. (2005) Molecular recycling within amyloid fibrils. *Nature*, **436**, 554–558.

109 Shankar, G.M., Leissring, M.A., Adame, A., Sun, X., Spooner, E.,

Masliah, E., Selkoe, D.J., Lemere, C.A., and Walsh, D.M. (2009) Biochemical and immunohistochemical analysis of an Alzheimer's disease mouse model reveals the presence of multiple cerebral Aβ assembly forms throughout life. *Neurobiol. Dis.*, **36**, 293–302.

110 Smith, J.F., Knowles, T.P.J., Dobson, C.M., MacPhee, C.E., and Welland, M.E. (2006) Characterization of the nanoscale properties of individual amyloid fibrils. *Proc. Natl. Acad. Sci. USA*, **103**, 15806–15811.

111 Merlini, G., Ascari, E., Amboldi, N., Bellotti, V., Arbustini, E., Perfetti, V., Ferrari, M., Zorzoli, I., Marinone, M.G., Garini, P., Diegolis, M., Trizio, D., and Ballinari, D. (1995) Interaction of the anthracycline 4'-iodo-4'-deoxydoxorubicin with amyloid fibrils: inhibition of amyloidogenesis. *Proc. Natl. Acad. Sci. USA*, **92**, 2959–2963.

112 Cardoso, I. and Saraiva, M.J. (2006) Doxycycline disrupts transthyretin amyloid: evidence from studies in a FAP transgenic mice model. *FASEB J.*, **20**, 234–239.

113 DeGroot, J. (2004) The AGE of the matrix: chemistry, consequence and cure. *Curr. Opin. Pharmacol.*, **4**, 301–305.

114 Hepler, R.W., Grimm, K.M., Nahas, D.D., Breese, R., Dodson, E.C., Acton, P., Keller, P.M., Yeager, M., Wang, H., Shughrue, P., Kinney, G., and Joyce, J.G. (2006) Solution state characterization of amyloid beta-derived diffusible ligands. *Biochemistry*, **45**, 15157–15167.

115 Danzer, K.M., Krebs, S.K., Wolff, M., Birk, G., and Hengerer, B. (2009) Seeding induced by α-synuclein oligomers provides evidence for spreading of α-synuclein pathology. *J. Neurochem.*, **111**, 192–203.

116 El-Agnaf, O.M., Salem, S.A., Paleologou, K.E., Curran, M.D., Gibson, M.J., Court, J.A., Schlossmacher, M.G., and Allsop, D. (2006) Detection of oligomeric forms of α-synuclein protein in human plasma as a potential biomarker for Parkinson's disease. *FASEB J.*, **20**, 419–425.

117 Lee, H.J., Patel, S., and Lee, S.J. (2005) Intravesicular localization and exocytosis of α-synuclein and its aggregates. *J. Neurosci.*, **25**, 6016–6024.

118 Campioni, S., Mannini, B., Pensalfini, A., Zampagni, M., Parrini, C., Evangelisti, E., Relini, A., Stefani, M., Dobson, C.M., Cecchi, C., and Chiti, F. (2010) The causative link between the structure of aberrant protein oligomers and their ability to cause cell dysfunction. *Nat. Chem. Biol.*, **6**, 140–147.

119 Picone, P., Carrotta, R., Montana, G., Nobile, M.R., San Biagio, P.L., and Di Carlo, M. (2009) Aβ oligomers and fibrillar aggregates induce different apoptotic pathways in LAN5 neuroblastoma cell cultures. *Biophys. J.*, **96**, 4200–4211.

120 Pountney, D.L., Voelcker, N.H., and Gai, W.P. (2005) Annular α-synuclein oligomers are potentially toxic agents in α-synucleinopathy. *Neurotox. Res.*, **7**, 59–67.

# 6
# Intracellular Amyloid β: a Modification to the Amyloid Hypothesis in Alzheimer's Disease

*Yan Zhang*

## 6.1
## Introduction

Alzheimer's disease (AD) is a progressive neurodegenerative disorder characterized by impaired cognition and memory. AD patients' brains are characterized by three major pathological hallmarks: extracellular senile plaques composed primarily of ~4 kDa amyloid β peptide (Aβ), intracellular neurofibrillary tangles (NFTs), and degenerating neurons [1]. Because of the extracellular location of senile plaques, extracellular Aβ (eAβ), especially $eAβ_{42}$, is thought to be one of the primary causes of AD [2–8]. The eAβ toxicity hypothesis is supported by evidence that fibrillar eAβ could be toxic to various cell lines and primary neurons in cultures [2, 9–17]. In addition, eAβ levels are increased in the brains and sera of AD patients [1]. Several possible mechanisms have been suggested to account for eAβ cytotoxicity, for example, upregulation of vulnerability of cells to secondary insults [12, 18], changes in calcium influx [19], activation of inflammation and microglia [20, 21], induction of tau hyperphosphorylation [22], induction of apoptosis [23–25], induction of lysosomal protease activity, and direct damage to membranes [26, 27]. Furthermore, an Aβ oligomeric fraction but not fibrils has been shown to impair long-term potentiation (LTP) formation [28]. More recently, Aβ oligomers isolated from AD transgenic model Tg2576 mouse brains have been reported to induce cognitive defects when transferred into wild-type mouse brains, verifying the role of eAβ in memory function [28, 29]. A mouse model that produces greater amounts of Aβ oligomers but not fibrils was shown to have impaired hippocampal synaptic plasticity and memory [30].

However, several lines of evidence suggest that eAβ may have reduced impact on AD pathology. For example, there are cognitively normal aged individuals with heavy plaque load [31–33]. Even at high micromolar concentrations, eAβ failed to demonstrate consistent cytotoxicity in rat PC12 cells, human IMR32 cells, and monkey cerebral cortical neurons [34–36]. Cultured human primary neurons are resistant to 10 μM of eAβ treatment [37, 38]. Hence eAβ deposition alone is inadequate to account for all AD pathology. A new modified amyloid hypothesis, the so-called "intracellular amyloid hypothesis," is attracting increasing attention [39]. This hypothesis states

*Lipids and Cellular Membranes in Amyloid Diseases*, First Edition. Edited by Raz Jelinek.
© 2011 Wiley-VCH Verlag GmbH & Co. KGaA. Published 2011 by Wiley-VCH Verlag GmbH & Co. KGaA.

that intracellular Aβ (iAβ) accumulation is the early, causative event in AD development. iAβ leads to cytotoxicity; the extracellular amyloid deposition observed in late stages of disease is the result of cell death and destruction [39]. The significance of iAβ is a complementary explanation for defective neuronal physiology and neuronal cell loss in AD.

## 6.2
## Evidence for the Presence of Intracellular Amyloid

### 6.2.1
### Detection of Intracellular Amyloid

iAβ is widely detected in neuronal cells and primary human neurons [40]. Significant accumulation of iAβ is detected in mild cognitive impairment (MCI), AD patient brains (Figure 6.1) [41–49], Down's syndrome (DS) patient brains [48, 50, 51], and the brain stem and the temporal lobe of young drug abusers [52]. Using various antibodies directed against the C-terminus of Aβ, iAβ immunoreactivity is specifically observed, prior to Aβ deposit formation, in most neurons of the CA areas of the hippocampus and the laminae III and V of the frontal and temporal cortices in early AD and juvenile DS brains [53]. Interestingly, iAβ is not prominently detected in non-AD neurons and the amount actually diminishes after amyloid deposition in neuropathologically confirmed AD patients [53].

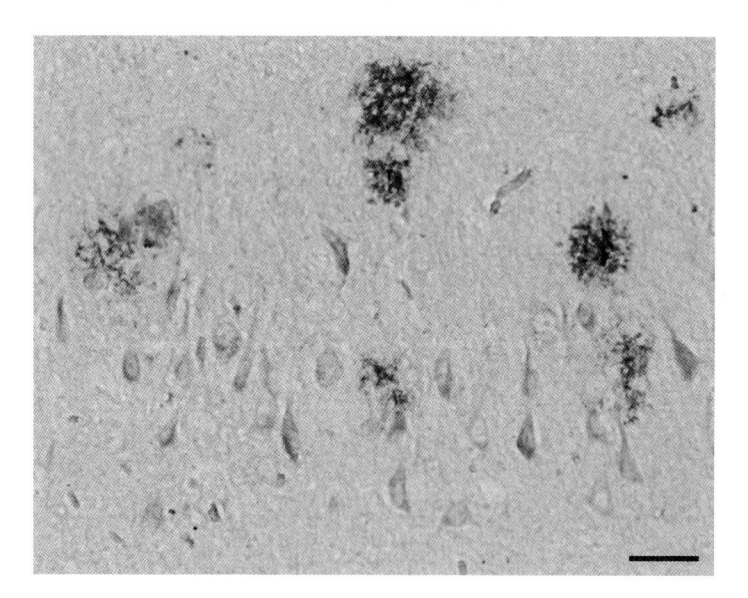

**Figure 6.1** Detection of intracellular Aβ in AD patient brain. Representative image of intraneuronal Aβ$_{42}$ and plaque labeling in human Alzheimer's disease brain using Aβ$_{42}$ specific antibody (G2-13, Millipore); citric buffer pretreatment was used. Bar = 100 μm. This image was kindly provided by Estibaliz Capetillo-Zarate and Gunnar Gouras, Weill Cornell Medical College.

iAβ is often observed in neurons of various transgenic AD mouse models [54], including inducible p25 [55], 5xFAD [56], and arcAβ [57]. Moreover, cultured primary neurons from *Drosophila* and cynomolgus monkey all demonstrated accumulation of iAβ$_{1-42}$ [58–61]. However, modifications and variations of the Aβ detection methods can affect the results of iAβ immunostaining. In fixed neurons, a specific antigen retrieval technique, such as "heating," enhances the immunolabeling of both anti-Aβ$_{42}$-N-terminal and -C-terminal antibodies, but has no effect on anti-Aβ$_{40}$ antibodies. In contrast, formic acid treatment abolishes the effect of "heating," suggesting that the amount of detectable iAβ may be underestimated in many studies using the conventional formic acid treatment in immunohistochemistry protocols [62]. Accumulation of iAβ appears to be an early change and occurs prior to neuronal loss and NFT appearance in transgenic AD mouse brains. In the amyloid precursor protein (APP)/PS1 double mutant mice, iAβ accumulation precedes NFT formation [63]. In APP mutant mice, synaptic loss and iAβ detection occur before the appearance of Aβ plaques [64–67]. In Tg2576 mice, iAβ accumulates in the outer membranes of multivesicular bodies multivesicular body (MVBs) and associates with synaptic dystrophy [48, 68], suggesting that iAβ may be an early event before neuronal degeneration.

Although Aβ is mainly produced by neurons, it is often detected in other types of cells. Glial cells have been shown to produce high levels of Aβ in normal human brains [11]. In cultured human brain cells, both astrocytes and microglia have been demonstrated to produce Aβ [69, 70]. Strikingly, in AD patient brains, iAβ immunoreactivity is detected in astrocytes [71, 72], especially for those present in the vicinity of the diffuse plaques [73]. In DS patient brains where Aβ plaques and high levels of Aβ are found, iAβ accumulation occurs in astrocytes before plaque formation [74]. Aβ-immunoreactive granules accumulating in astrocytes in aged non-demented human brains have also been reported [75]. Other than neurons and glial cells, iAβ is also detected in muscle cells of inclusion body myositis patients [76–78], another degenerative disease associated with Aβ deposition.

### 6.2.2
### Neurotoxicty of Intracellular Amyloid

Several pieces of evidence suggest that accumulation of iAβ may be related to cell death. In P19 cells overexpressing human APP751, an increase in iAβ is associated with a decrease in mitochondria membrane potential, a functional change that is related to free radical release and cell death [79]. Rat brains overexpressing human APP with iAβ accumulation have neuronal loss in the cortical regions, and inhibition of iAβ accumulation by γ-secretase inhibitor promotes cell survival [80]. In AD patient brains, iAβ deposits correlate with the degree of apoptotic cell death demonstrated by terminal deoxynucleotidyl transferase–biotin dUTP nick-end labeling (TUNEL) [81]. Microinjection of Aβ$_{1-42}$ induces significant cell death in human primary cultured neurons [82]. In contrast, microinjecting the control reversed sequence peptides or extracellularly expressing Aβ$_{1-42}$ cDNA constructs are unable to induce neuronal cell death [82]. The cellular toxicity of iAβ may be cell-type specific, because it induces cell death only in human primary neurons, and not in NT2, LaN1, or M17 cells [82]. It also appears that the Aβ oligomers but not fibrils may be the toxic species [82].

### 6.2.3
### Possible Mechanisms of Intracellular Amyloid Toxicity

Several molecular and cellular mechanisms may account for the iAβ neurotoxicity. Accumulation of iAβ may impair subcellular compartment structures, including inducing alterations in axonal structure and transport. In aged G-protein coupled receptor kinase 5 (GRK5)-deficient mice that show selective working memory impairment, iAβ is detected in the hippocampal neuronal cell bodies and axons [83]. In these mice, iAβ accumulation is also associated with clusters of axonal swellings [83]. In the double APP and PS1 mutant transgenic model, stable expression of human iAβ increases the number of Golgi apparatus elements, lysosomes, and lipofuscin bodies in the hippocampal area [84]. In the spinal cords of transgenic mice overexpressing human APP751 and PS1$^{M146L}$, upregulation of iAβ is associated with an age-dependent axonopathy with the formation of axonal spheroids and myelin ovoids [85]. Since iAβ accumulates in the MVBs in the APP mutant transgenic mice and in AD patient brains, iAβ may impair the MVB sorting pathway and inhibit the activities of proteasome and deubiquitinating enzymes [86].

In addition to inducing subcellular compartment structural changes, iAβ toxicity correlates with the regulation of several apoptotic factors and pathways. It is well known that oxidative DNA damage induces iAβ accumulation and p53 mRNA increase in the nuclei of the guinea-pig primary neurons [87]; both events are also commonly observed in the degenerating neurons in AD patient brains [41, 87]. Incidentally, iAβ activates p53 by direct interaction with the p53 promoter [87], which might lead to a subsequent cell death pathway such as Bax and caspase-6 [82]. There is also evidence showing that iAβ may suppress the Akt signaling pathway in rat primary cortical neurons and in the Tg2576 AD transgenic model [88] that may lead to glycogen synthase kinase 3β (GSK3β) activation and cell death [88]. In transgenic rats with significant accumulation of iAβ, a dramatic upregulation of mitogen-activated protein kinase extracellular signal-regulated kinase 2 (ERK2) associated with increased tau phosphorylation is observed [67]. Since ERK2 has been linked to cAMP response element (CRE)-directed gene expression regulation, protein synthesis and late LTP [89], accumulation of iAβ could possibly be linked to synaptic dysfunction, a prominent feature of AD pathology. Heat shock proteins (Hsps) are probably involved in Akt neuroprotection against iAβ toxicity, since small interference RNA of Hsp70 blocks Akt neuroprotection [88]. Hsp70 has also been shown to be protective against iAβ toxicity in human primary neurons [90].

## 6.3
## Sources of Intracellular Amyloid

Aβ is generated by the sequential enzymatic cleavage of APP, a type I transmembrane protein [91–93]. APP is initially synthesized and inserted into the endoplasmic reticulum (ER), and can be cleaved at the C terminus by α-secretase near the cell surface to generate a secreted fragment [6]. In addition to the α-secretase pathway, it can be sequentially cleaved by β- and γ-secretases to generate Aβ containing 39–43 amino acids [94, 95]. The majority of Aβ peptides is constituted by the 40 amino acid long A$β_{1-40}$; only 10% of the species are the 42 amino acid peptide A$β_{1-42}$. To date, the physiological functions of APP and Aβ are still not well understood. These cleavages also generate the soluble extracellular fragments secreted amyloid precursor protein (sAPP) and amyloid precursor protein intracellular domain (AICD) [96].

The exact mechanisms leading to iAβ accumulation in AD remain unclear. There are at least three hypothetical pathways that may result in iAβ and its accumulation. First, passive leakage along any component of the secretory pathway can lead membrane-bound Aβ to release to the cytosol (Figure 6.2). APP processing is thought to occur in the endoplasmic reticulum/intermediate compartment (ER/IC) [97, 98], medial Golgi saccules [99], and the trans-Golgi network (TGN) [59, 100–102]. Studies using rat hippocampal neurons overexpressing APP showed that the nuclear envelope and the ER are immunostained with

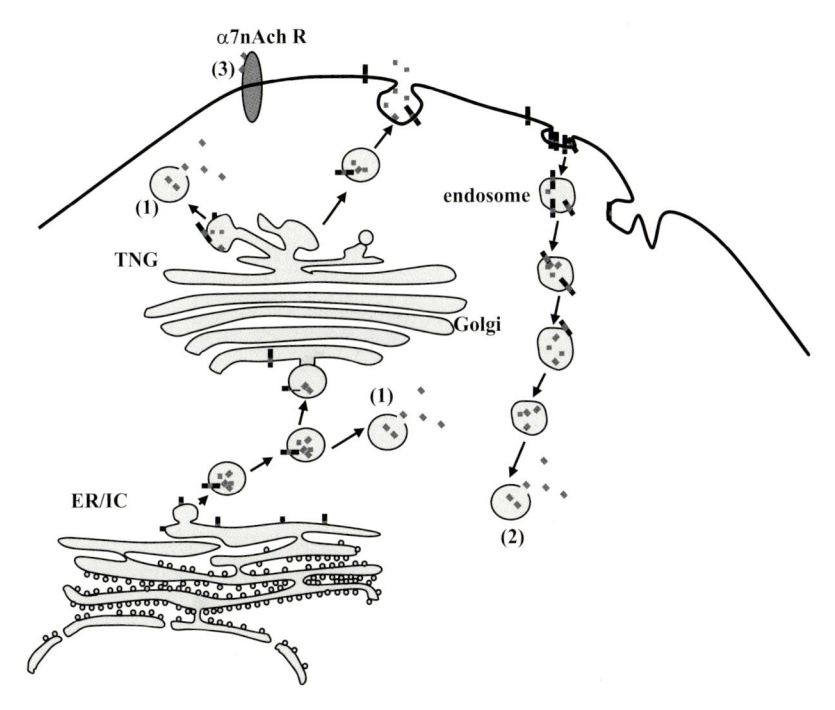

Figure 6.2 Origin and source of intracellular Aβ. APP can be cleaved by β-secretase and γ-secretase to form A$β_{42}$ in the ER, Golgi, and TGN. (1) Passive leakage along any component of the secretory pathway can lead membrane-bound Aβ to be released to the cytosol. (2) Aβ within the endosome/lysosome system can damage the lysosomal membrane and leak into the cytosol. (3) eAβ binds to surface receptors such as the α7AchR and is internalized into the cell.

monoclonal antibody specific for the A$\beta_{42}$ [103], indicating that the cleavage of APP by γ-secretase may occur within the ER. Furthermore, it has been shown that A$\beta_{x-40}$ is generated exclusively within the TGN and packaged into post-TGN secretary vesicles [59, 101], whereas A$\beta_{x-42}$ could be made and retained within the ER in an insoluble state, before entering the TGN and packaged into secretary vesicles [59]. When insoluble Aβ is produced in the ER, it can be recognized as a misfolded protein and exported back to the cytosol. The misfolded proteins are ubiquitinated and sent to the proteasomes for degradation [104]. Since the proteasome activity decreases with aging, inefficient degradation and clearance of Aβ can result in iAβ accumulation.

Second, Aβ can be generated in the endosomal/lysosomal system (Figure 6.1), because β-secretase [105, 106], presenilins [107, 108], and APP [109] can localize in the endosomes. This is of particular relevance to AD pathogenesis since abnormalities of the neuronal early endosomes are early pathologic events of sporadic AD and AD associated with DS [53, 110]. Respective overexpression of rab5 and MPR46, molecules that enhance the delivery of APP and lysosomal proteases to the early endosome, increases Aβ production [111, 112]. A double-immunostaining study showed that iAβ localizes primarily to rab5-positive endosomes in neurons of AD patient brains and is prominent in the enlarged endosomes, suggesting that the endosome is a primary site for Aβ production. It has been suggested that Aβ can increase the membrane permeability of lysosomes [58]. Therefore, the Aβ within the endosome/lysosome system may damage the lysosome membrane and leak into the cytosol. In addition, there is also evidence showing that autophagy-mediated Aβ generation is associated with both the turnover of APP-rich membrane systems and γ-secretase activity [113], providing another source for iAβ. Autophagy is a cell "self-digestion" process during cellular stress. Aβ in the lysosomes may cause leakage in the autophagic vacuole membrane, allowing Aβ to be released into the cytosol, hence with the subsequent formation of iAβ.

Third, Aβ is actively taken up by surface receptors such as the α7 nicotinic acetylcholine receptors (α7AchR) [46] (Figure 6.1). In addition, some molecular factors, including homocysteic acid [114] and cholesterol [115], can promote intracellular Aβ accumulation. Using a novel *in vitro* system, Cole *et al.* [115] distinguished the cholesterol-independent and cholesterol-dependent statin effects on APP processing and showed that low cholesterol levels would favor α-secretase cleavage whereas low isoprenoid levels would increase the accumulation of APP and amyloidogenic fragments in cellular compartments [115].

## 6.4
### Relationship Between Intracellular and Extracelluar Amyloid

eAβ can be a possible source of iAβ. The eAβ species can interact with cell surface receptors or the cell membrane to gain access into the cells, hence contributing to iAβ population. Aβ fibrils present as insoluble deposits can dissolve in soluble Aβ monomers, as demonstrated by the anti-amyloid antibody and amyloid vaccine therapies that cleared Aβ plaques [29]. The solubilized Aβ may subsequently gain access into the cells via receptor- or membrane-mediated mechanisms as described, if not degraded by the appropriate proteases such as insulin-dependent enzyme and neprilysin [116]. On the other hand, iAβ and its accumulation are detrimental to the neurons. It is possible that excess accumulation of iAβ may lead to cellular autolysis, releasing the iAβ into the extracellular matrix, contributing to amyloid centers for further deposition activities and formation of a "plaque" [45]. In this case, iAβ can be an origin of eAβ plaques. It is possible that the eAβ and iAβ present interact and feed back in a cyclic pathway.

## 6.5 Prevention of Intracellular Amyloid Toxicity

Both estrogens and androgens, at physiological concentrations, protect against iAβ cytotoxicity in human primary neurons [90]. The study also identifies Hsp70 as a key neuroprotectant [90]. Hsp70 alone is sufficient to rescue iAβ-induced cell death in human neurons, suggesting that Hsp70, a stress-induced chaperone, can attenuate iAβ toxicity [90]. In addition, a cell-permeable tricyclic pyrone, CP2, reduces iAβ toxicity and oligomeric complex of iAβ in neuroblastoma MC65 cells and in primary cortical neurons, suggesting that CP2 and its analogs may be promising anti-iAβ molecules [117]. Galanin has also been shown to be protective against iAβ toxicity in human primary neurons [118]. Galanin protection is associated with downregulation of Bax levels in neurons [118]. These studies targeting iAβ-induced neuronal loss indicate that iAβ toxicity is reversible. Whether these molecules can recover the neuronal function and improve the cognitive or behavioral performance remains to be evaluated. Neuronal loss is a relatively late event in neurodegeneration. Many changes, including synapse dysfunction, electrophysiological property alteration, and morphological atrophy, occur before neuronal loss. Intracellular Aβ can alter the electrophysiological properties of human primary neurons [119], which would subsequently lead to neuronal loss. Intracellular Aβ and its accumulation are an early event prior to senile plaque and NFT formation in transgenic AD mouse models; therefore, targeting may promise new targets and therapies for AD.

## 6.6 Concluding Remarks

The neurotoxicity and detection of eAβ in AD have been well documented for a long time. Recently, abundant evidence for the detection of iAβ in AD and other amyloid-related diseases, AD transgenic animal models, and confirmation of iAβ toxicity in human primary neurons, synaptic dysfunction, and cognitive impairment in transgenic animal models has also been presented. Hence both eAβ and iAβ are both critical components of the "amyloid hypothesis," supporting the critical role of Aβ in AD development [120]. The exact origin of iAβ and the molecular and cellular consequences of iAβ accumulation are speculative. Future studies need to be carried out to address the molecular sources and the role of iAβ in memory function and AD pathology. Such knowledge may contribute to our understanding of AD and allow better design of therapies and preventions.

## 6.7
## Disclosure Statement

The author declares no actual or potential conflicts of interest, including any financial, personal, or other relationships with other people or organizations, within 3 years of beginning this work that could inappropriately influence (bias) their work.

## Acknowledgments

This work was supported by the National Program of Basic Research sponsored by the Ministry of Science and Technology of China (2009CB941301), a Peking University President Research Grant, and a Ministry of Education Recruiting Research Grant.

## References

1 Price, D.L. and Sisodia, S.S. (1998) Mutant genes in familial Alzheimer's disease and transgenic models. *Annu. Rev. Neurosci.*, **21**, 479–505.

2 Yankner, B.A., Caceres, A., and Duffy, L.K. (1990) Nerve growth factor potentiates the neurotoxicity of beta amyloid. *Proc. Natl. Acad. Sci. USA*, **87**, 9020–9023.

3 Roses, A.D. (1996) The Alzheimer diseases. *Curr. Opin. Neurobiol.*, **6**, 644–650.

4 Scheuner, D. et al. (1996) Secreted amyloid beta-protein similar to that in the senile plaques of Alzheimer's disease is increased *in vivo* by the presenilin 1 and 2 and APP mutations linked to familial Alzheimer's disease. *Nat. Med.*, **2**, 864–870.

5 Yankner, B.A. (1996) Mechanisms of neuronal degeneration in Alzheimer's disease. *Neuron*, **16**, 921–932.

6 Sinha, S. and Lieberburg, I. (1999) Cellular mechanisms of beta-amyloid production and secretion. *Proc. Natl. Acad. Sci. USA*, **96**, 11049–11053.

7 De Strooper, B. and Annaert, W. (2000) Proteolytic processing and cell biological functions of the amyloid precursor protein. *J. Cell Sci.*, **113**, 1857–1870.

8 Wang, S.S., Rymer, D.L., and Good, T.A. (2001) Reduction in cholesterol and sialic acid content protects cells from the toxic effects of beta-amyloid peptides. *J. Biol. Chem.*, **276**, 42027–42034.

9 Kowall, N.W., Beal, M.F., Busciglio, J., Duffy, L.K., and Yankner, B.A. (1991) An *in vivo* model for the neurodegenerative effects of beta amyloid and protection by substance P. *Proc. Natl. Acad. Sci. USA*, **88**, 7247–7251.

10 Pike, C.J., Walencewicz, A.J., Glabe, C.G., and Cotman, C.W. (1991) In vitro aging of

beta-amyloid protein causes peptide aggregation and neurotoxicity. *Brain Res.*, **563**, 311–314.
11. Busciglio, J., Gabuzda, D.H., Matsudaira, P., and Yankner, B.A. (1993) Generation of beta-amyloid in the secretory pathway in neuronal and nonneuronal cells. *Proc. Natl. Acad. Sci. USA*, **90**, 2092–2096.
12. Behl, C., Davis, J., Lesley, R., and Schubert, D. (1994) Hydrogen peroxide mediates amyloid beta protein toxicity. *Cell*, **77**, 817–827.
13. Hoyer, S. (1994) Neurodegeneration, Alzheimer's disease, and beta-amyloid toxicity. *Life Sci.*, **55**, 1977–1983.
14. Lorenzo, A. and Yankner, B.A. (1994) Beta-amyloid neurotoxicity requires fibril formation and is inhibited by Congo Red. *Proc. Natl. Acad. Sci. USA*, **91**, 12243–12247.
15. Price, D.L., Sisodia, S.S., and Gandy, S.E. (1995) Amyloid beta amyloidosis in Alzheimer's disease. *Curr. Opin. Neurol.*, **8**, 268–274.
16. Lorenzo, A. and Yankner, B.A. (1996) Amyloid fibril toxicity in Alzheimer's disease and diabetes. *Ann. N. Y. Acad. Sci.*, **777**, 89–95.
17. Roher, A.E., Chaney, M.O., Kuo, Y.M., Webster, S.D., Stine, W.B., Haverkamp, L.J., Woods, A.S., Cotter, R.J., Tuohy, J.M., Krafft, G.A., Bonnell, B.S., and Emmerling, M.R. (1996) Morphology and toxicity of Abeta-(1–42) dimer derived from neuritic and vascular amyloid deposits of Alzheimer's disease. *J. Biol. Chem.*, **271**, 20631–20635.
18. Mattson, M.P., Barger, S.W., Cheng, B., Lieberburg, I., Smith-Swintosky, V.L., and Rydel, R.E. (1993) beta-Amyloid precursor protein metabolites and loss of neuronal $Ca^{2+}$ homeostasis in Alzheimer's disease. *Trends Neurosci.*, **16**, 409–414.
19. Ho, R., Ortiz, D., and Shea, T.B. (2001) Amyloid-beta promotes calcium influx and neurodegeneration via stimulation of L voltage-sensitive calcium channels rather than NMDA channels in cultured neurons. *J. Alzheimers Dis.*, **3**, 479–483.
20. Eikelenboom, P., Bate, C., Van Gool, W.A., Hoozemans, J.J., Rozemuller, J.M., Veerhuis, R., and Williams, A. (2002) Neuroinflammation in Alzheimer's disease and prion disease. *Glia*, **40**, 232–239.
21. Gasic-Milenkovic, J., Dukic-Stefanovic, S., Deuther-Conrad, W., Gartner, U., and Munch, G. (2003) Beta-amyloid peptide potentiates inflammatory responses induced by lipopolysaccharide, interferon-gamma and "advanced glycation endproducts" in a murine microglia cell line. *Eur. J. Neurosci.*, **17**, 813–821.
22. Ghribi, O., Herman, M.M., and Savory, J. (2003) Lithium inhibits Abeta-induced stress in endoplasmic reticulum of rabbit hippocampus but does not prevent oxidative damage and tau phosphorylation. *J. Neurosci. Res.*, **71**, 853–862.
23. Colurso, G.J., Nilson, J.E., and Vervoort, L.G. (2003) Quantitative assessment of DNA fragmentation and beta-amyloid deposition in insular cortex and midfrontal gyrus from patients with Alzheimer's disease. *Life Sci.*, **73**, 1795–1803.
24. Hashimoto, Y., Niikura, T., Chiba, T., Tsukamoto, E., Kadowaki, H., Nishitoh, H., Yamagishi, Y., Ishizaka, M., Yamada, M., Nawa, M., Terashita, K., Aiso, S., Ichijo, H., and Nishimoto, I. (2003) The cytoplasmic domain of Alzheimer's amyloid-beta protein precursor causes sustained apoptosis signal-regulating kinase 1/c-jun $NH_2$-terminal kinase-mediated neurotoxic signal via dimerization. *J. Pharmacol. Exp. Ther.*, **306**, 889–902.
25. Monsonego, A., Imitola, J., Zota, V., Oida, T., and Weiner, H.L. (2003) Microglia-mediated nitric oxide cytotoxicity of T cells following amyloid beta-peptide presentation to Th1 cells. *J. Immunol.*, **171**, 2216–2224.
26. Bahr, B.A. and Bendiske, J. (2002) The neuropathogenic contributions of lysosomal dysfunction. *J. Neurochem.*, **83**, 481–489.

27 Bendiske, J. and Bahr, B.A. (2003) Lysosomal activation is a compensatory response against protein accumulation and associated synaptopathogenesis – an approach for slowing Alzheimer disease? *J. Neuropathol. Exp. Neurol.*, **62**, 451–463.

28 Lesne, S., Koh, M.T., Kotilinek, L., Kayed, R., Glabe, C.G., Yang, A., Gallagher, M., and Ashe, K.H. (2006) A specific amyloid-beta protein assembly in the brain impairs memory. *Nature*, **440**, 352–357.

29 Haass, C. and Selkoe, D.J. (2007) Soluble protein oligomers in neurodegeneration: lessons from the Alzheimer's amyloid beta-peptide. *Nat. Rev. Mol. Cell Biol.*, **8**, 101–112.

30 Tomiyama, T., Matsuyama, S., Iso, H., Umeda, T., Takuma, H., Ohnishi, K., Ishibashi, K., Teraoka, R., Sakama, N., Yamashita, T., Nishitsuji, K., Ito, K., Shimada, H., Lambert, M.P., Klein, W.L., and Mori, H.A. (2010) A mouse model of amyloid beta oligomers: their contribution to synaptic alteration, abnormal tau phosphorylation, glial activation, and neuronal loss *in vivo*. *J. Neurosci.*, **30**, 4845–4856.

31 Terry, R.D., Masliah, E., and Salmon, D.P. (1991) Physical basis of cognitive alterations in Alzheimer's disease: synaptic loss is a major correlate of cognitive impairment. *Ann. Neurol.*, **30**, 572–580.

32 Gomez-Isla, T., Price, J.L., McKeel, D.W. Jr, Morris, J.C., Growdon, J.H., and Hyman, B.T. (1996) Profound loss of layer II entorhinal cortex neurons occurs in very mild Alzheimer's disease. *J. Neurosci.*, **16**, 4491–4500.

33 Gomez-Isla, T., Hollister, R., West, H., Mui, S., Growdon, J.H., Petersen, R.C., Parisi, J.E., and Hyman, B.T. (1997) Neuronal loss correlates with but exceeds neurofibrillary tangles in Alzheimer's disease. *Ann. Neurol.*, **41**, 17–24.

34 Busciglio, J., Lorenzo, A., and Yankner, B.A. (1992) Methodological variables in the assessment of beta amyloid neurotoxicity. *Neurobiol. Aging*, **13**, 609–612.

35 Podlisny, M.B., Stephenson, D.T., Frosch, M.P., Tolan, D.R., Lieberburg, I., Clemens, J.A., and Selkoe, D.J. (1993) Microinjection of synthetic amyloid beta-protein in monkey cerebral cortex fails to produce acute neurotoxicity. *Am. J. Pathol.*, **142**, 17–24.

36 Gschwind, M. and Huber, G. (1995) Apoptotic cell death induced by beta-amyloid 1–42 peptide is cell type dependent. *J. Neurochem.*, **65**, 292–300.

37 Mattson, M.P. (1992) Calcium as sculptor and destroyer of neural circuitry. *Exp. Gerontol.*, **27**, 29–49.

38 Paradis, E., Douillard, H., Koutroumanis, M., Goodyer, C., and LeBlanc, A. (1996) Amyloid beta peptide of Alzheimer's disease downregulates Bcl-2 and upregulates bax expression in human neurons. *J. Neurosci.*, **16**, 7533–7539.

39 Li, M., Chen, L., Lee, D.H., Yu, L.C., and Zhang, Y. (2007) The role of intracellular amyloid beta in Alzheimer's disease. *Prog. Neurobiol.*, **83**, 131–139.

40 Hayes, G.M., Howlett, D.R., and Griffin, G.E. (2002) Production of beta-amyloid by primary human foetal mixed brain cell cultures and its modulation by exogenous soluble beta-amyloid. *Neuroscience*, **113**, 641–646.

41 Nunomura, A., Tamaoki, T., Tanaka, K., Motohashi, N., Nakamura, M., Hayashi, T., Yamaguchi, H., Shimohama, S., Lee, H.G., Zhu, X., Smith, M.A., and Perry, G. (2010) Intraneuronal amyloid beta accumulation and oxidative damage to nucleic acids in Alzheimer disease. *Neurobiol. Dis.*, **37**, 731–737.

42 Chui, D.H., Tanahashi, H., Ozawa, K., Ikeda, S., Checler, F., Ueda, O., Suzuki, H., Araki, W., Inoue, H., Shirotani, K., Takahashi, K., Gallyas, F., and Tabira, T. (1999) Transgenic mice with Alzheimer presenilin 1 mutations show accelerated neurodegeneration without amyloid plaque formation. *Nat. Med.*, **5**, 560–564.

43 Gouras, G.K., Tsai, J., Naslund, J., Vincent, B., Edgar, M., Checler, F., Greenfield, J.P., Haroutunian, V., Buxbaum, J.D., Xu, H., Greengard, P., and Relkin, N.R. (2000) Intraneuronal

Abeta42 accumulation in human brain. *Am. J. Pathol.*, **156**, 15–20.

44 D'Andrea, M.R., Nagele, R.G., Wang, H.Y., Peterson, P.A., and Lee, D.H. (2001) Evidence that neurones accumulating amyloid can undergo lysis to form amyloid plaques in Alzheimer's disease. *Histopathology*, **38**, 120–134.

45 D'Andrea, M.R., Nagele, R.G., Gumula, N.A., Reiser, P.A., Polkovitch, D.A., Hertzog, B.M., and Andrade-Gordon, P. (2002) Lipofuscin and Abeta42 exhibit distinct distribution patterns in normal and Alzheimer's disease brains. *Neurosci. Lett.*, **323**, 45–49.

46 Nagele, R.G., D'Andrea, M.R., Anderson, W.J., and Wang, H.Y. (2002) Intracellular accumulation of beta-amyloid(1–42) in neurons is facilitated by the alpha7 nicotinic acetylcholine receptor in Alzheimer's disease. *Neuroscience*, **110**, 199–211.

47 Tabira, T., Chui, D.H., and Kuroda, S. (2002) Significance of intracellular Abeta42 accumulation in Alzheimer's disease. *Front. Biosci.*, **7**, a44–a49.

48 Takahashi, R.H., Milner, T.A., Li, F., Nam, E.E., Edgar, M.A., Yamaguchi, H., Beal, M.F., Xu, H., Greengard, P., and Gouras, G.K. (2002) Intraneuronal Alzheimer abeta42 accumulates in multivesicular bodies and is associated with synaptic pathology. *Am. J. Pathol.*, **161**, 1869–1879.

49 Wang, H.Y., D'Andrea, M.R., and Nagele, R.G. (2002) Cerebellar diffuse amyloid plaques are derived from dendritic Abeta42 accumulations in Purkinje cells. *Neurobiol. Aging*, **23**, 213–223.

50 Busciglio, J., Pelsman, A., Wong, C., Pigino, G., Yuan, M., Mori, H., and Yankner, B.A. (2002) Altered metabolism of the amyloid beta precursor protein is associated with mitochondrial dysfunction in Down's syndrome. *Neuron*, **33**, 677–688.

51 Mori, C., Spooner, E.T., Wisniewsk, K.E., Wisniewski, T.M., Yamaguch, H., Saido, T.C., Tolan, D.R., Selkoe, D.J., and Lemere, C.A. (2002) Intraneuronal Abeta42 accumulation in Down syndrome brain. *Amyloid*, **9**, 88–102.

52 Ramage, S.N., Anthony, I.C., Carnie, F.W., Busuttil, A., Robertson, R., and Bell, J.E. (2005) Hyperphosphorylated tau and amyloid precursor protein deposition is increased in the brains of young drug abusers. *Neuropathol. Appl. Neurobiol.*, **31**, 439–448.

53 Cataldo, A.M., Petanceska, S., Terio, N.B., Peterhoff, C.M., Durham, R., Mercken, M., Mehta, P.D., Buxbaum, J., Haroutunian, V., and Nixon, R.A. (2004) Abeta localization in abnormal endosomes: association with earliest Abeta elevations in AD and Down syndrome. *Neurobiol. Aging*, **25**, 1263–1272.

54 Van Broeck, B., Vanhoutte, G., Pirici, D., Van Dam, D., Wils, H., Cuijt, I., Vennekens, K., Zabielski, M., Michalik, A., Theuns, J., De Deyn, P.P., Van der Linden, A., Van Broeckhoven, C., and Kumar-Singh, S. (2008) Intraneuronal amyloid beta and reduced brain volume in a novel APP T714I mouse model for Alzheimer's disease. *Neurobiol. Aging*, **29**, 241–252.

55 Cruz, J.C., Kim, D., Moy, L.Y., Dobbin, M.M., Sun, X., Bronson, R.T., and Tsai, L.H. (2006) p25/cyclin-dependent kinase 5 induces production and intraneuronal accumulation of amyloid beta *in vivo*. *J. Neurosci.*, **26**, 10536–10541.

56 Oakley, H., Cole, S.L., Logan, S., Maus, E., Shao, P., Craft, J., Guillozet-Bongaarts, A., Ohno, M., Disterhoft, J., Van Eldik, L., Berry, R., and Vassar, R. (2006) Intraneuronal beta-amyloid aggregates, neurodegeneration, and neuron loss in transgenic mice with five familial Alzheimer's disease mutations: potential factors in amyloid plaque formation. *J. Neurosci.*, **26**, 10129–10140.

57 Knobloch, M., Konietzko, U., Krebs, D.C., and Nitsch, R.M. (2007) Intracellular Abeta and cognitive deficits precede beta-amyloid deposition in transgenic arcAbeta mice. *Neurobiol. Aging*, **28**, 1297–1306.

58 Yang, A.J., Chandswangbhuvana, D., Margol, L., and Glabe, C.G. (1998) Loss of endosomal/lysosomal membrane impermeability is an early event in amyloid Abeta1–42 pathogenesis. *J. Neurosci. Res.*, **52**, 691–698.

59 Greenfield, J.P., Tsai, J., Gouras, G.K., Hai, B., Thinakaran, G., Checler, F., Sisodia, S.S., Greengard, P., and Xu, H. (1999) Endoplasmic reticulum and trans-Golgi network generate distinct populations of Alzheimer beta-amyloid peptides. *Proc. Natl. Acad. Sci. USA*, **96**, 742–747.

60 Crowther, D.C., Kinghorn, K.J., Miranda, E., Page, R., Curry, J.A., Duthie, F.A., Gubb, D.C., and Lomas, D.A. (2005) Intraneuronal Abeta, non-amyloid aggregates and neurodegeneration in a *Drosophila* model of Alzheimer's disease. *Neuroscience*, **132**, 123–135.

61 Cuello, A.C. (2005) Intracellular and extracellular Abeta, a tale of two neuropathologies. *Brain Pathol.*, **15**, 66–71.

62 Ohyagi, Y., Tsuruta, Y., Motomura, K., Miyoshi, K., Kikuchi, H., Iwaki, T., Taniwaki, T., and Kira, J. (2007) Intraneuronal amyloid beta42 enhanced by heating but counteracted by formic acid. *J. Neurosci. Methods*, **159**, 134–138.

63 Wirths, O., Multhaup, G., Czech, C., Blanchard, V., Moussaoui, S., Tremp, G., Pradier, L., Beyreuther, K., and Bayer, T.A. (2001) Intraneuronal Abeta accumulation precedes plaque formation in beta-amyloid precursor protein and presenilin-1 double-transgenic mice. *Neurosci. Lett.*, **306**, 116–120.

64 Li, Y.P., Bushnell, A.F., Lee, C.M., Perlmutter, L.S., and Wong, S.K. (1996) Beta-amyloid induces apoptosis in human-derived neurotypic SH-SY5Y cells. *Brain Res.*, **738**, 196–204.

65 Masliah, E., Sisk, A., Mallory, M., Mucke, L., Schenk, D., and Games, D. (1996) Comparison of neurodegenerative pathology in transgenic mice overexpressing V717F beta-amyloid precursor protein and Alzheimer's disease. *J. Neurosci.*, **16**, 5795–5811.

66 Hsia, A.Y., Masliah, E., McConlogue, L., Yu, G.Q., Tatsuno, G., Hu, K., Kholodenko, D., Malenka, R.C., Nicoll, R.A., and Mucke, L. (1999) Plaque-independent disruption of neural circuits in Alzheimer's disease mouse models. *Proc. Natl. Acad. Sci. USA*, **96**, 3228–3233.

67 Echeverria, V., Ducatenzeiler, A., Alhonen, L., Janne, J., Grant, S.M., Wandosell, F., Muro, A., Baralle, F., Li, H., Duff, K., Szyf, M., and Cuello, A.C. (2004) Rat transgenic models with a phenotype of intracellular Abeta accumulation in hippocampus and cortex. *J. Alzheimers Dis.*, **6**, 209–219.

68 Takahashi, R.H., Almeida, C.G., Kearney, P.F., Yu, F., Lin, M.T., Milner, T.A., and Gouras, G.K. (2004) Oligomerization of Alzheimer's beta-amyloid within processes and synapses of cultured neurons and brain. *J. Neurosci.*, **24**, 3592–3599.

69 LeBlanc, A.C., Xue, R., and Gambetti, P. (1996) Amyloid precursor protein metabolism in primary cell cultures of neurons, astrocytes, and microglia. *J. Neurochem.*, **66**, 2300–2310.

70 LeBlanc, A.C., Papadopoulos, M., Belair, C., Chu, W., Crosato, M., Powell, J., and Goodyer, C.G. (1997) Processing of amyloid precursor protein in human primary neuron and astrocyte cultures. *J. Neurochem.*, **68**, 1183–1190.

71 Kurt, M.A., Davies, D.C., and Kidd, M. (1999) beta-Amyloid immunoreactivity in astrocytes in Alzheimer's disease brain biopsies: an electron microscope study. *Exp. Neurol.*, **158**, 221–228.

72 Oide, T., Kinoshita, T., and Arima, K. (2006) Regression stage senile plaques in the natural course of Alzheimer's disease. *Neuropathol. Appl. Neurobiol.*, **32**, 539–556.

73 Thal, D.R., Hartig, W., and Schober, R. (1999) Diffuse plaques in the molecular layer show intracellular A beta(8–17)-immunoreactive deposits in subpial astrocytes. *Clin. Neuropathol.*, **18**, 226–231.

74 Gyure, K.A., Durham, R., Stewart, W.F., Smialek, J.E., and Troncoso, J.C. (2001)

Intraneuronal abeta-amyloid precedes development of amyloid plaques in Down syndrome. *Arch. Pathol. Lab. Med.*, **125**, 489–492.

75 Funato, H., Yoshimura, M., Yamazaki, T., Saido, T.C., Ito, Y., Yokofujita, J., Okeda, R., and Ihara, Y. (1998) Astrocytes containing amyloid beta-protein (Abeta)-positive granules are associated with Abeta40-positive diffuse plaques in the aged human brain. *Am. J. Pathol.*, **152**, 983–992.

76 Askanas, V., Engel, W.K., and Alvarez, R.B. (1992) Light and electron microscopic localization of beta-amyloid protein in muscle biopsies of patients with inclusion-body myositis. *Am. J. Pathol.*, **141**, 31–36.

77 Sugarman, M.C., Yamasaki, T.R., Oddo, S., Echegoyen, J.C., Murphy, M.P., Golde, T.E., Jannatipour, M., Leissring, M.A., and LaFerla, F.M. (2002) Inclusion body myositis-like phenotype induced by transgenic overexpression of beta APP in skeletal muscle. *Proc. Natl. Acad. Sci. USA*, **99**, 6334–6339.

78 Dalakas, M.C. (2006) Sporadic inclusion body myositis – diagnosis, pathogenesis and therapeutic strategies. *Nat. Clin. Pract. Neurol.*, **2**, 437–447.

79 Grant, S.M., Shankar, S.L., Chalmers-Redman, R.M., Tatton, W.G., Szyf, M., and Cuello, A.C. (1999) Mitochondrial abnormalities in neuroectodermal cells stably expressing human amyloid precursor protein (hAPP751). *Neuroreport*, **10**, 41–46.

80 Kienlen-Campard, P., Miolet, S., Tasiaux, B., and Octave, J.N. (2002) Intracellular amyloid-beta 1–42, but not extracellular soluble amyloid-beta peptides, induces neuronal apoptosis. *J. Biol. Chem.*, **277**, 15666–15670.

81 Chui, D.H., Dobo, E., Makifuchi, T., Akiyama, H., Kawakatsu, S., Petit, A., Checler, F., Araki, W., Takahashi, K., and Tabira, T. (2001) Apoptotic neurons in Alzheimer's disease frequently show intracellular Abeta42 labeling. *J. Alzheimers Dis.*, **3**, 231–239.

82 Zhang, Y., McLaughlin, R., Goodyer, C., and LeBlanc, A. (2002) Selective cytotoxicity of intracellular amyloid beta peptide1–42 through p53 and Bax in cultured primary human neurons. *J. Cell Biol.*, **156**, 519–529.

83 Suo, Z., Cox, A.A., Bartelli, N., Rasul, I., Festoff, B.W., Premont, R.T., and Arendash, G.W. (2007) GRK5 deficiency leads to early Alzheimer-like pathology and working memory impairment. *Neurobiol. Aging*, **28**, 1873–1888.

84 Lopez, E.M., Bell, K.F., Ribeiro-da-Silva, A., and Cuello, A.C. (2004) Early changes in neurons of the hippocampus and neocortex in transgenic rats expressing intracellular human a-beta. *J. Alzheimers Dis.*, **6**, 421–431, discussion 443–429.

85 Wirths, O., Weis, J., Szczygielski, J., Multhaup, G., and Bayer, T.A. (2006) Axonopathy in an APP/PS1 transgenic mouse model of Alzheimer's disease. *Acta Neuropathol. (Berl.)*, **111**, 312–319.

86 Almeida, C.G., Takahashi, R.H., and Gouras, G.K. (2006) Beta-amyloid accumulation impairs multivesicular body sorting by inhibiting the ubiquitin–proteasome system. *J. Neurosci.*, **26**, 4277–4288.

87 Ohyagi, Y., Asahara, H., Chui, D.H., Tsuruta, Y., Sakae, N., Miyoshi, K., Yamada, T., Kikuchi, H., Taniwaki, T., Murai, H., Ikezoe, K., Furuya, H., Kawarabayashi, T., Shoji, M., Checler, F., Iwaki, T., Makifuchi, T., Takeda, K., Kira, J., and Tabira, T. (2005) Intracellular Abeta42 activates p53 promoter: a pathway to neurodegeneration in Alzheimer's disease. *FASEB J.*, **19**, 255–257.

88 Magrane, J., Rosen, K.M., Smith, R.C., Walsh, K., Gouras, G.K., and Querfurth, H.W. (2005) Intraneuronal beta-amyloid expression downregulates the Akt survival pathway and blunts the stress response. *J. Neurosci.*, **25**, 10960–10969.

89 Allsopp, T.E., McLuckie, J., Kerr, L.E., Macleod, M., Sharkey, J., and Kelly, J.S. (2000) Caspase 6 activity initiates caspase 3 activation in cerebellar granule cell apoptosis. *Cell Death Differ.*, **7**, 984–993.

90 Zhang, Y., Champagne, N., Beitel, L.K., Goodyer, C.G., Trifiro, M., and LeBlanc, A. (2004) Estrogen and

androgen protection of human neurons against intracellular amyloid beta1–42 toxicity through heat shock protein 70. *J. Neurosci.*, **24**, 5315–5321.

91 Reinhard, C., Hebert, S.S., and De Strooper, B. (2005) The amyloid-beta precursor protein: integrating structure with biological function. *EMBO J.*, **24**, 3996–4006.

92 Heese, K. and Akatsu, H. (2006) Alzheimer's disease – an interactive perspective. *Curr. Alzheimer Res.*, **3**, 109–121.

93 Vetrivel, K.S. and Thinakaran, G. (2006) Amyloidogenic processing of beta-amyloid precursor protein in intracellular compartments. *Neurology*, **66**, S69–73.

94 Glenner, G.G. and Wong, C.W. (1984) Alzheimer's disease: initial report of the purification and characterization of a novel cerebrovascular amyloid protein. *Biochem. Biophys. Res. Commun.*, **120**, 885–890.

95 Ha, C. and Park, C.B. (2006) *Ex situ* atomic force microscopy analysis of beta-amyloid self-assembly and deposition on a synthetic template. *Langmuir*, **22**, 6977–6985.

96 Pellegrini, L., Passer, B.J., Tabaton, M., Ganjei, J.K., and D'Adamio, L. (1999) Alternative, non-secretase processing of Alzheimer's beta-amyloid precursor protein during apoptosis by caspase-6 and -8. *J. Biol. Chem.*, **274**, 21011–21016.

97 Chyung, A.S., Greenberg, B.D., Cook, D.G., Doms, R.W., and Lee, V.M. (1997) Novel beta-secretase cleavage of beta-amyloid precursor protein in the endoplasmic reticulum/intermediate compartment of NT2N cells. *J. Cell Biol.*, **138**, 671–680.

98 Cook, D.G., Forman, M.S., Sung, J.C., Leight, S., Kolson, D.L., Iwatsubo, T., Lee, V.M., and Doms, R.W. (1997) Alzheimer's A beta(1–42) is generated in the endoplasmic reticulum/intermediate compartment of NT2N cells. *Nat. Med.*, **3**, 1021–1023.

99 Sudoh, S., Kawamura, Y., Sato, S., Wang, R., Saido, T.C., Oyama, F., Sakaki, Y., Komano, H., and Yanagisawa, K. (1998) Presenilin 1 mutations linked to familial Alzheimer's disease increase the intracellular levels of amyloid beta-protein 1–42 and its N-terminally truncated variant(s) which are generated at distinct sites. *J. Neurochem.*, **71**, 1535–1543.

100 Peraus, G.C., Masters, C.L., and Beyreuther, K. (1997) Late compartments of amyloid precursor protein transport in SY5Y cells are involved in beta-amyloid secretion. *J. Neurosci.*, **17**, 7714–7724.

101 Xu, H., Sweeney, D., Wang, R., Thinakaran, G., Lo, A.C., Sisodia, S.S., Greengard, P., and Gandy, S. (1997) Generation of Alzheimer beta-amyloid protein in the trans-Golgi network in the apparent absence of vesicle formation. *Proc. Natl. Acad. Sci. USA*, **94**, 3748–3752.

102 Tomita, S., Kirino, Y., and Suzuki, T. (1998) A basic amino acid in the cytoplasmic domain of Alzheimer's beta-amyloid precursor protein (APP) is essential for cleavage of APP at the alpha-site. *J. Biol. Chem.*, **273**, 19304–19310.

103 Hartmann, H., Busciglio, J., Baumann, K.H., Staufenbiel, M., and Yankner, B.A. (1997) Developmental regulation of presenilin-1 processing in the brain suggests a role in neuronal differentiation. *J. Biol. Chem.*, **272**, 14505–14508.

104 VanSlyke, J.K. and Musil, L.S. (2002) Dislocation and degradation from the ER are regulated by cytosolic stress. *J. Cell Biol.*, **157**, 381–394.

105 Selkoe, D.J., Yamazaki, T., Citron, M., Podlisny, M.B., Koo, E.H., Teplow, D.B., and Haass, C. (1996) The role of APP processing and trafficking pathways in the formation of amyloid beta-protein. *Ann. N. Y. Acad. Sci.*, **777**, 57–64.

106 Huse, J.T., Pijak, D.S., Leslie, G.J., Lee, V.M., and Doms, R.W. (2000) Maturation and endosomal targeting of beta-site amyloid precursor protein-cleaving enzyme. The Alzheimer's disease beta-secretase. *J. Biol. Chem.*, **275**, 33729–33737.

107 Lah, J.J., Heilman, C.J., Nash, N.R., Rees, H.D., Yi, H., Counts, S.E., and Levey, A.I. (1997) Light and electron microscopic localization of presenilin-1

in primate brain. *J. Neurosci.*, **17**, 1971–1980.

108 Lah, J.J. and Levey, A.I. (2000) Endogenous presenilin-1 targets to endocytic rather than biosynthetic compartments. *Mol. Cell. Neurosci.*, **16**, 111–126.

109 Nordstedt, C., Caporaso, G.L., Thyberg, J., Gandy, S.E., and Greengard, P. (1993) Identification of the Alzheimer beta/A4 amyloid precursor protein in clathrin-coated vesicles purified from PC12 cells. *J. Biol. Chem.*, **268**, 608–612.

110 Cataldo, A.M., Barnett, J.L., Pieroni, C., and Nixon, R.A. (1997) Increased neuronal endocytosis and protease delivery to early endosomes in sporadic Alzheimer's disease: neuropathologic evidence for a mechanism of increased beta-amyloidogenesis. *J. Neurosci.*, **17**, 6142–6151.

111 Mathews, P.M., Guerra, C.B., Jiang, Y., Grbovic, O.M., Kao, B.H., Schmidt, S.D., Dinakar, R., Mercken, M., Hille-Rehfeld, A., Rohrer, J., Mehta, P., Cataldo, A.M., and Nixon, R.A. (2002) Alzheimer's disease-related overexpression of the cation-dependent mannose 6-phosphate receptor increases Abeta secretion: role for altered lysosomal hydrolase distribution in beta-amyloidogenesis. *J. Biol. Chem.*, **277**, 5299–5307.

112 Grbovic, O.M., Mathews, P.M., Jiang, Y., Schmidt, S.D., Dinakar, R., Summers-Terio, N.B., Ceresa, B.P., Nixon, R.A., and Cataldo, A.M. (2003) Rab5-stimulated up-regulation of the endocytic pathway increases intracellular beta-cleaved amyloid precursor protein carboxyl-terminal fragment levels and Abeta production. *J. Biol. Chem.*, **278**, 31261–31268.

113 Nixon, R.A. (2006) Autophagy in neurodegenerative disease: friend, foe or turncoat? *Trends Neurosci.*, **29**, 528–535.

114 Hasegawa, T., Ukai, W., Jo, D.G., Xu, X., Mattson, M.P., Nakagawa, M., Araki, W., Saito, T., and Yamada, T. (2005) Homocysteic acid induces intraneuronal accumulation of neurotoxic Abeta42: implications for the pathogenesis of Alzheimer's disease. *J. Neurosci. Res.*, **80**, 869–876.

115 Cole, S.L., Grudzien, A., Manhart, I.O., Kelly, B.L., Oakley, H., and Vassar, R. (2005) Statins cause intracellular accumulation of amyloid precursor protein, beta-secretase-cleaved fragments, and amyloid beta-peptide via an isoprenoid-dependent mechanism. *J. Biol. Chem.*, **280**, 18755–18770.

116 El-Amouri, S.S., Zhu, H., Yu, J., Gage, F.H., Verma, I.M., and Kindy, M.S. (2007) Neprilysin protects neurons against Abeta peptide toxicity. *Brain Res.*, **1152**, 191–200.

117 Maezawa, I., Hong, H.S., Wu, H.C., Battina, S.K., Rana, S., Iwamoto, T., Radke, G.A., Pettersson, E., Martin, G.M., Hua, D.H., and Jin, L.W. (2006) A novel tricyclic pyrone compound ameliorates cell death associated with intracellular amyloid-beta oligomeric complexes. *J. Neurochem.*, **98**, 57–67.

118 Cui, J., Chen, Q., Yue, X., Jiang, X., Gao, G.F., Yu, L.C., and Zhang, Y. (2010) Galanin protects against intracellular amyloid toxicity in human primary neurons. *J. Alzheimers Dis.*, **19**, 529–544.

119 Hou, J.F., Cui, J., Yu, L.C., and Zhang, Y. (2009) Intracellular amyloid induces impairments on electrophysiological properties of cultured human neurons. *Neurosci. Lett.*, **462**, 294–299.

120 Walsh, D.M. and Selkoe, D.J. (2007) A beta oligomers – a decade of discovery. *J. Neurochem.*, **101**, 1172–1184.

# 7
# Lipid Rafts Play a Crucial Role in Protein Interactions and Intracellular Signaling Involved in Neuronal Preservation Against Alzheimer's Disease

*Raquel Marin*

## 7.1
### Lipid Rafts: Keys to Signaling Platforms in Neurons

Intracellular signal transduction is initiated by a plethora of protein interactions, including receptors, kinases, and channels that are crucial for the correct neuronal communication and functions, and whose modifications determine cognitive and neurological impairments. An important concept is that the cognitive decline that occurs with normal aging and is exacerbated in neuropathologies is mainly related to functional changes in signal transduction cascades and cellular communication that modify neuronal responses, rather than to morphological modifications which are not evident during aging [1]. Although a set of cell surface proteins are found in liquid disordered regions of the plasma membrane, a large fraction of signaling proteins are located in liquid-ordered domains, or lipid rafts, which are the preferential locations for these proteins, due to the particular physico-chemical properties of these microdomains. In this regard, Lisanti *et al.* [2] were the first to suggest the "caveolae/raft signaling hypothesis," that is, the compartmentalization of proteins involved in transduction signals to provide a mechanism for the regulation and interaction between different intracellular pathways [2]. These macromolecular complexes may be considered specialized signaling platforms, or "signalosomes." In particular in neurons, signaling molecules preferentially located in lipid rafts include transmembrane proteins and lipid-modified proteins, in addition to intracellular signaling intermediates, such as trimeric and small GTPases, Src tyrosine kinases (STKs) family, lipid second messengers, and a variety of cytosolic signal transducers [3, 4], known to participate in neuronal growth, differentiation, preservation, and survival.

Furthermore, proteins in these microdomains also interact with resident lipids, suggesting that specific lipids may also take part in the processes developed by signaling molecules. In fact, lipid rafts are currently considered to be dynamic microenvironments where proteins and lipids can move and interact with different kinetics, changing their size and composition in response to a variety of intra- or extracellular stimuli that may ultimately favor specific protein interactions and

*Lipids and Cellular Membranes in Amyloid Diseases*, First Edition. Edited by Raz Jelinek.
© 2011 Wiley-VCH Verlag GmbH & Co. KGaA. Published 2011 by Wiley-VCH Verlag GmbH & Co. KGaA.

signaling cascades [5]. Therefore, the intrinsic composition and distribution of lipid hallmarks of rafts that modulate membrane fluidity, such as cholesterol, gangliosides, and polyunsaturated fatty acids (PUFAs), may affect the movement of proteins and presumably alter their function and signal transduction. Hence many raft intrinsic proteins preferentially contain lipid-modified structures that may contribute to their stabilization and correct functioning. In this sense, one of the earlier discoveries was the localization in these domains of glycosylphosphatidylinositol (GPI)-anchored proteins, that generally have saturated acyl chains and are preferentially anchored in the outer leaflet of the cell membrane. Although the GPI anchor does not completely cross the plasma membrane, it appears that it is crucial to initiate signaling events, probably through its association with other transmembrane proteins involved in intracellular signaling [6, 7]. Among the GPI-anchored proteins involved in neuropathology, one of the better characterized is the cellular prion protein ($PrP^c$) implicated in the pathogenesis of prion disease [8]. Prion disease is an amyloid disease characterized by the formation within neurons and other brain cells of protein plaques leading to cell death, which involves the conformational modification of normal $PrP^c$ in a pathogenic scrapie form, $PrP^{Sc}$ [9]. $PrP^c$ is constitutively expressed in neurons as a GPI-anchored protein localized in lipid rafts, and depletion in cholesterol but not sphingolipids affects its distribution in these microdomains [10]. Interestingly, there is an increasing body of evidence that the lipid raft environment plays a direct role in the conversion of $PrP^c$ into $PrP^{Sc}$ [9]. The downstream signaling of $PrP^c$ is dependent on its localization with respect to rafts, which induces the activation of some STKs, possibly Lyn, Src, Lck, or Fyn [11]. Thus, $PrP^c$ may be part of a multimolecular signaling complex which may be important in neuronal function [12].

Additional lipid modifications of signaling proteins inserted in rafts take place by binding to alternative saturated-chain lipids, such as palmitoylation and myristoylation. These modifications are found in, among others, STKs [13], scaffolding proteins [14], steroid receptors [15, 16], and GPI-anchored proteins [17], which may contribute to their stabilization and correct functioning in these microstructures. In addition, hallmark proteins of lipid rafts such as caveolin and flotillin undergo palmitoylation [18] and, in the case of caveolin, this requirement allows the coupling of Cav-1 to c-Src tyrosine kinase [14]. This evidence suggests that the lipid-modified nature of proteins integrated in lipid rafts may serve not only to target them to these domains but also to modulate the protein interactions that occur within rafts.

Together with GPI-anchored proteins, scaffolding proteins are the most abundant integral molecules of lipid rafts. They represent a particular group that have the intrinsic capacity to form lipid shells around themselves including, apart from caveolins and flotillins [19–21], the proteolipid MAL [22], stomatin [19], and some transmembrane proteins, such as presenilin-1 [23]. These structural proteins not only provide stabilizing scaffolds for lipid raft maintenance, but also participate in vesicular trafficking and signal transduction that modulate the final cellular response. Numerous demonstrations have concluded that the members of the caveolin family (caveolin 1, 2, and 3) serve to compartmentalize specific signaling molecules within lipid rafts, or caveolae, with the prospect of rapidly and selectively modulating cell signaling events, thereby proposing a "caveolae signaling hypothesis" [2]. In this regard, caveolins are known to regulate a variety of key signaling elements, including G-proteins, STKs, and some components of PI3K and MAPK pathways [24, 25]. Although the structures of caveolae and caveolin isoforms have not been fully elucidated in the nervous system, evidence suggests that neuronal lipid rafts may serve as docking points for numerous cell surface receptors which are recruited to this microdomain when bound to their specific ligands, activating numerous intracellular processes related to neuronal functioning. Among these receptors known to interact with caveolins are membrane estrogen receptors (ERs), which are regulated by interactions with caveolins [26]. Recent evidence suggests that, at least caveolin-1 (Cav-1) may play a crucial role in membrane ER function in the brain, which in turn determines many nervous system activities related to its preservation and maintenance [27, 28]. As further discussed below, we have observed the localization of Cav-1 in the cytoplasm and neuronal processes of different populations of human cortex and hippocampus, where it partially co-localizes with estrogen receptor α (ERα). Also, Cav-1 has been claimed to be involved in Aβ processing, suggesting that Aβ generation depends on the interactions of Cav-1 with the amyloid precursor protein (APP) [29]. Furthermore, Cav-1 expression is increased in senescent cells [30] and Alzheimer's disease (AD) brains [31], suggesting an involvement of this resident protein of lipid rafts in brain degeneration.

Flotillins belong to the so-called SPFH (stomatin/prohibiting/flotillin/HflK/C) protein family forming specialized rafts that, similarly to caveolae, provide stable platforms for the assembly of multiprotein complexes [21]. Flotillins not only are important for coordinated recruitment of the machinery for regulation of cytoskeletal remodeling [32] but have also been linked to the pathogenesis of AD. Indeed, flotillins appear to be upregulated in the cortex of patients with AD [33], where they accumulated at sites of amyloid β peptide (Aβ) production and secretion. However, further studies are required to elucidate fully the role of flotillin-1 in the progression of AD pathology.

A large body of evidence has demonstrated that lipid rafts are also platforms for neurotrophic signaling under the control of neurotrophins and glial-derived neurotrophic factor (GDNF)-family ligands, which are essential for synaptic transmission, axon guidance, and cell adhesion [3]. Src kinase activity is one of the main signaling proteins required to elicit GDNF bioactivity related to neurite outgrowth and neuronal survival [34]. Numerous receptor tyrosine kinases are located in lipid rafts, including TrkA [35], insulin receptor (IR) [36], epidermal growth factor receptor (EGFR) [24], and platelet-derived growth factor receptor (PDGFR) [37]. Accordingly, we have found the enrichment of insulin growth factor-1 receptor (IGF-1R) in lipid rafts from human cortex and hippocampus, in a complex with ERα and caveolin-1, suggesting that this receptor may also take part in multimolecular complexes in lipid rafts. Figure 7.1 illustrates a representative experiment demonstrating, by immunoprecipitation and immunoblotting in lipid raft fractions from human cerebral cortex, the association of IGF-1R with the complex formed by Cav-1, ERα, and voltage-dependent anion channel (VDAC).

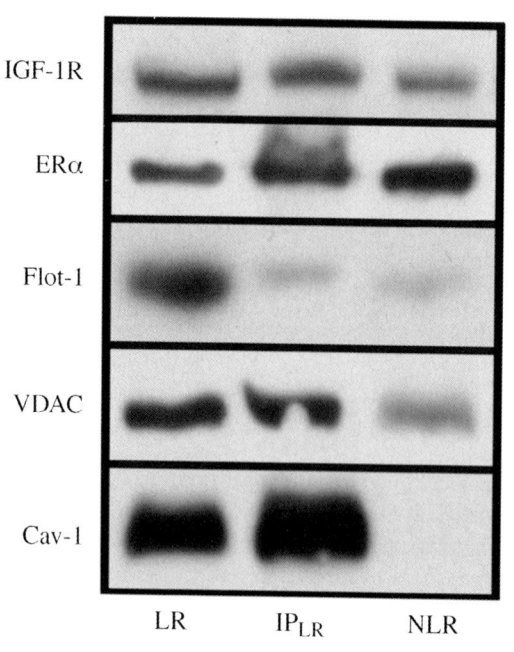

**Figure 7.1** Association of IGF-1R with caveolin-1 (Cav-1), ERα, and VDAC in lipid rafts of human cerebral cortex. Protein extracts from lipid rafts were used to immunoprecipitate with a specific anti-caveolin-1 antibody (IP$_{LR}$). The resultant precipitated protein immunocomplexes were subjected to SDS-PAGE (sodium dodecyl sulfate polyacrylamide gel electrophoresis), and immunoblotted with antibodies directed to ERα, VDAC, IGF-1R, and flotillin-1 (Flot-1), the last as a control of lipid raft purity. As immunoblotting control, total lipid rafts (LR) and non-lipid rafts (NLR) were also run. The figure illustrates a representative immunoblot assay after incubation with the different antibodies, observing a high degree of coprecipitation of IGF-1R with caveolin-1 in lipid raft fractions.

Emerging evidence also indicates that such rafts are important for neuronal synaptic transmission [3], and different neurotransmitter receptors and ion channels, for example, the voltage-gated $K^+$ channel Kv2.1 [38], nicotinic acetylcholine receptor (nAChR) [39], and GABA$_b$R receptor [40], are biochemically located in lipid rafts. Localization of ion channels to these microstructures appears to vary depending on the specific channel [38], a fact that modifies channel properties. In this context, our recent work on neuronal cell lines and human and mouse brain cortex and hippocampus have demonstrated that the pro-apoptotic plasma membrane VDAC is located in lipid rafts in physical contact with Cav-1 and ERα [41, 42], a fact that might be relevant in AD neuropathology, as discussed in the following sections.

Hence lipid rafts not only represent structural components of neuronal membranes, but also integrate protein signaling platforms which are crucial for the development of neuronal physiological activities related to neuroprotection.

## 7.2
### Estrogen Receptors Are Part of Signaling Platforms in Neuronal Rafts

Estrogen develops some crucial actions in the regulation of neuronal differentiation, synaptic plasticity, induction of neuronal survival, and regional neurogenesis in the adult [43, 44] that ultimately affect mood and cognitive processes [45, 46]. As largely demonstrated in the brain and other tissues, estrogen actions are initiated through hormone binding to two different estrogen receptors encoded by different genes, ERα and estrogen receptor β (ERβ), being the magnitude and type of responses determined by the relative population of specific ERs and their space–temporal expression [47]. Numerous pieces of evidence in different neural and non-neural tissues indicate a major role of ERs, ERα in particular, through different mechanisms of action that may coexist in the same cell. These mechanisms take place partly through the activation of a canonical (nuclear) ER nuclear estrogen receptor (nER) that may interact with different coregulators to induce the transcription of specific genes (termed the "classical mechanism" [48]). In this context, it has been demonstrated in the past that estrogen neuroprotection against Aβ-induced toxicity is partially due to gene transactivation via a classical ERα-mediated mechanism [49, 50]. More recent reports have shown that estrogens can also interact with specific ERs located in close contact with the neuronal membrane membrane estrogen receptor (mERs) to activate different, highly coordinated, signaling pathways that take place within a few minutes (termed "alternative mechanisms," [48, 51]). Even though the molecular nature of mERs has not been fully elucidated, the generally held view maintains that endogenous mERs may be produced by the same transcript as its classical nuclear counterpart [52]. However, the scenario may be even more complex in neurons because, apart from ERα and ERβ, additional cytoplasmic and plasma mER-like receptors have been reported, including G protein-coupled receptors (GPR30), a functional ERα variant with lower molecular weight (ER-46), and other distinct receptors such as ER-X [27, 53]. Even though the role of these additional ERs in the brain has not been elucidated, GPR30 is known to mediate rapid cellular signaling [54], and is expressed in different brain regions, including cortical, hippocampal, and striatal areas [55], suggesting a role for this receptor in neuronal survival. Moreover, ER-X is a membrane-related ER found in neocortical explants that shows some homologies with ERα, and has also been related to neuroprotection [56]. The presence of neuronal membrane-related ERs similar to classical ERα has given rise to some controversy about the manner in which a hydrophilic molecule lacking transmembrane domains may be inserted into the plasma membrane, in order to interact rapidly with their natural molecular targets. This apparent paradox has been partially solved by the recent finding of mERs present in neuronal lipid rafts where the receptor may be stabilized by its binding to raft resident molecules. Thus, ER-X is highly enriched in neuronal rafts where the receptor has been implicated in functional membrane signaling events. Moreover, we have provided the first evidence of a raft-located ERα in murine septal and hippocampal neurons, and in human frontal cortex and hippocampus ([41, 57]). In these membrane fractions and also in microsomal fractions from different mouse brain areas [58], ERα was found to interact physically with Cav-1 and with other proteins, such as a plasmalemmal voltage-dependent anion channel (pl-VDAC) and the insulin growth factor-1 receptor, which may be part of this molecular complex at the neuronal membrane [57]. It is important to underline that currently available evidence has demonstrated that the ER and IGF-1R pathways cross-talk with each other to promote neuroprotective events, and that this association may be affected by aging [59]. Thus, the ERα–IGF-1R complex may be differentially activated by estrogens and IGF, the natural ligand of IGF-1R, and may operate by adapting the neuronal response to the changes in the extracellular levels of these ligands by using different components of their signaling machinery. In this macromolecular complex, caveolin-1 may be the pivotal anchoring protein that provides stability for the integration and functionality of ERα, facilitating its associations with other signaling proteins in the raft microstructure.

A plausible possibility is that components of caveolin-related "signalosome" may bind to the caveolar scaffolding domain (CSD) present in this protein, through a sequence motif which is conserved in different signaling proteins known to interact with this anchoring marker [24]. In agreement with this hypothesis, using several bioinformatic tools, we have found a consensus sequence susceptible to binding to CSD in both the ligand binding domain of ERα and in the second intracellular loop of the mitochondrial VDAC structure which is conserved in mouse and human sequences [58]. However, further investigation is required to confirm this possibility as the topology of pl-VDAC has still not been elucidated, although this membrane channel is claimed to be an isoform of mitochondrial VDAC, that has incorporated a leader sequence for its trafficking to the plasma membrane [60]. Hence ERs involved in plasma membrane actions in the brain appear to be multifaceted and versatile molecules that may be integrated in particular membrane microdomains where they may be differentially activated by estradiol and other extracellular signals. This may be crucial in the estrogen mechanisms developed by neurons to achieve neuroprotection, and it might be at the basis of the failure of estrogen replacement therapies supplied in post-menopausal periods, when the amount of membrane ERs may be decreased or absent. Indeed, changes in the intracellular distribution of both ERα and ERβ have been observed in different areas of AD patients, with an increase in nuclear ERs as compared with healthy controls of similar age [61, 62].

## 7.3
### Role of Lipid Raft ERα–VDAC Interactions in Neuronal Preservation Against Aβ Toxicity

Substantial literature based on clinical studies has reported the putative beneficial effects of estrogen on cognitive functioning during menopausal periods [63]. Compelling support for these data has been provided by an impressive amount of evidence on estrogen-mediated neuronal preservation against AD, Parkinson's disease, cerebral ischemia, and other neuropathologies, mainly in brain areas involved in memory and cognition (reviewed in [44, 45, 53]). In particular in AD, the most common form of dementia in the elderly, some experimental studies in cultured neurons using paradigms that evoke the neuropathological events of this disease support the view that estrogen can protect against different aspects of Aβ-mediated toxicity, such as reducing Aβ accumulation, which is a critical parameter in AD progression, regulating Aβ clearance and activating anti-apoptotic mechanisms [53, 64]. In support of this concept, several clinical studies have concluded that there are beneficial effects of estrogen therapy against the progression of AD [63].

These estrogen actions may take place through genomic and non-genomic actions that may converge into the preservation of nerve cells. Thus, emerging data have indicated that rapid estrogen membrane-related actions may participate in preventing cell degeneration against Aβ neurotoxicity [65–68]. Using murine septal SN56 cells, we have characterized a non-genomic mechanism of estrogen action with the participation of a membrane ERα, which triggers the intracellular and transient activation of the Raf-1/MEK/MAPK cascade that elicits neuroprotection against Aβ treatment. Activation of this signaling pathway appears to be a main strategy developed by estrogen, and a variety of data on different cultured neurons has indicated the involvement of mER in MAPK signaling, resulting in cell preservation against different injuries [27, 69]. Additionally, in *in vivo* experiments using ovariectomized rats, a moderate activation of MAPK after a short exposure to estrogen has been assessed in specific cognitive brain regions [70], suggesting that estrogen response in neurons may be dependent on cell type.

Moreover, other signaling pathways may also be involved in the control of neuronal survival, including the PI3K/Akt pathway [53, 64], although very little is known about its participation in protection against Aβ-induced neurotoxicity [71, 72]. Interestingly, a role of PI3K signaling in Aβ-induced memory loss has recently been demonstrated [73], evoking the relevance of this pathway in AD pathogenesis and neuronal death prevention. Additionally, it has been determined that glycogen synthase kinase-3 (GSK-3) activation is a critical step in the detrimental events in AD [74]. In this context, it has been evidenced that estrogen can cause the phosphorylation and hence inactivation of GSK-3β, and thereby reduction of neuronal death, by a mechanism mediated by ER [75]. This interaction may affect important processes that occur in AD, such as levels of phospho-tau phosphorylation and consequently neurofibrillary tangle formation. Overall, these data indicate that estrogens acting at the plasma membrane may trigger different intracellular transduction pathways that may converge in neuronal survival.

Part of the strategies developed by estrogen to achieve neuroprotection against AD may involve the regulation of putative modulators of Aβ-induced toxicity, such as the VDAC. The VDAC, traditionally known as a mitochondrial porin, is known to participate in transmembrane fluxes of ions and metabolites, and in apoptosis regulation, through direct interaction with apoptogenic factors leading to caspase activation and cytochrome *c* release [76]. Moreover, since the first discovery of the presence of a VDAC present at the plasma membrane of human lymphocytes (plasmalemmal voltage-dependent anion channel (pl-VDAC)) [77], numerous pieces of evidence in the last two decades have also demonstrated the presence of VDAC at the plasma membrane of different cellular types [60, 78], including immortalized neurons ([41, 58, 79, 80]). Additionally, VDAC has been found to be a resident protein of lipid rafts in neurons from both mouse and human cognitive areas, such as frontal cortex, septum, and hippocampus [41, 42, 81], suggesting that VDAC location in neuronal rafts may be a general phenomenon. Although the exact role of this channel at the plasma membrane is still unclear, pl-VDAC has been claimed to participate in different aspects of cellular homeostasis and cell volume regulation [78, 82, 83] and, in particular in neurons, it may participate in the extrinsic apoptotic pathway [84]. Related to the potential role of pl-VDAC in neuronal apoptosis, we have recently demonstrated the participation of membrane VDAC in Aβ-induced cytotoxicity, as evidenced by using anti-VDAC antibodies to inactivate the porin in the presence of the amyloid [41]. In support of this concept, we have observed by immunofluorescence in human brain tissue sections from AD patients a strong VDAC immunoreactivity in dystrophic neuritic processes of amyloid plaques and neurofibrillary tangles, two main anatomopathological parameters of AD injury. A similar pattern of VDAC distribution has also been reproduced in aged APP/PS1 transgenic mice [85], which is accepted as a model of early onset of AD. Indeed, VDAC has been shown to increase substantially its accumulation in lipid rafts from AD brains, as compared with rafts from healthy brains of similar age [42], suggesting that the porin present in these membrane fractions may be involved in some aspects of AD neuropathology. Therefore, one could hypothesize that Aβ may physically interact with VDAC in lipid rafts, thereby contributing to the channel opening, ultimately leading to intracellular apoptosis. In line with this possibility, binding of this peptide to gangliosides, lipid raft components, is thought to induce the assembly of Aβ proteins involved in the formation of senile plaques [86–88]. An alternative possibility is that VDAC might be activated by membrane depolarization as a result of an increase in intracellular $Ca^{2+}$ levels provoked by Aβ interacting with the neuronal membrane. In agreement with this hypothesis, some data have indicated that Aβ peptide may induce cellular toxicity by regulating $Ca^{2+}$ homeostasis, based on its property to activate $Ca^{2+}$ channels [89]. It should also be mentioned that a large number of studies have proposed that cell exposure to Aβ peptide results in unregulated flux of $Ca^{2+}$ through the plasma membrane, upon disruption of plasma membrane integrity. In addition, several studies have highlighted the importance of the specific interaction of Aβ (and other protein amyloids) with glutamate receptors (AMPA and NMDA). Notably, a rise in intracellular $Ca^{2+}$ induced by Aβ decreases the availability of AMPA receptors at the synapses, thereby affecting synaptic plasticity [90, 91]. Aβ also affects NMDA receptor activation, culminating in intracellular $Ca^{2+}$ overload, which disrupts neuronal transmission [92].

A suggested mechanism by which estrogens control VDAC activation may be via post-translational modifications of the porin. Prior electrophysiological studies in neuroblastoma cells have demonstrated that triphenylethylene antiestrogens (i.e., tamoxifen and toremifene) activate MaxiCl⁻ channels, the electrophysiological correlates of VDAC. MaxiCl⁻ channel activation and voltage-dependent opening can be prevented by extracellular application of estrogen through a process that involves the channel phosphorylation [93]. In agreement with this, we have observed by 2-D electrophoresis and Western blotting in septal and hippocampal cells that estrogen maintains pl-VDAC in a phosphorylated status whereas tamoxifen provokes this channel dephosphorylation, indicating an antagonist effect of estrogen and antiestrogen in VDAC post-translational regulation (unpublished results). Also, an example in epithelial cells has suggested that estrogens can directly modulate VDAC expression [94]. In support of this concept, data in the temporal, frontal, and occipital cortex of AD brains have indicated changes in VDAC levels, observing post-mortem differences in its phosphorylation pattern that may be related to synaptic loss [95]. It should also be mentioned that this porin has been found to be present in a nitrated form in hippocampus of AD brains, as an alternative isoform modification, and this may produce an irreversible dysfunction of the channel [96]. Overall, these data indicate that VDAC regulation may be crucial in the pathogenesis of AD, which may take place through the modulation of its post-translational modifications. Moreover, part of the alternative mechanisms of estrogen related neuroprotection may involve the regulation of VDAC phosphorylation, in order to maintain the channel in a closing inactivated status.

## 7.4
### Disruption of ERα–VDAC Complex in AD Brains

It has been extensively speculated that changes in the lipid composition of lipid rafts may contribute to AD pathology [97]. Thus, multiple lines of investigation have demonstrated the role of cholesterol, one of the major lipid constituents of lipid rafts, in amyloidogenic processing of APP [9, 98], suggesting a dynamic interaction of APP with lipid rafts. Apart from cholesterol, gangliosides are other lipid raft components that appear to be involved in Aβ peptide formation and processing [29], and have also been observed to be modified in lipid rafts of AD patients [99]. In addition, other classes of lipids such as phospholipids have been claimed either to mediate or to modulate key pathological processes associated with AD in relation to phospholipase D activity. Moreover, although less characterized, PUFAs may also play an important role in lipid raft stability, and their deficiency has been associated with AD pathology [100]. In particular, docosahexanoic acid (DHA) has been shown to be highly enriched in neuronal membrane phospholipids [101]. Indeed, we have recently demonstrated that AD lipid rafts obtained from brain cortex at late stages exhibited significant reductions in DHA compared with age-matched controls [102]. These abnormally low levels of PUFA are in consonance with previous observations in whole membranes from different brain areas of AD patients [103], and are correlated with reduced unsaturation and peroxidizability indexes [102]. Overall, these observations suggest that changes in brain lipid composition are important determinants of AD progression.

Taking these findings into account, it can be suggested that anomalies in the lipid composition of lipid rafts may presumably result in a profound modification of the physico-chemical properties of these structures, such as an increase in membrane viscosity and rigidity, which may largely affect the activities and interactions of raft resident proteins [102]. As an example of this hypothesis, we have recently checked whether the presence of ERα and VDAC at the plasma membrane may be altered in lipid rafts from cortical areas at late stages of AD pathology [42], as a correlate with the observed aberrant amounts of DHA in these areas [102]. The results demonstrated that pl-VDAC increased its concentration and interaction with caveolin-1 in lipid rafts of AD patients, whereas ERα levels in these fractions were reduced. In fact, and in agreement with previous data [104], ERα was mostly observed in astrocytes, suggesting a role of this receptor in estrogen protective effects related to these glial cells. Therefore, it is conceivable that anomalies in the composition of lipid rafts may interfere in pl-VDAC–mERα interactions and consequent modulation (i.e., phosphorylation) of the porin by estrogens, thus contributing to reducing the defenses facing Aβ-induced toxicity. These disrupted interactions may also affect other proteins participating in this signaling complex, such as IGF-1R. Figure 7.2 shows a schematic hypothesis of the proposed disruption in the interaction of these molecules in neuronal lipid rafts as a consequence of altered physico-chemical properties of these microstructures following AD neuropathology. Further studies are worth pursuing in order to elucidate the abnormal lipid raft dynamics of these molecules in neurons affected by AD.

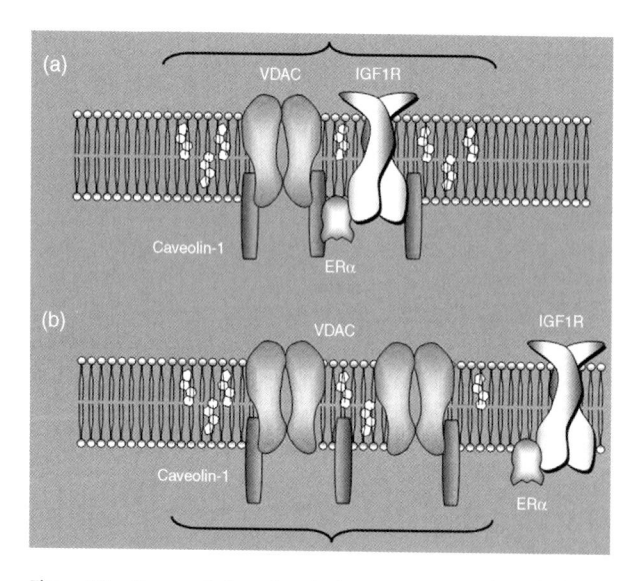

**Figure 7.2** Proposed disruption in the interaction of caveolin-1, VDAC, ERα, and IGF-1R in neuronal lipid rafts as a consequence of alterations in these microstructures related to AD neuropathology. In normal circumstances (a), VDAC and ERα may interact with caveolin-1, as a pivotal marker of this complex. IGF-1R may also be part of this signaling platform, through interactions with ERα. In the pathological scenario (b), where lipid rafts might have an aberrant structure as a result of the altered lipid composition, at least ERα, and probably also IGF-1R, may be dissociated from this complex in lipid rafts, whereas VDAC maintains and even increases its interaction with caveolin-1.

## 7.5
## Future Studies

There is general agreement about the importance of lipid–protein and protein–protein interactions in lipid rafts as crucial parameters for the integrity of these membrane microstructures, which may be profoundly modified in different neuropathologies including AD. Raft protein interactions in dynamic signaling platforms may have a pivotal role in the regulation of distinct cellular responses directed at neuronal preservation. Among the relevant protein interactions related to neuroprotection may be VDAC–ERα association together with caveolin-1, as a part of a signaling complex involved in preservation against AD. Therefore, on the one hand, the involvement of plasmalemmal VDAC in the modulation of Aβ toxicity probably related to the channel activation (opening) and, on the other hand, the demonstrated effects of estrogen, through its binding to membrane ERα, in neuronal preservation against Aβ injury, lead us to hypothesize whether the receptor may be crucial in the maintenance of VDAC inactivation (closing). pl-VDAC regulation through its phosphorylation status might be a key factor for neuroprotection and, therefore, studies on the potential post-transductional modifications of this porin at the neuronal membrane may be at the basis of estrogen mechanisms leading to brain preservation. It is enticing to speculate that disruption of VDAC association with ERα observed in AD may induce irreversible post-transductional changes in this channel, such as nitration, thereby contributing to oxidative stress and lipid raft impairment. Future work on these interactive actions may contribute to the development of novel strategies on early diagnoses to prevent this devastating disease.

## Acknowledgments

This work was supported by grants SAF2007-66148-C02-01/02 (Spanish Ministry of Education and Science). The author thanks Dr. Mario Diaz for helpful comments and graphic design.

## References

1. Gallagher, M. (2003) Aging and hippocampal/cortical circuits in rodents. *Alzheimer Dis. Assoc. Disord.*, **17**, S45–S47.
2. Lisanti, M.P., Scherer, P.E., Tang, Z., and Sargiacomo, M. (1994) Caveolae, caveolin and caveolin-rich membrane domains: a signalling hypothesis. *Trends Cell Biol*, **4** (7), 231–235.
3. Tsui-Pierchala, B., Encinas, M., Milbrandt, J., and Jonson, E.M. (2002) Lipid rafts in neuronal signaling and function. *Trends Neurosci.*, **25** (8), 412–417.
4. Ma, D.W. (2007) Lipid mediators in membrane rafts are important determinants of human health and disease. *Appl. Physiol. Nutr. Metab.*, **32** (3), 341–350.
5. Lingwood, D. and Simons, K. (2010) Lipid rafts as a membrane-organizing principle. *Science*, **327**, 46–50.

6 Robinson, P.J. (1997) Signal transduction via GPI-anchored membrane proteins. *Adv. Exp. Med. Biol.*, **419**, 365–370.

7 Jones, D.R. and Varela-Nieto, I. (1998) The role of glycosylphosphatidylinositol in signal transduction. *Int. J. Biochem. Cell Biol.*, **30** (3), 313–326.

8 Chesebro, B., Trifilo, M., Race, R., Meade-White, K., Teng, C., LaCasse, R. et al. (2005) Anchorless prion protein results in infectious amyloid disease without clinical scrapie. *Science*, **308** (5727), 1435–1439.

9 Taylor, D.R. and Hooper, N.M. (2007) Role of lipid rafts in the processing of the pathogenic prion and Alzheimer's amyloid-beta proteins. *Semin. Cell Dev. Biol.*, **18** (5), 638–648.

10 Sarnataro, D., Campana, V., Paladino, S., Stornaiuolo, M., Nitsch, L., and Zurzolo, C. (2004) PrP$^c$ association with lipid rafts in the early secretory pathway stabilizes its cellular conformation. *Mol. Biol. Cell*, **15**, 4031–4042.

11 Mattei, V., Garofalo, T., Misasi, R., Cifcella, A., Manganelli, V., Lucania, G., Pavan, A., and Sorice, M. (2004) Prion protein is a component of the multimolecular signaling complex involved in T cell activation. *FEBS Lett.*, **560**, 14–18.

12 Russelakis-Carneiro, M., Hetz, C., Maundrell, K., and Soto, C. (2004) Prion replication alters the distribution of synaptophysin and caveolin 1 in neuronal lipid rafts. *Am. J. Pathol.*, **165** (5), 1839–1848.

13 Sato, I., Obata, Y., Kasahara, K., Nakayama, Y., Fukumoto, Y., Yamasaki, T., Yokohama, K.K., Saito, T., and Yamaguchi, N. (2009) Differential trafficking of Src, Lyn, Yes and Fyn is specified by the state of palmitoylation in the SH4 domain. *J. Cell Sci.*, **122** (7), 965–975.

14 Lee, H., Woodman, S.E., Engelman, J.A., Volonte, D., Galbiati, F., Kaufman, H.L., Lublin, D.M., and Lisanti, M.P. (2001) Palmitoylation of caveolin-1 at a single site (Cys-156) controls its coupling to the c-Src tyrosine kinase. *J. Biol. Chem.*, **276** (37), 35150–35158.

15 Marino, M. and Ascenzi, P. (2006) Steroid hormone rapid signaling: the pivotal role of S-palmitoylation. *IUBMB Life*, **58** (12), 716–719.

16 Pedram, A., Razandi, M., Sainson, R.C., Kim, J.K., Hughes, C.C., and Levin, E.R. (2007) A conserved mechanism for steroid receptor translocation to the plasma membrane. *J. Biol. Chem.*, **282** (31), 22278–22288.

17 Murakami, Y., Siripanyapinyo, U., Hong, Y., Kang, J.Y., Ishihara, S., Nakakuma, H., Maeda, Y., and Kinoshita, T. (2003) PIG-W is critical for inositol acylation but not for flipping of glycosylphosphatidylinositol-anchor. *Mol. Biol. Cell*, **14**, 4285–4295.

18 Morrow, I.C., Rea, S., Martin, S., Prior, I.A., Prohaska, B., Hancock, J.F., James, D.E., and Parton, R.G. (2002) Flotillin-1/Reggie-2 traffics to surface raft domains via a novel Golgi-independent pathway. *J. Biol. Chem.*, **277** (50), 48834–48841.

19 Salzer, U. and Prohaska, R. (2001) Stomatin, flotillin-1 and flotillin-2 are major integral proteins of erythrocyte lipid rafts. *Blood*, **97**, 1141–1143.

20 Parton, R.G. (2003) Caveolae-form ultrastructure to molecular mechanisms. *Nat. Rev. Mol. Cell Biol.*, **4**, 162–167.

21 Langhorst, M.F., Reuter, A., and Stuermer, C.A.O. (2005) Scaffolding microdomains and beyond: the function of reggie/flotillin proteins. *Cell. Mol. Life Sci.*, **62**, 2228–2240.

22 Cheong, K.H., Zacchetti, D., Schneeberger, E.E., and Simons, K. (1999) VIP17/MAL, a lipid-raft associated protein, is involved in apical transport in MDCK cells. *Proc. Natl. Acad. Sci. USA*, **96**, 6241–6248.

23 Eckert, G.P. and Müller, W.E. (2009) Presenilin 1 modifies lipid raft composition of neuronal membranes. *Biochem. Biophys. Res. Commun.*, **382** (4), 673–677.

24 Couet, J., Li, S., Okamoto, T., Ikezu, T., and Lisanti, M.P. (1997) Identification of peptide and protein ligands for the caveolae-associated proteins. *J. Biol. Chem.*, **272**, 6525–6533.

25 Arcaro, A., Aubert, M., Espinosa del Hierro, M.E., Khanzada, U.K.,

Angelidou, S., Tetely, T.D., Bittermann, A.G., Frame, M.C., and Seckl, M.J. (2007) Critical role for lipid raft-associated Src kinases in activation of PI3K-Akt signalling. *Cell Signal.*, **19**, 1081–1092.

26 Boulware, M.I., Kordasiewicz, J., and Mermelstein, P.G. (2007) Caveolin proteins are essential for distinct effects of membrane estrogen receptors in neurons. *J. Neurosci.*, **27** (37), 9941–9950.

27 Toran-Allerand, C.D. (2004) Minireview: a plethora of estrogen receptors in the brain: where will it end? *Endocrinology*, **145**, 1069–1074.

28 Luoma, J.I., Boulware, M.I., and Mermelstein, P.G. (2008) Caveolin proteins and estrogen signalling in the brain. *Mol. Cell. Endocrinol.*, **290** (1–2), 8–13.

29 Ehehalt, R., Keller, P., Haass, C., Thiele, C., and Simons, K. (2003) Amyloidogenic processing of the Alzheimer beta-amyloid precursor protein depends on lipid rafts. *J. Cell Biol.*, **160** (1), 113–123.

30 Kang, M.J., Chung, Y.H., Hwang, Ch., Murata, M., Fujimoto, T., Mook-Jung, I.H., Cha, Ch.I., and Park, W.Y. (2006) Caveolin-1 upregulation in senescent neurons alters amyloid precursor protein processing. *Exp. Mol. Med*, **38** (2), 126–133.

31 Gaudreault, S.B., Dea, D., and Poirier, J. (2004) Increased caveolin-1 expression in Alzheimer's disease brain. *Neurobiol. Aging*, **25**, 753–759.

32 Langhorst, M.F., Jaeger, F.A., Mueller, S., Hartman, L.S., Luxenhofer, G., and Stuermer, C.A.O. (2008) Reggies/flotillins regulate cytoskeletal remodelling during neuronal differentiation via CAP/ponsin and Rho GTPases. *Eur. J. Cell Biol.*, **87**, 921–931.

33 Kokubo, H., Lemere, C.A., and Yamaguchi, H. (2000) Localization of flotillins inhuman brain and their accumulation with the progression of Alzheimer's disease pathology. *Neurosci. Lett.*, **290**, 93–96.

34 Encinas, M., Tansey, M.G., Tsui-Pierchala, B.A., Comella, J.X., Milbrandt, J., and Johnson, E.M. (2001) c-Src is required for glial cell line-derived neurotrophic factor (GDNF) family ligand-mediated neuronal survival via a phosphatidylinositol-3 kinase (PI-3K)-dependent pathway. *J. Neurosci.*, **21** (5), 1464–1472.

35 Limpert, A.S., Karlo, J.C., and Landreth, G.E. (2007) Nerve growth factor stimulates the concentration of TrkA within lipid rafts and extracellular signal-regulated kinase activation through c-Cbl-associated protein. *Mol. Cell Biol.*, **27** (16), 5686–5698.

36 Yamamoto, M., Toya, Y., Schewencke, C., Lisanti, M.P., Myers, M.G., and Ishikawa, Y. (1998) Caveolin is an activator of insulin receptor signalling. *J. Biol. Chem.*, **273**, 26962–26968.

37 Liu, J., Oh, P., Horner, T., Rogers, R.A., and Schnitzer, J.E. (1997) Organized endothelial cell surface signal transduction in caveolae distinct from glycosylphosphatidylinositol-anchored protein microdomains. *J. Biol. Chem.*, **272** (11), 7211–7222.

38 Martens, J.R., Navarro-Polanco, R., Coppock, E.A., Nishiyama, A., Parshley, L., Grobaski, T.D., and Tamkun, M.M. (2000) Differential targeting of shaker-like potassium channels to lipid rafts. *J. Biol. Chem.*, **275**, 7443–7446.

39 Stetzkowski-Marden, F., Recouvreur, M., Camus, G., Cartaud, A., Marchand, S., and Cartaud, J. (2006) Rafts are required for acetylcholine receptor clustering. *J. Mol. Neurosci.*, **30** (1–2), 37–38.

40 Becher, A., White, J.H., and McIlhinney, R.A. (2001) The γ-aminobutyric acid receptor B, but not the metabotropic glutamate receptor type-1, associates with lipid rafts in the rat cerebellum. *J. Neurochem.*, **79**, 787–795.

41 Marin, R., Ramirez, C.M., Gonzalez, M., Gonzalez-Muñoz, E., Zorzano, A., Camps, M., Alonso, R., and Diaz, M. (2007) Voltage-dependent anion channel (VDAC) participates in amyloid beta-induced toxicity and interacts with plasma membrane estrogen receptor alpha in septal and hippocampal neurons. *Mol. Membr. Biol.*, **24** (2), 148–160.

42 Ramirez, C.M., Gonzalez, M., Diaz, M., Alonso, R., Ferrer, I., Santere, G., Puig, B., Meyer, G., and Marin, R. (2009) VDAC

and ERα interaction in caveolae from human cortex is altered in Alzheimer's disease. *Mol. Cell. Neurosci.*, **42**, 172–183.

43 Maggi, A., Ciana, P., Belcredito, S., and Vegeto, E. (2004) Estrogens in the nervous system: mechanisms and nonreproductive functions. *Annu. Rev. Physiol.*, **66**, 291–313.

44 Suzuki, S., Brown, C.M., and Wise, P.M. (2006) Mechanisms of neuroprotection by estrogen. *Endocrine*, **29**, 209–215.

45 Garcia-Segura, L.M., Azcoitia, I., and DonCarlos, L.L. (2001) Neuroprotection by estradiol. *Prog. Neurobiol.*, **63**, 29–60.

46 Woolley, C.S. (2007) Acute effects of estrogen on neuronal physiology. *Annu. Rev. Pharmacol. Toxicol.*, **47**, 657–680.

47 Rissman, E.F. (2008) Roles of oestrogen receptors alpha and beta in behavioural neuroendocrinology: beyond Ying/Yang. *J. Neuroendocrinol.*, **20** (6), 873–879.

48 Nadal, A., Diaz, M., and Valverde, M.A. (2001) The estrogen trinity: membrane, cytosolic and nuclear effects. *News Physiol. Sci.*, **16**, 251–255.

49 Kim, H., Bang, O.Y., Jun, M.W., Ha, S.D., Hong, H.S., Huh, K., Kim, S.U., and Moo-Jung, I. (2001) Neuroprotective effects of estrogen against beta-amyloid toxicity are mediated by estrogen receptors in cultured neuronal cells. *Neurosci. Lett.*, **302**, 58–62.

50 Marin, R., Guerra, B., Hernandez-Jimenez, J.G., Kang, X.L., Fraser, J.D., Lopez, F.J., and Alonso, R. (2003) Estradiol prevents amyloid-β peptide-induced cell death in a cholinergic cell line via modulation of a classical estrogen receptor. *Neuroscience*, **121**, 917–926.

51 Vasudevan, N. and Pfaff, D.W. (2008) Non-genomic actions of estrogens and their interaction with genomic actions in the brain. *Front. Neuroendocrinol.*, **29**, 238–257.

52 Watson, C.S. and Gametchu, B. (1999) Membrane-initiated steroid actions and the proteins that mediate them. *Proc. Soc. Exp. Biol. Med.*, **22**, 9–19.

53 Marin, R., Guerra, B., Alonso, M., Ramirez, C.M., and Diaz, M. (2005) Estrogen activates classical and alternative mechanisms to orchestrate neuroprotection. *Curr. Neurovasc. Res.*, **2**, 287–301.

54 Revankar, C.M., Cimino, D.F., Sklar, L.A., Arterburn, J.B., and Prossnitz, E.R. (2005) A transmembrane intracellular estrogen receptor mediates rapid cell signaling. *Science*, **307**, 1625–1630.

55 Brailiou, E., Dun, S.L., Brailou, G.C., Mizou, K., Sklar, L.A., Oprea, T.I., Prossnitz, E.R., and Dun, N.J. (2007) Distribution and characterization of estrogen receptor G protein-coupled receptor 30 in the rat central nervous system. *J. Endocrinol.*, **193**, 311–321.

56 Toran-Allerand, C.D., Guan, X., MacLusky, N.J., Hovath, T.L., Diano, S., Singh, M., Connolly, E.S., Nethrapalli, I.S., and Tinnikov, A.A. (2002) ER-X: a novel, plasma membrane-associated, putative estrogen receptor that is regulated during developmentand alter ischemic brain injury. *J. Neurosci.*, **22**, 8391–8401.

57 Marin, R., Diaz, M., Alonso, R., Sanz, A., Arévalo, M.A., and Garcia-Segura, L.M. (2009) Role of estrogen receptor α in membrane-initiated signaling in neural cells: interaction with IGF-1 receptor. *J. Steroid Biochem. Mol. Biol.*, **114**, 2–7.

58 Marin, R., Ramirez, C.M., Morales, A., Gonzalez, M., Alonso, R., and Diaz, M. (2008) Modulation of Aβ-induced neurotoxicity by estrogen receptor alpha and other associated proteins in lipid rafts. *Steroids*, **73**, 992–996.

59 Garcia-Segura, L.M., Diz-Chaves, Y., Perez-Martin, M., and Darnaudéry, M. (2007) Estradiol, insulin-like growth factor-1 and brain aging. *Psychoneuroendocrinology*, **32**, 557–561.

60 Buettner, R., Papoutsoglou, G., Scemes, E., Spray, D.C., and Dermietzel, R. (2000) Evidence for secretory pathway localization of a voltage-dependent anion channel isoform. *Proc. Natl. Acad. Sci. USA*, **97**, 3201–3206.

61 Ishunina, T.A. and Swaab, D.F. (2003) Increased neuronal metabolic activity and estrogen receptors in the vertical limb of the diagonal band of Broca in Alzheimer's disease: relation to sex and aging. *Exp. Neurol.*, **183**, 159–172.

62 Hestiantoro, A. and Swaab, D.F. (2004) Changes in estrogen receptor-α and -β in the infundibular nucleus of the human hypothalamus are related to the occurrence of Alzheimer's disease neuropathology. *J. Neurosci. Res.*, **60**, 321–327.

63 Sherwin, B.B. and Henry, J.F. (2008) Brain aging modulates the neuroprotective effects of estrogen on selective aspects of cognition in women: a critical review. *Front. Neuroendocrinol.*, **29**, 88–113.

64 Correia, S.C., Santos, R.X., Cardoso, S., Carvalho, C., Santos, M.S., Oliveira, C.T., and Moreira, P.I. (2010) Effects of estrogen in the brain: is it a neuroprotective agent in Alzheimer's disease? *Curr. Aging Sci.*, **3** (2), 113–126.

65 Bi, R., Broutman, G., Roy, M.R., Thompson, R.F., and Baudry, M. (2000) The tyrosine kinase and mitogen-activated protein kinase pathways mediate multiple effects of estrogen in hippocampus. *Proc. Natl. Acad. Sci. USA*, **97**, 3602–3607.

66 Manthey, D., Heck, S., Engert, S., and Behl, C. (2001) Estrogen induces a rapid secretion of amyloid β precursor protein via the mitogen-activated protein kinase pathway. *Eur. J. Biochem.*, **268**, 4285–4291.

67 Fitzpatrick, J.L., Mize, A.L., Wade, C.B., Harris, J.A., Shapiro, R.A., and Dorsa, D.M. (2002) Estrogen-mediated neuroprotection against β-amyloid toxicity requires expression of estrogen receptor α or β and activation of the MAPK pathway. *J. Neurochem.*, **82**, 674–682.

68 Marin, R., Guerra, B., Morales, A., Díaz, M., and Alonso, R. (2003) An oestrogen membrane receptor participates in estradiol actions for the prevention of amyloid-β peptide$_{1-40}$-induced toxicity in septal-derived cholinergic SN56 cells. *J. Neurochem.*, **85**, 1180–1189.

69 Singer, C.A., Figueroa-Masot, X.A., Batchelor, R.H., and Dorsa, D.M. (1999) The mitogen-activated protein kinase pathway mediates estrogen neuroprotection after glutamate toxicity in primary cortical neurons. *J. Neurosci.*, **19**, 2455–2463.

70 Bryant, D.N., Sheldahl, L.C., Marriot, L.K., Shapiro, R.A., and Dorsa, D.M. (2006) Multiple pathways transmit neuroprotective effects of gonadal steroids. *Endocrine*, **29** (2), 199–207.

71 Zhang, L., Rubinow, D.R., Xaing, G., Li, B.S., Chang, Y.H., Maric, D., Marker, J.L., and Ma, W. (2001) Estrogen protects against beta-amyloid-induced neurotoxicity in rat hippocampal neurons by activation of Akt. *Neuroreport*, **12**, 1919–1923.

72 Zhao, L., Yao, J., Mao, Z., Cheng, S., Wang, Y., and Brinton, R.D. (2010) 17beta-estradiol regulates insulin-degrading enzyme expression via an ERbeta/PI3-K pathway in hippocampus: relevance to Alzheimer's prevention. *Neurobiol. Aging*, doi: 10.1016/j.neurobiolaging.2009.12.010.

73 Chiang, H.Ch., Wang, L., Xie, Z., Yau, A., and Zhong, Y.I. (2010) PI3 kinase signaling is involved in Aβ-induced memory loss in *Drosophila*. *Proc. Natl. Acad. Sci. USA*, **107** (15), 7060–7065.

74 Takashima, A. (2006) GSK-3 is essential in the pathogenesis of Alzheimer's disease. *J. Alzheimers Dis.*, **9** (3), 309–317.

75 Goodenough, S., Schleusner, D., Pietrzik, C., Skutella, T., and Behl, C. (2005) Glycogen synthase kinase 3β links neuroprotection by 17β-estradiol to key Alzheimer processes. *Neuroscience*, **132**, 581–589.

76 Tsujimoto, Y. and Shimizu, S. (2007) Role of the mitochondrial membrane permeability transition in cell death. *Apoptosis*, **12**, 835–840.

77 Thinnes, F.P., Gota, H., Kayser, H., Benz, R., Schmidt, W.E., Kratzin, H.D., and Hilschmann, N. (1989) Identification of human porins. I. Purification of a porin from human B-lymphocytes (Porin 31HL) and the topochemical proof of its expression on the plasmalemma of the progenitor cells. *Biol. Chem. Hoppe Seyler*, **370**, 1253–1264.

78 De Pinto, V., Messina, A., Lane, D.J.R., and Lawen, A. (2010) Voltage-dependent anion-selective channel (VDAC) in the

plasma membrane. *FEBS Lett.*, **584** (9), 1793–1799.

79 Bahamonde, M.I. and Valverde, M.A. (2003) Voltage-dependent anion channel localizes to the plasma membrane and peripheral but not perinuclear mitochondria. *Pflugers Arch.*, **446**, 309–313.

80 Elinder, F., Akanda, N., Tofighi, R., Shimizu, S., Tsujimoto, Y., Orrenius, S., and Ceccatelli, S. (2005) Opening of plasma membrane voltage-dependent anion channels (VDAC) precedes caspase activation in neuronal apoptosis induced by toxic stimuli. *Cell Death Differ.*, **12**, 1134–1140.

81 Bàthori, G., Parolini, I., Tombola, F., Szabò, I., Messina, A., Oliva, M. et al. (1999) Porin is present in the plasma membrane where it is concentrated in caveolae and caveolae-related domains. *J. Biol. Chem.*, **274**, 29607–29612.

82 Okada, S.F., O'Neal, W.K., Huang, P., Nicholas, R.A., Ostrowski, L.E., Craigen, W.J., Lazarowski, E.R., and Boucher, R.C. (2004) Voltage-dependent anion channel-1 (VDAC-1) contributes to ATP release and cell volume regulation in murine cells. *J. Gen. Physiol.*, **124**, 513–526.

83 Thinnes, F.P. (2009) Neuroendocrine differentiation of LNCaP cell suggests: VDAC in the cell membrane is involved in the extrinsic apoptotic pathway. *Mol. Genet. Metab.*, **97**, 241–243.

84 Akanda, N., Tofighi, R., Brask, J., Tamm, C., Elinder, F., and Ceccatelli, S. (2008) Voltage-dependent anion channels (VDAC) in the plasma membrane play a critical role in apoptosis in differentiated hippocampal neurons but not in neural stem cells. *Cell Cycle*, **7**, 3225–3234.

85 Ferrer, I. (2009) Altered mitochondria, energy metabolism, voltage-dependent anion channel, and lipid rafts converge to exhaust neurons in Alzheimer's disease. *J. Bioenerg. Biomembr.*, **41**, 425–431.

86 Kakio, A., Nishimoto, S., Kozutsumi, Y., and Matsuzaki, K. (2003) Formation of a membrane-active form of amyloid β-protein in raft-like model membranes. *Biochem. Biophys. Res. Commun.*, **303**, 514–518.

87 Yamamoto, N., Hirabayashi, Y., Amari, M., Yamaguchi, H., Romanov, G., Van Nostrand, W.E., and Yanagisawa, K. (2005) Assembly of hereditary amyloid-beta protein variants in the presence of favorable gangliosides. *FEBS Lett.*, **579** (10), 2185–2190.

88 Zampagni, M., Evangelisti, E., Cascella, R., Liguri, G., Becatti, M., Pensalfini, A. et al. (2010) Lipid rafts are primary mediators of amyloid oxidative attack on plasma membrane. *J. Mol. Med.*, **88**, 597–608.

89 Pollard, H.B., Rojas, E., and Arispe, N. (1993) A new hypothesis for the mechanism of amyloid toxicity, based on the calcium channel activity of amyloid beta protein (A beta P) in phospholipid bilayer membranes. *Ann. N. Y. Acad. Sci.*, **695**, 165–168.

90 Hsieh, H., Boehm, J., Sato, C., Iwatsubo, T., Tomita, T., Sisodia, S., and Malinow, R. (2006) AMPAR removal underlies Aβ-induced synaptic depression and dendritic spine loss. *Neuron*, **52**, 831–843.

91 Liu, S.J., Gasperini, R., Foa, L., and Small, D.H. (2010) Amyloid-β decreases cell-surface AMPA receptors by increasing intracellular calcium and phosphorylation of GluR2. *J. Alzheimer's Dis.*, **21**, doi: 10.3233/JAD-2010-091654.

92 Parameshwaran, K., Dhanasekaran, M., and Suppiramaniam, V. (2008) Amyloid beta peptides and glutamatergic synaptic dysregulation. *Exp. Neurol.*, **210**, 7–13.

93 Diaz, M., Bahamonde, M.I., Lock, H., Muñoz, F.J., Hardy, S.P., Posas, F., and Valverde, M.A. (2001) Okadaic acid-sensitive activation of Maxi-Cl$^-$ channels by triphenylethylene antioestrogens in C1300 mouse neuroblastoma cells. *J. Physiol.*, **536**, 79–88.

94 Tagaki-Morishita, Y., Yamada, N., Sugihara, A., Iwasaki, T., Tsujimura, T., and Terada, N. (2003) Mouse uterine epithelial apoptosis is associated with the expression of mitochondrial voltage-dependent anion channels, release of cytochrome *c* from mitochondria, and the ratio of Bax to Bcl-2 or Bcl-X. *Biol. Reprod.*, **68**, 1178–1184.

95 Yoo, B.C., Fountoulakis, M., Cairns, N., and Lubec, G. (2001) Changes of

voltage-dependent anion-selective channel proteins VDAC1 and VDAC2 brain levels in patients with Alzheimer's disease and Down syndrome. *Electrophoresis*, **22**, 172–179.

96 Sultana, R., Poon, H.F., Cai, J., Pierce, W.M., Merchant, M., Klein, J.B. *et al.* (2006) Identification of nitrated proteins in Alzheimer's disease brain using a redox proteomics approach. *Neurobiol. Dis.*, **22**, 76–87.

97 Michel, V. and Bakovic, M. (2007) Lipid rafts in health and disease. *Biol. Cell*, **99**, 129–140.

98 Vetrivel, K.S. and Thinakaran, G. (2010) Membrane rafts in Alzheimer's disease beta-amyloid production. *Biochim. Biophys. Acta*, **1801**, 860–867.

99 Molander-Melin, M., Blennow, M., Bogdanovic, N., Dellheden, B., Mansson, J.E., and Fredman, P. (2005) Structural membrane alterations in Alzheimer brains found to be associated with regional disease development, increased density of gangliosides GM1 and GM2 and loss of cholesterol in detergent-resistant membrane domains. *J. Neurochem.*, **92**, 171–182.

100 Young, G. and Conquer, J. (2005) Omega-3 fatty acids and neuropsychiatric disorders. *Reprod. Nutr. Dev.*, **45**, 1–28.

101 Lane, R.M. and Farlow, M.R. (2005) Lipid homeostasis and apolipoprotein E in the development and progression of Alzheimer's disease. *J. Lipid Res.*, **46**, 949–968.

102 Martin, V., Fabelo, N., Santpere, G., Puig, B., Marin, R., Ferrer, I., and Diaz, M. (2010) Lipid alterations in lipid rafts from Alzheimer's disease human brain cortex. *J. Alzheimer's Dis.*, **19**, 489–502.

103 Plourde, M., Fortier, M., Vandal, M., Tremblay-Mercier, J., Freemantle, E., Bégin, M., Pifferi, F., and Cunnane, S.C. (2007) Unresolved issues in the link between docosahexanoic acid and Alzheimer's disease. *Prostaglandins Leukot. Essent. Fatty Acids*, **77**, 301–308.

104 Garcia-Ovejero, D., Veiga, S., Garcia-Segura, L.M., and DonCarlos, L.L. (2002) Glial expression of estrogen and androgen receptors after rat brain injury. *J. Comp. Neurol.*, **450**, 256–271.

# 8
# Alzheimer's Disease as a Membrane-Associated Enzymopathy of β-Amyloid Precursor Protein (APP) Secretases

*Saori Hata, Yuhki Saito, and Toshiharu Suzuki*

## 8.1
## Introduction

The aged population is rapidly increasing in advanced countries. Putting policies in place that encourage healthy living is important for preserving an active and productive society in the context of a decreasing work force and the presence of fewer younger individuals. One of the major difficulties in preserving a healthy, independent lifestyle in the context of aging is dementia, which impairs neurocognition. Increased numbers of senile individuals with dementia can burden a nation with higher medical and nursing care costs, which in turn can dampen social and economic activity. Therefore, overcoming "senile dementia" that results from neurodegenerative disease is an important issue in advanced nations with large populations of aged individuals, and is vital for maintaining the personal health of seniors.

Alzheimer's disease (AD), which was first described by Alois Alzheimer at the beginning of the twentieth century [1], is now the most common form of dementia among age-dependent neurodegenerative diseases and is accompanied by amyloidosis. During the last 30 years, clinical and basic AD research has progressed and increased our understanding of the molecular mechanisms of AD pathogenesis. In particular, the isolation and characterization of amyloid β (Aβ) peptide, which participates in the formation of the senile/amyloid plaques in AD patients, and the identification of genes that cause familial Alzheimer's disease (FAD) contributed to our understanding of AD pathogenesis [2]. Currently, two types of genes have been identified that are associated with FAD. One of these is the gene for the β-amyloid precursor protein (APP), and the other is the presenilin (PS) genes presenilin 1 (PS1) and presenilin 2 (PS2) [3]. APP, a type I membrane protein, undergoes primary proteolytic cleavage by α- or β-secretase at its extracellular/luminal juxtamembrane region, and then undergoes secondary cleavage by γ-secretase within its transmembrane region. Consecutive cleavage by β- and γ-secretases generates Aβ, whereas cleavage by α- and γ-secretases generates the shorter amyloidolytic peptide p3 (Figure 8.1c). The generation and oligomerization of Aβ is believed to play a central

*Lipids and Cellular Membranes in Amyloid Diseases*, First Edition. Edited by Raz Jelinek.
© 2011 Wiley-VCH Verlag GmbH & Co. KGaA. Published 2011 by Wiley-VCH Verlag GmbH & Co. KGaA.

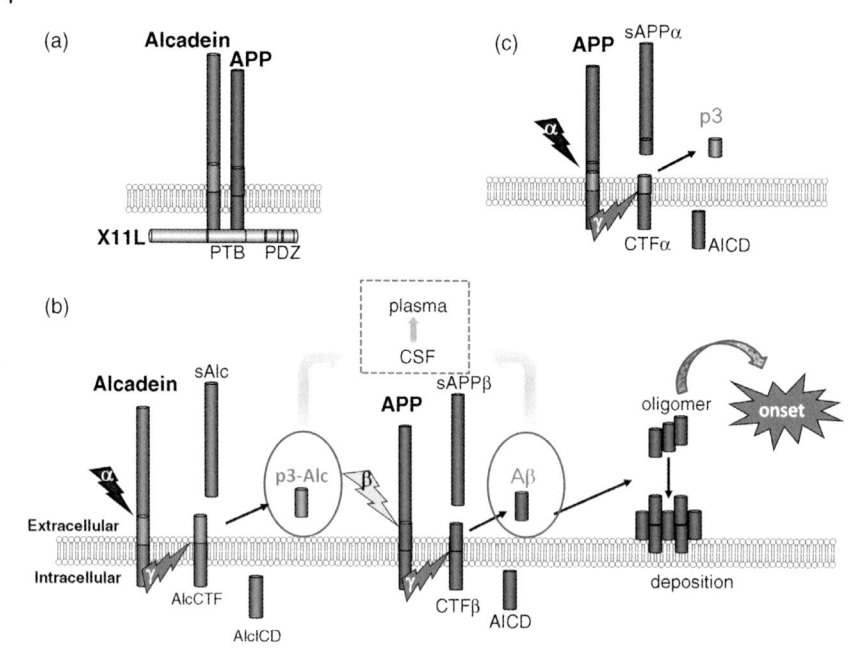

**Figure 8.1** Schematic diagram showing the metabolism of APP and Alc. Alc and APP are subject to similar proteolytic metabolism by secretases. When both proteins form a tripartite complex including the X11L protein, which mediates the cytoplasmic interaction between APP and Alc, the proteins are stable (a) [9]. In the absence of X11L, APP and Alc are first cleaved in a similar fashion at their primary cleavage sites [9]. Alc is cleaved by the APP α-secretase, generating sAlc and AlcCTF. AlcCTF is further cleaved by γ-secretase to generate p3-Alc and AlcICD. In contrast to Aβ, p3-Alc is not prone to aggregation and the generation of amyloidosis. APP is cleaved by (b) β- or (c) α-secretase, and then secondarily cleaved by γ-secretase. Cleavage of APP by the combination of β- and γ-secretase generates the Aβ peptide, which appears to induce neurotoxicity by forming oligomers that deposit in the brain as amyloid plaques (b), while the combination of α- and γ-cleavages generates the amyloidolytic p3 peptide, which is quickly degraded and cannot be detected in CSF and blood samples (c).

role in AD pathogenesis, especially in the induction of neuronal dysfunction [4]. Interaction of APP with the cytoplasmic adapter protein X11-like (X11L) regulates the amyloidogenic cleavage of APP [5]. Mutations in the APP and PS genes that have been detected in FAD patients induce alterations in the quality and quantity of Aβ generation. Thus, pathogenic mutations in both of these genes are a major causal factor of FAD etiology. However, FAD patients comprise only a fraction of the total number of AD patients. The majority of AD cases are sporadic (sporadic Alzheimer's disease (SAD)), which are not associated with mutations in the APP and PS genes. Despite this, SAD patients show a pathology that is almost identical with that of the FAD patients, except that onset of SAD occurs relatively late in life. Due to these similarities, there has been speculation that qualitative and quantitative alterations in Aβ generation may occur in a fixed percentage of SAD patients, even in the absence of PS and APP gene mutations. Therefore, important questions remain as to the molecular mechanisms by which Aβ generation is altered in the absence of causative gene mutations. Recent progress in research suggests that these mechanisms may involve the dysfunction of APP secretases and/or alterations in their functional context, which may lead to a type of enzymopathy.

APP is a secretase substrate, and therefore it is reasonable to expect that mutations in APP may affect the catalytic efficiency of proteolytic enzymes such as secretases. In fact, the Swedish APP gene mutation, which alters the N-terminus near the APP β-cleavage site, increases β-site cleavage of APP and thus increases Aβ generation [6]. PS is the catalytic subunit of the membrane-associated γ-secretase complex, which is composed of PS1 or PS2, nicastrin (NCT), anterior pharynx-defective 1 (APH-1), and presenilin enhancer 2 (PEN-2). FAD-linked mutations in PS affect γ-site cleavage, resulting in the generation of more pathogenic Aβ species such as $A\beta_{42}$. In general, FAD mutations in PS are known to increase the ratio of $A\beta_{42}$ production to $A\beta_{40}$, which is the major product of amyloidogenic cleavage of APP by β- and γ-secretases and is less pathogenic than $A\beta_{42}$ [7]. Hence the relationship between FAD mutations in causative genes and the production of neurotoxic Aβ is relatively simple from the standpoint of enzymology, as it reflects the relationship between a substrate and its enzyme.

In contrast to FAD pathogenesis, the pathogenic mechanisms involved in SAD are thought to be more complicated. One of these mechanisms may involve secretase activity, whereas another may affect Aβ-degrading enzymes in the extracellular milieu, such as neprilysine- and/or insulin-degrading enzymes. In other cases, some pathogenic alterations in the neurons themselves may affect the oligomerization of Aβ or alter the intracellular trafficking and distribution of APP and its secretases. Because almost all of the players involved in APP metabolism, APP, α-secretase (ADAM10 and/or ADAM17), β-secretase (BACE), and the components of the γ-secretase complex (PS1 or PS2, NCT, APH-1 and PEN-2), are membrane-associated proteins, alteration of the lipid components and/or the membrane status, membrane-associated intracellular protein trafficking, and the consequent protein distribution in neurons are all likely to affect APP metabolism directly and/or indirectly [5, 8]. Indeed, among the apolipoprotein E isoforms that play an important role in lipid efflux in neurons, apoE4 is known as a significant risk factor for age-related SAD, epidemiologically. Furthermore, impairment of membrane vesicle transport in terminally differentiated neurons, which often have a long axon and highly branched dendrites, increases neurotoxic Aβ generation.

In this chapter, we first focus on the involvement of lipid rafts in the molecular regulation of β-cleavage of APP. Lipid rafts are cholesterol- and sphingolipid-abundant detergent resistant membrane (DRM) domains and are rich in enzymatically active β- and γ-secretases. Increased localization of APP in these membrane microdomains can result from dysfunction of a component of the APP interactome and/or alterations in membrane lipid components. This increase in APP membrane localization can result in increased amyloidogenic β-cleavage of APP and the production of Aβ, even if β-secretase enzymatic activity itself has not been affected and remains appropriately regulated. Second, we discuss γ-secretase enzymopathies that can occur independently of PS mutations. γ-Secretase hydrolyzes peptide bonds within type I membrane proteins in regions of the membrane that are poor in water. Altered circumstances and/or conditions can change the catalytic properties of the γ-secretase complex enzymatic reaction and result in slippage at the cleaving sites or change the cleavage efficiency. Such enzymopathies facilitate quantitative and qualitative changes in Aβ generation and can trigger the onset of SAD, regardless in the absence of PS gene mutations.

Because Aβ can form aggregates, it is difficult to estimate net changes in Aβ generation by testing Aβ quantity and quality in cerebrospinal fluid (CSF) and blood. Indeed, the level of Aβ$_{42}$, which is thought to be a significant cause of AD pathogenesis, gradually decreases as AD progresses, possibly due to its aggregation and deposition in the brain where it forms amyloid or senile plaques. Hence changes in the quality and quantity of pathogenic Aβ are not suitable indicators of γ-secretase dysfunction (enzymopathy), although these metrics can be used as an index of AD pathology. In the second part of this chapter, we therefore discuss alcadein (Alc) and its metabolic fragment p3-Alc and the potential for using these components to measure γ-secretase enzymopathy [9, 10]. p3-Alc is formed following cleavage of Alc by α- and γ-secretases in a fashion similar to APP metabolism in the brain. In contrast to Aβ, however, p3-Alc does not aggregate and is detectable in the CSF and blood (Figure 8.1). Therefore, qualitative and quantitative alterations in p3-Alc in the CSF and blood could reflect a net alteration in Aβ quality and quantity. Moreover, changes in p3-Alc in SAD patients who do not carry PS mutations could indicate a γ-secretase enzymopathy. Based on these ideas, basic biochemical analyses of human samples were performed. The results suggest that in a fixed population of SAD subjects, the onset of SAD is triggered by γ-secretase dysfunction (see 8.3). Identification and classification of the fundamental pathogenic factor(s) that underlie pathogenic Aβ generation in SAD are critical in selecting the most effective treatment among the many AD drugs already developed, including those that target β-secretase, γ-secretase, Aβ aggregation, Aβ degradation, and lipid metabolism, among others.

### 8.1.1
#### Cholesterol and Alzheimer's Disease Pathogenesis

Cholesterol is an essential structural component of the plasma membrane of every cell and is a major component of lipid rafts together with sphingolipids. The central nervous system (CNS) accounts for 2% of the total body mass but contains a quarter of total body cholesterol [11]. Several lines of evidence indicate that cholesterol homeostasis is clinically associated with AD pathogenesis. First, high cholesterol and low-density lipoprotein (LDL) levels in plasma correlate with Aβ load in the brains of AD patients. Second, high cholesterol levels in plasma in later middle age increase the risk of AD pathogenesis. Third, patients treated with statins, inhibitors of 3-hydroxymethyl 3-glutaryl-coenzyme A (HMG-CoA) reductase, showed a reduced incidence of AD. Fourth, cholesterol loading and depletion studies demonstrated a correlation between cholesterol levels and the efficiency of Aβ production and deposition in cultured cells and a transgenic mice model of AD. Moreover, various animal experiments have shown a positive correlation between plasma cholesterol levels and cerebral Aβ amounts. Finally, a cholesterol transport inhibitor, U18666A, and mutations in NPC1 and NPC2, which lead to endosomal cholesterol accumulation, neuronal dysfunction, and death in Niemann–Pick type C disease, increase the generation of $A\beta_{40}$ and $A\beta_{42}$ [12]. These observations strongly suggest that cholesterol homeostasis, intracellular trafficking, and distribution of cholesterol are relevant in APP processing and AD pathogenesis [13].

## 8.1.2
### ApoE, Lipoprotein Receptors and Alzheimer's Disease

Although *APP*, *PS1*, and *PS2* have been identified as the genes responsible for FAD [3], these mutations do not account for the vast majority of SAD cases. A major effort to determine the genetic causes of SAD uncovered apolipoprotein E4 (apoE-ε4), a component of lipoprotein particles that transports lipids such as cholesterol in the blood and other body fluids, as a risk factor associated with SAD [14]. Although apoE is mainly produced and secreted by the liver, it is also produced in the brain by astrocytes and microglia and is believed to deliver cholesterol to neurons, where it is required for development and maintenance of neuronal plasticity and function. In humans, three allelic variations in the *ApoE* gene exist at a single locus on chromosome 19 (ε2, ε3, and ε4). These alleles encode three protein isoforms (apoE2, apoE3, and apoE4) of 299 amino acids that differ at only two residues [E2 (Cys112, Cys158), E3 (Cys112, Arg158), and E4 (Arg112, Arg158)] but have different biological properties. ApoE4 appears to be less efficient than the other isoforms in promoting cholesterol efflux from neurons and astrocytes, and has a decreased effect on cholesterol uptakes when compared with apoE2 and apoE3. These data suggested that apoE isoform differences could influence AD pathogenesis by affecting cholesterol homeostasis. ApoE also appears to be involved in the binding and clearance of Aβ, with apoE3 and apoE2 being more effective than apoE4. The interaction of ApoE3 with oligomeric Aβ and apoE receptors also reduces the neurotoxicity of Aβ oligomers to a greater extent than interaction with ApoE4. In other studies, apoE4 has been found to affect signaling pathways and to show reduced protection against oxidative stress, and is associated with greater cholinergic dysfunction in AD [15].

ApoE appears to be necessary for amyloid deposition, as apoE knockout (KO) mice crossed to AD model mice have much less Aβ deposition compared with mice expressing either apoE3 or apoE4. These studies show that apoE assists Aβ deposition and that apoE4 may increase AD risk by increasing amyloid deposition levels. In addition, the cholesterol transporter ATP-binding cassette sub-family A member 1 (ABCA1) is a crucial regulator of apoE levels and lipidation in the brain. An ABCA1 deficiency leads to the loss of about 80% of the apoE in the brain, and the residual 20% is poorly lipidated. This poorly lipidated apoE is associated with increased amyloid burden in mouse models of AD, implying that apoE lipidation by ABCA1 affects amyloid deposition or clearance. On the other hand, over-expression of ABCA1 promotes apoE lipidation and eliminates amyloid plaque formation. Therefore, apoE lipid binding capacity may be highly relevant to AD pathogenesis [3], suggesting that the status of lipids in brain neurons can affect amyloidosis, including Aβ generation and aggregation.

The main ApoE receptors expressed in the CNS are members of the LDLR gene family. Neurons obtain cholesterol through the uptake of apoE–cholesterol complexes via LDLR family receptors. ApoE receptors, such as LDLR-related protein 1 (LRP1/LRP/CD91), LRP1B (LRP-DIT), LRP2 (megalin), ApoER2 (LRP8), and the sortilin-related receptor (SOR1/LR11/SorLA), can also bind and internalize apoE-containing lipoproteins. This family of ApoE receptor proteins also interacts with APP and regulates APP metabolism and trafficking, as described below.

### 8.1.2.1 LRP1 and LRP1B

LRP1 is highly expressed in the brain, where it is cleaved by furin in the trans Golgi network (TGN) and then by BACE1 and γ-secretase. LRP1 binds to APP though the cytoplasmid adapter protein FE65, which promotes the endocytosis of APP via clathrin-coated pits, delivers APP to late endosomal compartments, and enhances Aβ generation. LRP1 mediates Aβ clearance *in vitro* by binding either to Aβ itself or to Aβ complexes. Moreover, several polymorphisms within the LRP1 gene have been correlated with an increased risk of AD [16, 17]. ApoE4, a genetic risk factor for SAD, enhances LRP1-dependent APP endocytosis and Aβ generation. These findings suggest that LRP1 could play a pivotal role in AD pathogenesis. Studies of LRP1 neuron-specific KO mice have shown that LRP1 is involved in synaptic transmission and motor function in the CNS. Other studies have demonstrated that LRP1 neuron-specific KO mice have lower brain cholesterol levels and increased apoE levels, suggesting that LRP1 is necessary for neuronal uptake of cholesterol–apoE complexes [3].

LRP1B shares 59% amino acid identity with LRP1, but its effects on APP metabolism are different from those of LRP1. LRP1B suppresses the endocytosis of APP from the plasma membrane and decreases amyloidogenic metabolism of APP. This diversity in LRP1 and LRP1B function may be attributed to differences in the endocytotic rate of these molecules (LRP1 $t_{1/2} < 0.5$ min, LRP1B $t_{1/2} < 8$ min) [18].

### 8.1.2.2 SorLA/LR11

SorLA/LR11 is a ~250 kDa membrane receptor that is highly expressed in the CNS and contains a vacuolar protein sorting 10 protein (vps10p) domain. SorLA/LR11 is homologous to a yeast receptor that transports proteins between the late Golgi and the endosome. Furthermore, homology analyses suggest that mammalian SorLA predominantly localizes to intracellular compartments and is more likely to be involved in intracellular rather than endocytic transport. SorLA/LR11 interacts directly with APP and the two proteins colocalize in Golgi compartments and in the endosome. Direct interaction between SorLA/LR11 and APP blocks the interaction of APP with β-secretase, effectively reducing APP cleavage in post-Golgi compartments. In contrast to LRP1 and LRP1B, SorLA does not affect the modulation of the rate of APP internalization from the cell membrane. SorLA/LR11 expression levels are reduced in the brains of AD patients, and the generation of Aβ is also increased in the brains of SorLA/LR11 KO mice [18]. In a transgenic expression study, SorLA/LR11 altered the APP intracellular distribution and reduced Aβ levels in cultured cells [19]. Thus, SorLA/LR11 acts as a sorting molecule that prevents APP processing and Aβ generation.

### 8.1.2.3 ApoER2/LRP8
ApoER2 is mainly expressed in the hippocampus and cerebral cortex and has a pivotal role in neuronal migration during brain development. ApoER2 acts as a receptor for apoE, reelin, and F-spondin. Like APP, ApoER2 also undergoes proteolytic processing by α- and γ-secretases and binds Fe65, X11s, the JNK interacting proteins (JIPs) Dab1 and Dab2. The expression levels of APP and Aβ production increased in the frontal cortex of ApoER2 KO mice. Furthermore, a genetic polymorphism in the *LRP8* (*APOER2*) gene has been associated with AD [20].

## 8.1.3
### Lipid Rafts and Alzheimer's Disease

Lipid rafts are dynamic and highly ordered membrane microdomains that are rich in cholesterol and sphingolipids, such as seramide, gangliosides, glycerophospholipids, and sterols. Lipid rafts are estimated to have an average diameter of 50 nm, but several classes of lipid rafts that vary in size and duration can exist in a cell [21]. Lipid rafts are formed in the Golgi and are transported from the Golgi to the plasma membrane [22, 23]. In contrast, retrograde transport vesicles (from the Golgi to the ER) are composed of very little cholesterol and sphingolipids [23]. Lipid rafts exist in the plasma membrane and are endocyted from the membrane through the endocytic pathway and recycled back into the plasma membrane or returned to the Golgi apparatus [24]. Lipid rafts serve as platforms for cell signaling, pathogen entry, cell adhesion, and protein sorting and trafficking, and are biochemically defined as the DRM fraction [22, 25]. Aβ generation and aggregation occur in lipid rafts, and both BACE1 and γ-secretase activity is higher in lipid rafts than in other membrane regions. These data suggest that lipid rafts play a key role in APP processing and AD pathogenesis.

### 8.1.3.1 APP-Cleaving Enzyme and Lipid Rafts
Less than 10% of full-length APP associates with lipid rafts. In contrast, lipid rafts are the primary site of APP C-terminal fragment (CTF) accumulation. BACE1, the APP β-secretase, generates the β CTF form by cleaving the APP ectodomain (Figure 8.1b). This amyloidogenic cleavage occurs in cholesterol- and sphingolipid-enriched DRM domains and, indeed, high levels of BACE colocalize with APP in DRMs [26, 27]. BACE1 mutant proteins with an attached GPI anchor group localize at higher levels

in DRMs than wild-type BACE1 and cleave APP more efficiently [26]. The lipid composition of plasma membrane modulates BACE1 activity [28]. Hence the degree of BACE1 localization in lipid rafts relative to non-raft regions that are composed of different lipids at distinct ratios is important for amyloidogenic APP metabolism.

Acyl modification, such as *S*-palmitoylation and *N*-myristoylation, is a post-translational modification that targets transmembrane- or membrane-attached proteins to lipid rafts [29]. BACE1 undergoes *S*-palmitoylation at four cysteine residue (Cys474/478/482/485) [30, 31]. Specific mutations in BACE1 completely abolish *S*-palmitoylation, preventing association with DRMs. This mutant form of BACE1, however, functions as a β-secretase and allows Aβ generation from APP. These data suggest that BACE1 *S*-palmitoylation is important for its localization to lipid rafts under physiological conditions, but is dispensable for BACE1 activity. One caveat to these experiments, however, is that the DRM fractions were biochemically prepared using several different kinds of detergents, such as CHAPS, Lubrol WX, and Triton X-100. These data must therefore be viewed in the light of the fact that distinct classes of DRMs are isolated using different kind of detergents.

### 8.1.3.2 APP, X11 Family Proteins, and Lipid Rafts

*S*-Palmitoylation sites are present in APP secretase but not in APP. What determines the distribution of APP and APP CTFs between DRMs and non-DRM regions? X11/Mint family proteins may be one key group of regulators that modulate APP distribution between DRMs and non-DRMs. X11 family proteins (X11s) consist of X11/X11α/Mint1, X11L/X11β/Mint2, and X11L2/X11γ/Mint3, and are encoded by separate genes on human chromosomes 9, 15, and 19, respectively. X11s contain an evolutionarily conserved central phosphotyrosine binding/interaction (PTB/PI) domain and two C-terminal PDZ domains, which were first discovered to share the domain within post-synaptic density protein 95 (PSD95), *Drosophila* disc large tumor suppressor (Dlg), and zonula occludens-1 protein (ZO-1) [32]. The PTB/PI domain and the PDZ domain are well-characterized modular protein–protein interaction domains that are found in many adapter proteins and signaling molecules. X11 proteins interact with various kinds of proteins, such as APP, Alc, apoER2, munc18, KIF17, kalirin, hyperpolarization-activated cyclic nucleotide gated (HCN) channel, and arfs, through their PTB/PI and PDZ domains. The interaction of X11L with APP stabilizes APP metabolism and suppresses Aβ generation *in vitro* (Figure 8.1a) [33–35]. This suppressive effect is enhanced by co-expression of Alc, which results in the formation of an APP–X11L–Alc trimeric complex. X11 and X11L also suppress APP metabolism and Aβ generation *in vivo* [36–38]. X11 or X11L transgenic mice crossed to APPswe transgenic mice showed reduced amyloid deposits and decreased levels of Aβ$_{40}$ and Aβ$_{42}$ in the brain compared with APPswe transgenic mice. In addition, endogenous APP metabolism was enhanced in the brain of X11L KO mice [36]. However, the molecular mechanisms involved in X11- and X11L-mediated suppression of APP amyloidogenic metabolism remained unclear. To address this issue, X11 and X11L double KO mice were generated, and the amount of APP CTF and Aβ were quantified. In the brain of mice lacking X11

(a) X11+/+, X11L+/+ neuron      (b) X11-/-, X11L-/- neuron

DRM (lipid raft)

APP    sAPP    CTFβ    Aβ    X11s    β-secretase   γ-secretase

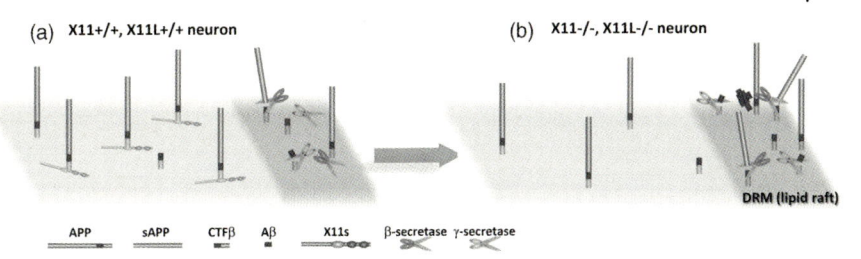

**Figure 8.2** Role of the X11 proteins, X11 and X11L, in regulating lipid raft–DRM association and the amyloidogenic β-cleavage of APP. The X11 proteins X11 and X11L are neuron-specific adapter proteins that regulate APP metabolism at several steps in the APP secretory pathway. In the brain, X11L is expressed widely and colocalizes with APP rather than X11 in neurons [33, 39]. X11s associate with APP outside DRMs (light zone), and prevent translocation of APP into DRMs (dark zone). Following dissociation from X11s, APP can spontaneously enter DRM regions, which are rich in active β- and γ-secretases (a). In neurons lacking X11s, APP can more easily access DRM regions, increasing the chance that APP will be cleaved by β- and γ-secretases (b). In some SAD cases, the malfunction of X11L and/or a weak association between X11L to APP in the context of aging may facilitate the amyloidogenic processing of APP [39]. Altered lipid composition in the membrane may affect the size and/or numbers of DRM regions and increase the chance that APP may enter these areas, which can also stimulate the generation of neurotoxic Aβ. In either case, β-secretase activity itself does not change, and therefore is not enzymopathic, but the pathogenic metabolism of APP is nonetheless increased.

and/or X11L, the levels of CTFβ and Aβ were increased relative to wild-type levels. In addition, X11L suppressed APP metabolism more strongly than X11. The absence of X11s resulted in more APP and APP CTF translocation into DRMs and enhanced colocalization of APP or APP CTF with BACE1 in DRMs [39] (Figure 8.2). We also showed that cytoplasmic X11s could be recovered in membrane fractions and that they mainly localized to non-DRMs. In addition, other independent groups have shown that APP localizes in membrane microdomains with the X11–munc18–syntaxin1 complex, and these microdomains could be distinguished from BACE1 microdomains [40]. Phosphorylation of munc18 by cdk5 causes disassociation of APP-X11–mun18–syntaxin1 complexes and shifts APP to BACE1-containing microdomains. Moreover, neuronal hyperactivity, which promotes Aβ generation, enhances APP movement into BACE1 microdomains in a cdk5-dependent manner. These data indicate that APP associates with X11s outside DRMs to prevent APP translocation into lipid rafts where amyloidogenic metabolism of APP occurs (Figure 8.2).

Dysfunction of X11s in aged neurons may therefore contribute to SAD etiology. The dysfunction of X11s could lead to a weakening of the association between X11s and APP, resulting in greater translocation of APP into DRMs. Alterations in the lipid composition of membranes may enlarge lipid raft areas or increase the number of lipid rafts, which could enhance APP translocation into DRMs. These qualitative alterations in X11s and/or lipid metabolism could result in increased β-cleavage of APP, even if β-secretase itself is not enzymopathic.

## 8.2
### Intramembrane-Cleaving Enzyme of Type I Membrane Proteins

#### 8.2.1
#### γ-Secretase

Aβ peptides are generated through sequential processing of the APP by β- and γ-secretases (Figure 8.1). APP is initially cleaved by β-secretase (BACE), releasing sAPPβ and generating CTFβ. CTFβ is further cleaved by the γ-secretase complex to generate Aβ peptides (A$\beta_{40}$ and A$\beta_{42}$) and the APP intracellular domain (AICD) fragment. In an alternative pathway, APP is cleaved by α-secretase within the Aβ domain, and the resulting CTFα fragment is also cleaved by γ-secretase to release the nonamyloidogenic peptide p3 (Figure 8.1b). The γ-secretase complex is composed of PS, NCT, APH-1, and PEN-2. Genetic mutations in PS1 and PS2 are responsible for familial early-onset AD [3]. More than 190 mutations have been identified [41], and these mutations result in an increase in the ratio of A$\beta_{42}$ to A$\beta_{40}$, which might affect the formation of neurotoxic oligomers and lead to AD [4].

In addition to the proteolysis of APP, γ-secretase cleaves various type I transmembrane proteins, including APP family members (APLP1, APLP2), cell fate determination molecules (Notch, Jagged), adhesion molecules (N-cadherin, E-cadherin, CD44 and nectin-1α), the regulatory β2 subunit of voltage-gate sodium channels, neurotrophin receptor (p75), LDL receptor-related protein and receptor tyrosine kinase (ErbB4) [42]. The substrates for γ-secretase are generally derived from large precursor proteins that undergo an ectodomain shedding event before being cleaved by γ-secretase [43]. The enzymes that catalyze the ectodomain shedding of γ-secretase substrates include members of the disintegrin and metalloprotease (ADAM) family, β-site APP cleaving enzymes (BACE1), and matrix metalloproteases (MMPs).

Components of the γ-secretase complex are assembled in a stepwise, well-ordered fashion during membrane trafficking through the endoplasmic reticulum (ER) and Golgi apparatus [44]. Uncomplexed PS is metabolically unstable and is rapidly degraded during trafficking, but PS is stable in high molecular weight complexes containing NCT and APH-1, and the addition of PEN-2 is thought to activate γ-secretase. NCT has also been suggested to function as a γ-secretase–substrate receptor [45].

#### 8.2.2
#### γ-Secretase and Cholesterol

There is growing evidence that cholesterol is linked to the production of Aβ and the progression of AD. Several epidemiological studies have suggested that the use of statins, cholesterol synthesis inhibitors (HMG-CoA reductase blockers), may reduce the risk of developing dementia, including AD [46, 47], as described in Section 8.1. *In vitro* studies have demonstrated that cholesterol depletion inhibits Aβ production in hippocampal neurons [48, 49] and increases α-secretase activity in cultured cells [50]. *In vivo* studies have also verified that cholesterol synthesis inhibitors can reduce Aβ generation both in guinea pigs [49] and in an AD mouse model [51]. In contrast, high-cholesterol diets increased Aβ levels in rabbits and in an AD mouse model [52]. Hence, in addition to the ApoE4 allele AD risk factor [14, 15], cholesterol may also impact the development of AD. Lowering cholesterol may inhibit amyloidogenic processing of APP, possibly as a result of lipid raft disruption, as described in Section 8.1.

Several reports have examined the relationship between γ-secretase and cholesterol using DRMs. Both endoproteolytic fragments of PS1, the PS1 CTF and the PS1 NTF (N-terminal fragment), localize in DRM fractions [53, 54]. The other γ-secretase components, NCT, Aph-1, and Pen-2, are also enriched in DRM fractions, and the presence of all four core components is required for the complex to become raft-associated and functionally active [31, 55]. Thus, γ-secretase activity is found in the buoyant, cholesterol-rich DRM fraction [56], and active γ-secretase has also been shown to localize in DRMs isolated from human brain [57]. Interestingly, unlike APP, CTFs derived from Notch1, Jagged2, Deleted in Colorectal Carcinoma (DCC), and N-cadherin remain largely detergent soluble, indicative of their spatial segregation in non-raft domains [58].

## 8.3
### Alcadein Processing by γ-Secretase in Alzheimer's Disease

### 8.3.1
#### Alcadein as a γ-Secretase Substrate in Neurons

Alc family proteins are evolutionary conserved, type I membrane proteins that are predominantly expressed in neuronal tissues [34]. The first Alc family member was identified as a binding protein for the neuron-specific adapter X11L and was also identified as a postsynaptic $Ca^{2+}$-binding membrane protein called calsyntenin [59]. In humans, Alc family proteins consist of Alcα1 (971 amino acids), Alcα2 (981 amino acids), Alcβ (956 amino acids), and Alcγ (955 amino acids). Alcα, Alcβ, and Alcγ are encoded by independent genes, whereas Alcα1 and Alcα2 are splice variants derived from the Alcα gene. Evolutionarily conserved Alcs contain two cadherin motifs at the extracellular/luminal domain, an Asn–Pro (NP) sequence for X11 and X11L binding, a WD motif (the consensus, D/E–W–D–D–S–A/T–L–T/S amino acid sequence in Alc) for kinesin light chain (KLC) binding, and a highly acidic sequence in the cytoplasmic region [60].

Alc interacts directly with KLC and functions as a kinesin-1 cargo receptor, which mediates anterograde transport of membrane vesicles [60]. Because APP also functions as a kinesin-1 cargo receptor, Alc-containing vesicles can compete for kinesin-1 with APP-containing vesicles, and disruption of APP containing vesicle transport increases Aβ generation [60]. Thus, Alc has been suggested to play an important role in vesicular transport and to contribute to AD pathogenesis.

A genome-wide human study of genes involved in hippocampus-dependent episodic memory alterations resulted in the identification of CLSTN2 (the gene encoding Alcγ/calsyntenin2) [61]. Single nucleotide polymorphisms (SNPs) in the Alcγ gene have been associated with episodic memory performance [62]. These reports suggest that Alc may play an important role in learning and memory.

In neurons, both Alc and APP bind X11L in the cytoplasm, leading to Alc and APP co-association [9, 34]. Formation of this APP–X11L–Alc complex stabilizes the metabolism of both APP and Alc, which is discussed in the next paragraph. In AD brains, Alcα localizes in dystrophic neurites together with APP [34]. Overall, these observations strongly indicate a correlation between AD and Alc, and additional analysis of Alc may contribute to increase our understanding of AD pathogenesis as an enzymopathy.

### 8.3.2
### Alcadein Processing and γ-Secretase Dysfunction

Formation of the APP–X11L–Alc complex stabilizes the metabolism of both APP and Alc in cells [34]. Dissociation of X11L from this tripartite complex induces the coordinated cleavage of APP and Alc, resulting in the generation of Aβ and p3 from APP and p3-Alc from Alc. In addition, their respective cytoplasmic domain fragments, AICD and AlcICD, are released [9] (Figure 8.1).

All three members of the Alc protein family (Alcα, Alcβ, Alcγ) are subject to primary cleavage at the extracellular juxtamembrane region by APP α-secretase, which results in the release of their large ectodomains (sAlc, corresponding to the APP ectodomain sAPP) [10]. The membrane tethered C-terminal fragment of Alc (AlcCTF) is subsequently cleaved by the γ-secretase complex, liberating p3-Alc and AlcICD. Thus, the proteolytic processing of Alc family proteins is very similar to that of APP.

The primary protease of Alc family proteins is ADAM10, which has been identified as the APP α-secretase. In ADAM 10 $-/-$ cells, the release of sAlc and sAPP was similarly decreased. In addition, sAlc increased in a dose-dependent manner in response to the expression of exogenous ADAM17 [10]. Hence ADAM17 may cleave Alc proteins in addition to APP. In contrast, studies suggest that BACE1, which is thought to be the APP β-secretase [63], is unlikely to be involved in the primary cleavage of Alcα, Alcβ, and Alcγ [10]. These results indicate that both Alc and APP are primarily cleaved by the same enzyme, α-secretase (ADAM10 or ADAM17), following individual or coordinated liberation of these proteins from X11L [10].

p3-Alc peptides are released into cell-conditioned media and into human CSF by intramembrane cleavage of the Alc CTF. Similarly to the secretion of APP-generated $A\beta_{40}$ and $A\beta_{42}$, which possess different C-termini, several p3-Alc species are secreted that result from cleavage at different γ-secretase sites. In human CSF, the major p3-Alcα released is p3-Alcα35 (composed of 35 amino acids), the major p3-Alcβ is p3-Alcβ37, and the major p3-Alcγ is p3-Alcγ31 (Figure 8.3). p3-Alcα2N+35 is the major secreted species in conditioned media harvested from HEK293 cells expressing Alcα,

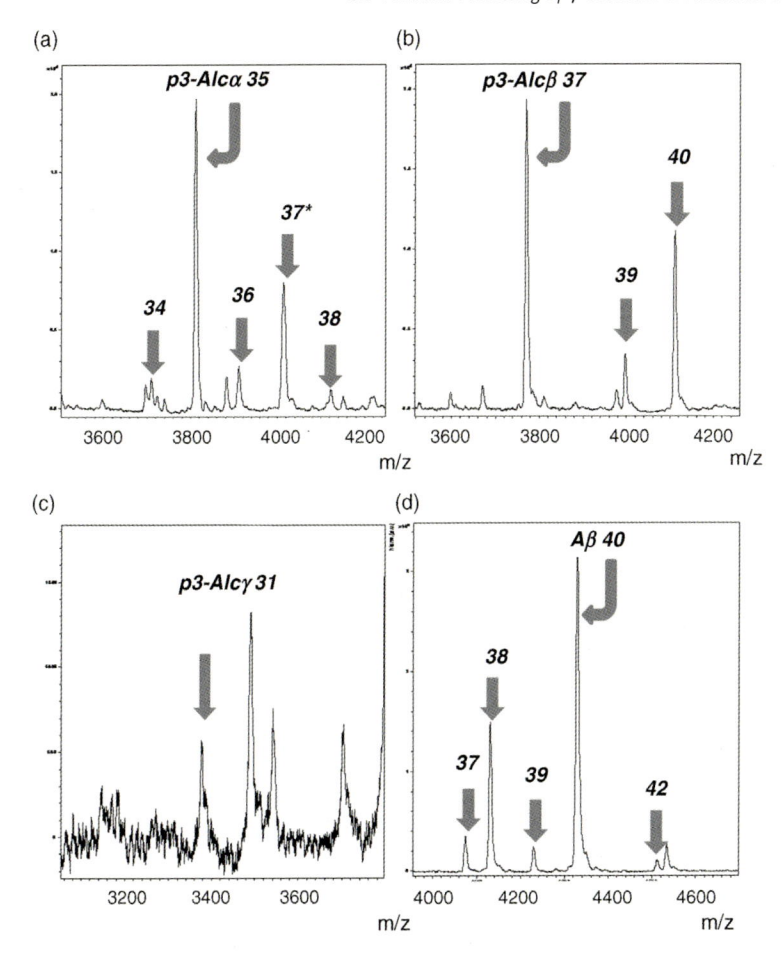

**Figure 8.3** Profiles of p3-Alc and Aβ species in human CSF. The major and minor species of (a) p3-Alcα, (b) p3-Alcβ, (c) p3-Alcγ, and (d) Aβ in human CSF were separated using a combination of immunoprecipitation and MALDI/TOF-MS (matrix-assisted laser desorption/ionization time-of-flight mass spectrometry). The major p3-Alc species derived from Alcα, Alcβ, and Alcγ are p3-Alcα35, p3-Alcβ37, and p3-Alcγ31, respectively [10, 64], and the major Aβ species derived from APP is Aβ$_{40}$. The minor species in CSF possess an identical N-terminus but have a different C-termini due to altered cleavage by γ-secretase. The amino acid sequences of p3-Alc and Aβ are shown in Figure 8.4.

and this species possesses the same C-terminus as p3-Alcα 35. The amino acid sequences of p3-Alcs and Aβ are indicated in Figure 8.4.

FAD-linked PS1 mutations are known to alter the cleavage of APP and increase the ratio of Aβ$_{42}$ (minor species) to Aβ$_{40}$ (major species). These mutations also alter the secretion ratio of minor p3-Alc species to major species (p3-Alcα2N+38/2N+35, p3-Alcβ 40/37, p3-Alcγ 34/31) [10]. This suggests that Alc may be a useful tool to detect γ-secretase dysfunction. Hence in clinical situations, analyzing not only Aβ but also p3-Alc, which is not prone to aggregation, may be useful to detect γ-secretase

## APP695

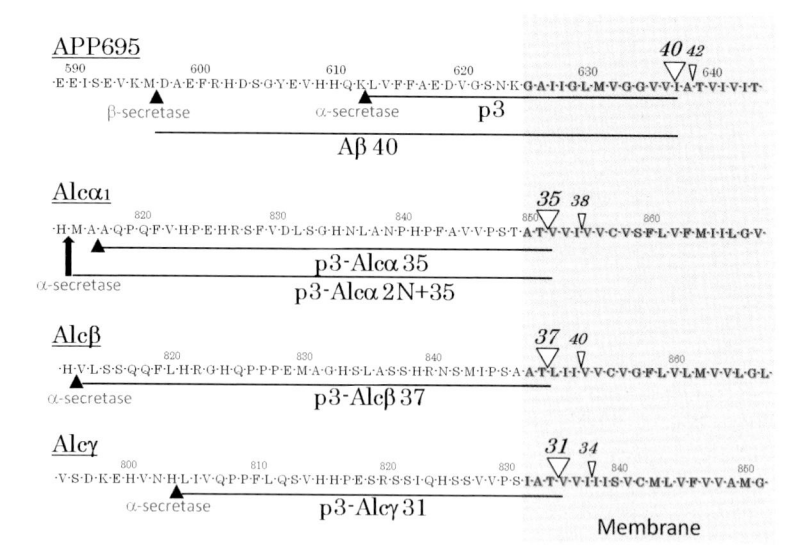

```
     590              600              610              620              630    40 42  640
                                                                                 ▽▽
·E·I·S·E·V·K·M·D·A·E·F·R·H·D·S·G·Y·E·V·H·H·Q·K·L·V·F·F·A·E·D·V·G·S·N·K·G·A·I·I·G·L·M·V·G·G·V·V·I·A·T·V·I·V·I·T·
                ▲                              ▲
           β-secretase                    α-secretase        p3
                                  ──────────────────────────────────────
                                                  Aβ 40
```

## Alcα1

```
                820              830              840    35  38       860
                                                         ▽   ▽
·H·M·A·A·Q·P·Q·F·V·H·P·E·H·R·S·F·V·D·L·S·G·H·N·L·A·N·P·H·P·F·A·V·V·P·S·T·A·T·V·V·I·V·V·C·V·S·F·L·V·F·M·I·I·L·G·V·
  ↑       ▲
α-secretase                                  ──────────────────
                                                  p3-Alcα 35
                                    ────────────────────────────
                                          p3-Alcα 2N+35
```

## Alcβ

```
                820              830              840    37  40       860
                                                         ▽   ▽
·H·V·L·S·S·Q·Q·F·L·H·R·G·H·Q·P·P·P·E·M·A·G·H·S·L·A·S·S·H·R·N·S·M·I·P·S·A·A·T·L·I·I·V·V·C·V·G·F·L·V·L·M·V·V·L·G·L·
        ▲
 α-secretase                                     ──────────────────
                                                    p3-Alcβ 37
```

## Alcγ

```
        800              810              820              830   31 34       850
                                                                 ▽  ▽
·V·S·D·K·E·H·V·N·H·L·I·V·Q·P·P·F·L·Q·S·V·H·H·P·E·S·R·S·S·I·Q·H·S·S·V·V·P·S·I·A·T·V·V·I·I·S·V·C·M·L·V·F·V·V·A·M·G·
        ▲
 α-secretase                                     ──────────────
                                                    p3-Alcγ 31
                                                                              Membrane
```

**Figure 8.4** Amino acid sequence of p3-Alc and Aβ. The amino acid sequences of the major p3-Alc species (p3-Alcα35, p3-Alcβ37, and p3-Alcγ31) along with the sequence of the major Aβ$_{40}$ and the p3 peptide, are indicated by the underlined letters. The major primary α- and β-cleavage sites are indicated by closed arrowheads, and the major secondary γ-cleavage sites are indicated by open arrowheads. In Alcα, the closed arrow indicates the primary α-cleavage that is the dominant site cleaved in cultured cells [10], which generates p3-Alcα35 with two extra amino acids at the N-terminus (p3-Alcα2N+35). The amino acid numbers are from human APP695, Alcα1, Alcβ and Alcγ [10, 64].

dysfunction. In studies using cells expressing FAD-linked PS1 mutants, the Aβ$_{42}$/Aβ$_{40}$ ratio and the p3-Alcα(minor/major) ratio altered covariantly. We observed a similar trend in the CSF of SAD subjects [64], suggesting a pathogenic malfunction of the γ-secretase complex in sporadic AD patients similar to that observed in FAD-linked PS mutation pathogenesis.

Taken together, these studies indicate that (i) Alc and APP form a complex and are coordinately metabolized; (ii) Alc and APP both function as cargo receptors of kinesin-1; (iii) Alcs largely colocalize with APP in healthy neurons and also in brains of AD patients; (iv) Alc and APP are proteolytically processed by the same proteases in a highly similar fashion; and (v) p3-Alc, the metabolic product of Alc, may reveal pathogenic malfunction of γ-secretase in SAD. The relationship between lipid composition and AD will be further clarified by additional studies on Alc metabolism and p3-Alc generation.

## References

1 Alzheimer, A., Stelzmann, R.A., Schnitzlein, H.N., and Murtagh, F.R. (1995) An English translation of Alzheimer's 1907 paper, "Uber eine eigenartige Erkankung der Hirnrinde". *Clin. Anat.*, **8** (6), 429–431.

2 Perry, G., Avila, J., Kinoshita, J., and Smith, M.A. (eds.) (2006) *Alzheimer's Disease; a Century of Scientific and Clinical Research*, IOS Press, Amsterdam.

3 Goedert, M. and Spillantini, M.G. (2006) A century of Alzheimer's disease. *Science*, **314** (5800), 777–781.

4 Shankar, G.M., Li, S., Methta, T.H., Garcia-Munoz, A., Shepardson, N.E., Smith, I., Brett, F.M., Farrell, M.A., Rowan, M.J., Lemere, C.A., Regan, C.M., Walsh, D.M., Sabatini, B.L., and Selkoe, D.J. (2008) Amyloid-β protein dimers isolated directly from Alzheimer's brain impair synaptic plasticity and memory. *Nat. Med.*, **14** (8), 837–842.

5 Suzuki, T. and Nakaya, T. (2008) Regulation of amyloid β-protein precursor by phosphorylation and protein interactions. *J. Biol. Chem.*, **283** (44), 29633–29637.

6 Citron, M., Oltesdorf, T., Haass, C., McConlogue, L., Hung, A.Y., Seubert, P., Vigo-Pelfrey, C., Lieberburg, I., and Selkoe, D.J. (1992) Mutation of the β-amyloid precursor protein in familial Alzheimer's disease β-protein production. *Nature*, **360** (6405), 672–674.

7 Bergmans, B.A. and De Strooper, B. (2010) Gamma-secretases: from cell biology to therapeutic strategies. *Lancet Neurol.*, **9** (2), 215–226.

8 Thinakaran, G. and Koo, E. (2008) Amyloid precursor protein trafficking, processing, and function. *J. Biol. Chem.*, **283** (44), 29615–29619.

9 Araki, Y., Miyagi, N., Kato, N., Yoshida, T., Wada, S., Nishimura, M., Komano, H., Yamamoto, T., De Strooper, B., Yamamoto, K., and Suzuki, T. (2004) Coordinated metabolism of alcadein and amyloid β-protein precursor regulates FE65-dependent gene transactivation. *J. Biol. Chem.*, **279** (23), 24343–24354.

10 Hata, S., Fujishige, S., Araki, Y., Kato, N., Araseki, M., Nishimura, M., Hartmann, D., Saftig, P., Fahrenholz, F., Taniguchi, M., Urakami, K., Akatsu, H., Martins, R.N., Yamamoto, K., Maeda, M., Yamamoto, T., Nakaya, T., Gandy, S., and Suzuki, T. (2009) Alcadein cleavages by APP α- and γ-secretases generate small peptides p3-Alcs indicating Alzheimer disease-related γ-secretase dysfunction. *J. Biol. Chem.*, **284** (152), 36024–36033.

11 Dietschy, J.M. and Turley, S.D. (2004) Thematic review series: brain lipids. Cholesterol metabolism in the central nervous system during early development and in the mature animal. *J. Lipid. Res.*, **45** (8), 1375–1397.

12 Martins, J.I., Berger, T., Sharman, J.M., Verdile, G., Fuller, J.S., and Martins, N.R. (2009) Cholesterol metabolism and transport in the pathogenesis of Alzheimer's disease. *J. Neurochem.*, **111** (6), 1275–1308.

13 Puglielli, L., Tanzi, R.E., and Kovacs, D.M. (2003) Alzheimer's disease: the cholesterol connection. *Nat. Neurosci.*, **6** (4), 345–351.

14 Corder, E.H., Saunders, A.M., Strittmatter, W.J., Schmechel, D.E., Gaskell, P.C., Small, G.W., Roses, A.D., Haines, J.L., and Pericak Vance, M.A. (1993) Gene dose of apolipoprotein E type 4 allele and the risk of Alzheimer's disease in late onset families. *Science*, **261** (5123), 921–923.

15 Bu, G. (2009) Apolipoprotein E and its receptors in Alzheimer's disease: pathways, pathogenesis and therapy. *Nat. Rev. Neurosci.*, **10** (5), 333–344.

16 Kang, D.E., Pietrzik, C.U., Baum, L., Chevallier, N., Merriam, D.E., Kounnas, M.Z., Wagner, S.L., Troncoso, J.C., Kawas, C.H., Katzman, R., and Koo, E.H. (2000) Modulation of amyloid beta-protein clearance and Alzheimer's disease susceptibility by the LDL receptor-related protein pathway. *J. Clin. Invest.*, **106** (9), 1159–1166.

17 Kang, D.E., Saitoh, T., Chen, X., Xia, Y., Masliah, E., Hansen, L.A., Thomas, R.G., Thal, L.J., and Katzman, R. (1997) Genetic association of the low-density lipoprotein receptor-related protein gene (LRP), an apolipoprotein E receptor, with late-onset Alzheimer's disease. *Neurology*, **49**, 56–61.

18 Andersen, M.O. and Willnow, E.T. (2006) Lipoprotein receptors in Alzheimer's disease. *Trends Neurosci.*, **29** (12), 687–694.

19 Offe, K., Dodson, S.E., Shoemaker, J.T., Fritz, J.J., Gearing, M., Levey, A.I., and Lah, J.J. (2006) The lipoprotein receptor

LR11 regulates amyloid beta production and amyloid precursor protein traffic in endosomal compartments. *J. Neurosci.*, **26** (5), 1596–1603.

20 Bu, G. and Marzolo, M. (2009) Lipoprotein receptors and cholesterol in APP trafficking and proteolytic processing, implications for Alzheimer's disease. *Semin. Cell Dev. Biol.*, **20** (2), 191–200.

21 Hancock, J.F. (2006) Lipid rafts: contentious only from simplistic standpoints. *Nat. Rev. Mol. Cell. Biol.*, **7** (6), 456–462.

22 Brown, D.A. and London, E. (1998) Functions of lipid rafts in biological membranes. *Annu. Rev. Cell. Dev. Biol.*, **14**, 111–136.

23 Brugger, B., Sandhoff, R., Wegehingel, S., Gorgas, K., Malsam, J., Helms, J.B., Lehmann, W.D., Nickel, W., and Wieland, F.T. (2000) Evidence for segregation of sphingomyelin and cholesterol during formation of COPI-coated vesicles. *J. Cell. Biol.*, **151** (3), 507–518.

24 Mukherjee, S. and Maxfield, F.R. (2000) Role of membrane organization and membrane domains in endocytic lipid trafficking. *Traffic*, **1** (3), 203–211.

25 Helms, J.B. and Zurzolo, C. (2004) Lipids as targeting signals: lipid rafts and intracellular trafficking. *Traffic*, **5** (4), 247–254.

26 Cordy, J.M., Hussain, I., Dingwall, C., Hooper, N.M., and Turner, A.J. (2003) Exclusively targeting beta-secretase to lipid rafts by GPI-anchor addition up-regulates beta-site processing of the amyloid precursor protein. *Proc. Natl. Acad. Sci. USA*, **100** (20), 11735–11740.

27 Ehehalt, R., Keller, P., Haass, C., Thiele, C., and Simons, K. (2003) Amyloidogenic processing of the Alzheimer beta-amyloid precursor protein depends on lipid rafts. *J. Cell. Biol.*, **160** (1), 113–123.

28 Kalvodova, L., Kahya, N., Schwille, P., Ehehalt, R., Verkade, P., Drechsel, D., and Simons, K. (2005) Lipids as modulators of proteolytic activity of BACE: involvement of cholesterol, glycosphingolipids, and anionic phospholipids *in vitro*. *J. Biol. Chem.*, **280** (44), 36815–36823.

29 Melkonian, K.A., Ostermeyer, A.G., Chen, J.Z., Roth, M.G., and Brown, D.A. (1999) Role of lipid modifications in targeting proteins to detergent-resistant membrane rafts. Many raft proteins are acylated, while few are prenylated. *J. Biol. Chem.*, **274** (6), 3910–3917.

30 Benjannet, S., Elagoz, A., Wickham, L., Mamarbachi, M., Munzer, J.S., Basak, A., Lazure, C., Cromlish, J.A., Sisodia, S., Checler, F., Chretien, M., and Seidah, N.G. (2001) Post-translational processing of beta-secretase (beta-amyloid-converting enzyme) and its ectodomain shedding. The pro- and transmembrane/cytosolic domains affect its cellular activity and amyloid-beta production. *J. Biol. Chem.*, **276** (14), 10879–10887.

31 Vetrivel, K.S., Meckler, X., Chen, Y., Nguyen, P.D., Seidah, N.G., Vassar, R., Wong, P.C., Fukata, M., Kounnas, M.Z., and Thinakaran, G. (2009) Alzheimer disease Abeta production in the absence of S-palmitoylation-dependent targeting of BACE1 to lipid rafts. *J. Biol. Chem.*, **284** (6), 3793–3803.

32 Miller, C.C., McLoughlin, D.M., Lau, K.F., Tennant, M.E., and Rogelj, B. (2006) The X11 proteins, Abeta production and Alzheimer's disease. *Trends Neurosci.*, **29** (5), 280–285.

33 Tomita, S., Ozaki, T., Taru, H., Oguchi, S., Takeda, S., Yagi, Y., Sakiyama, S., Kirino, Y., and Suzuki, T. (1999) Interaction of a neuron-specific protein containing PDZ domains with Alzheimer's amyloid precursor protein. *J. Biol. Chem.*, **274** (4), 2243–2254.

34 Araki, Y., Tomita, S., Yamaguchi, H., Miyagi, N., Sumioka, A., Kirino, Y., and Suzuki, T. (2003) Novel cadherin-related membrane proteins, alcadeins, enhance the X11-like protein-mediated stabilization of amyloid beta-protein precursor metabolism. *J. Biol. Chem.*, **278** (49), 49448–49458.

35 Taru, H. and Suzuki, T. (2009) Regulation of the physiological function and metabolism of AβPP by AβPP binding

proteins. *J. Alzheimers Dis.*, **18** (2), 253–265.

36 Sano, Y., Syuzo Takabatake, A., Nakaya, T., Saito, Y., Tomita, S., Itohara, S., and Suzuki, T. (2006) Enhanced amyloidogenic metabolism of the amyloid beta-protein precursor in the X11L-deficient mouse brain. *J. Biol. Chem.*, **281** (49), 37853–37860.

37 Lee, J.H., Lau, K.F., Perkinton, M.S., Standen, C.L., Shemilt, S.J., Mercken, L., Cooper, J.D., McLoughlin, D.M., and Miller, C.C. (2003) The neuronal adaptor protein X11alpha reduces Abeta levels in the brains of Alzheimer's APPswe Tg2576 transgenic mice. *J. Biol. Chem.*, **278** (47), 47025–47029.

38 Lee, J.H., Lau, K.F., Perkinton, M.S., Standen, C.L., Rogelj, B., Falinska, A., McLoughlin, D.M., and Miller, C.C. (2004) The neuronal adaptor protein X11beta reduces amyloid beta-protein levels and amyloid plaque formation in the brains of transgenic mice. *J. Biol. Chem.*, **279** (47), 49099–49104.

39 Saito, Y., Sano, Y., Vassar, R., Gandy, S., Nakaya, T., Yamamoto, T., and Suzuki, T. (2008) X11 proteins regulate the translocation of amyloid beta-protein precursor (APP) into detergent-resistant membrane and suppress the amyloidogenic cleavage of APP by beta-site-cleaving enzyme in brain. *J. Biol. Chem.*, **283** (51), 35763–35771.

40 Sakurai, T., Kaneko, K., Okuno, M., Wada, K., Kashiyama, T., Shimizu, H., Akagi, T., Hashikawa, T., and Nukina, N. (2008) Membrane microdomain switching: a regulatory mechanism of amyloid precursor protein processing. *J. Cell. Biol.*, **183** (2), 339–352.

41 Cruts, M. and Brouwers, N. Alzheimer Disease and Frontotemporal Dementia Mutation Database, http://www.molgen.ua.ac.be/ADMutations/.

42 Beel, A.J. and Sanders, C.R. (2008) Substrate specificity of gamma-secretase and other intramembrane proteases. *Cell. Mol. Life Sci.*, **65** (9), 1311–1334.

43 Struhl, G. and Adachi, A. (2000) Requirements for presenilin-dependent cleavage of notch and other transmembrane proteins. *Mol. Cell*, **6** (3), 625–636.

44 Takasugi, N., Tomita, T., Hayashi, I., Tsuruoka, M., Niimura, M., Takahashi, Y., Thinakaran, G., and Iwatsubo, T. (2003) The role of presenilin cofactors in the gamma-secretase complex. *Nature*, **422** (6930), 438–441.

45 Shah, S., Lee, S.F., Tabuchi, K., Hao, Y.H., Yu, C., LaPlant, Q., Ball, H., Dann, C.E. III, Sudhof, T., and Yu, G. (2005) Nicastrin functions as a gamma-secretase-substrate receptor. *Cell*, **122** (3), 435–447.

46 Jick, H., Zornberg, G.L., Jick, S.S., Seshadri, S., and Drachman, D.A. (2000) Statins and the risk of dementia. *Lancet*, **356** (9242), 1627–1631.

47 Wolozin, B., Kellman, W., Ruosseau, P., Celesia, G.G., and Siegel, G. (2000) Decreased prevalence of Alzheimer disease associated with 3-hydroxy-3-methylglutaryl coenzyme A reductase inhibitors. *Arch. Neurol.*, **57** (10), 1439–1443.

48 Simons, M., Keller, P., De Strooper, B., Beyreuther, K., Dotti, C.G., and Simons, K. (1998) Cholesterol depletion inhibits the generation of beta-amyloid in hippocampal neurons. *Proc. Natl. Acad. Sci. USA*, **95** (11), 6460–6464.

49 Fassbender, K., Simons, M., Bergmann, C., Stroick, M., Lutjohann, D., Keller, P., Runz, H., Kuhl, S., Bertsch, T., von Bergmann, K., Hennerici, M., Beyreuther, K., and Hartmann, T. (2001) Simvastatin strongly reduces levels of Alzheimer's disease beta-amyloid peptides Abeta 42 and Abeta 40 *in vitro* and *in vivo*. *Proc. Natl. Acad. Sci. USA*, **98** (10), 5856–5861.

50 Kojro, E., Gimpl, G., Lammich, S., Marz, W., and Fahrenholz, F. (2001) Low cholesterol stimulates the nonamyloidogenic pathway by its effect on the alpha-secretase ADAM 10. *Proc. Natl. Acad. Sci. USA*, **98** (10), 5815–5820.

51 Refolo, L.M., Pappolla, M.A., LaFrancois, J., Malester, B., Schmidt, S.D., Thomas-Bryant, T., Tint, G.S., Wang, R., Mercken, M., Petanceska, S.S., and Duff, K.E. (2001) A cholesterol-lowering drug reduces beta-amyloid pathology in a

transgenic mouse model of Alzheimer's disease. *Neurobiol. Dis.*, **8** (5), 890–899.

52 Refolo, L.M., Malester, B., LaFrancois, J., Bryant-Thomas, T., Wang, R., Tint, G.S., Sambamurti, K., Duff, K., and Pappolla, M.A. (2000) Hypercholesterolemia accelerates the Alzheimer's amyloid pathology in a transgenic mouse model. *Neurobiol. Dis.*, **7** (4), 321–331.

53 Lee, S.J., Liyanage, U., Bickel, P.E., Xia, W., Lansbury, P.T. Jr., and Kosik, K.S. (1998) A detergent-insoluble membrane compartment contains A beta *in vivo*. *Nat. Med.*, **4** (6), 730–734.

54 Parkin, E.T., Hussain, I., Karran, E.H., Turner, A.J., and Hooper, N.M. (1999) Characterization of detergent-insoluble complexes containing the familial Alzheimer's disease-associated presenilins. *J. Neurochem.*, **72** (4), 1534–1543.

55 Urano, Y., Hayashi, I., Isoo, N., Reid, P.C., Shibasaki, Y., Noguchi, N., Tomita, T., Iwatsubo, T., Hamakubo, T., and Kodama, T. (2005) Association of active gamma-secretase complex with lipid rafts. *J. Lipid Res.*, **46** (5), 904–912.

56 Wahrle, S., Das, P., Nyborg, A.C., McLendon, C., Shoji, M., Kawarabayashi, T., Younkin, L.H., Younkin, S.G., and Golde, T.E. (2002) Cholesterol-dependent gamma-secretase activity in buoyant cholesterol-rich membrane microdomains. *Neurobiol. Dis.*, **9** (1), 11–23.

57 Hur, J.Y., Welander, H., Behbahani, H., Aoki, M., Franberg, J., Winblad, B., Frykman, S., and Tjernberg, L.O. (2008) Active gamma-secretase is localized to detergent-resistant membranes in human brain. *FEBS J.*, **275** (6), 1174–1187.

58 Vetrivel, K.S., Cheng, H., Kim, S.H., Chen, Y., Barnes, N.Y., Parent, A.T., Sisodia, S.S., and Thinakaran, G. (2005) Spatial segregation of gamma-secretase and substrates in distinct membrane domains. *J. Biol. Chem.*, **280** (27), 25892–25900.

59 Vogt, L., Schrimpf, S.P., Meskenaite, V., Frischknecht, R., Kinter, J., Leone, D.P., Ziegler, U., and Sonderegger, P. (2001) Calsyntenin-1, a proteolytically processed postsynaptic membrane protein with a cytoplasmic calcium-binding domain. *Mol. Cell. Neurosci.*, **17** (1), 151–166.

60 Araki, Y., Miyagi, N., Kato, N., Yoshida, T., Wada, S., Nishimura, M., Komano, H., Yamamoto, T., De Strooper, B., Yamamoto, K., and Suzuki, T. (2007) The novel cargo alcadein induces vesicle association of kinesin-1 motor components and activates axonal transport. *EMBO J.*, **26** (6), 1475–1486.

61 Papassotiropoulos, A., Stephan, D.A., Huentelman, M.J., Hoerndli, F.J., Craig, D.W., Pearson, J.V., Huynh, K.D., Brunner, F., Corneveaux, J., Osborne, D., Wollmer, M.A., Aerni, A., Coluccia, D., Hanggi, J., Mondadori, C.R., Buchmann, A., Reiman, E.M., Caselli, R.J., Henke, K., and de Quervain, D.J. (2006) Common Kibra alleles are associated with human memory performance. *Science*, **314** (5798), 475–478.

62 Preuschhof, C., Heekeren, H.R., Li, S.C., Sander, T., Lindenberger, U., and Backman, L. (2010) KIBRA and CLSTN2 polymorphisms exert interactive effects on human episodic memory. *Neuropsychologia*, **48** (2), 402–408.

63 Cole, S.L. and Vassar, R. (2008) The role Main_Text of amyloid precursor protein processing by BACE1, the β-secretase, in Alzheimer disease pathophysiology. *J. Biol. Chem.*, **283** (4), 29621–29625.

64 Hata, S., Fujishige, S., Araki, Y., Taniguchi, M., Urakami, K., Peskind E., Akatsu, H., Araseki, M., Yamamoto, K., Martins, N. R., Maeda, M., Nishimura, M., Levey, A., Chung, K. A., Montine, T., Leverenz, J., Fagan, A., Goate, A., Bateman, R., Holtzman, D. M., Yamamoto, T., Nakaya, T., Gandy, S. and Suzuki, T. [2011] Alternative γ-secretase processing of g-secretase substrates in common forms of mild cognitive impairment and Alzheimer disease: Evidence for γ-secretase dysfunction. *Annal Neurol.* in press (Accepted manuscript online: 1 DEC 2010 03:29PM EST | DOI: 10.1002/ana.22343).

# 9
# Impaired Regulation of Glutamate Receptor Channels and Signaling Molecules by β-Amyloid in Alzheimer's Disease

*Zhen Yan*

## 9.1
### Introduction

A fundamental feature of Alzheimer's disease (AD) is the accumulation of β-amyloid (Aβ), a peptide generated from the amyloid precursor protein (APP). Although it is still unclear how Aβ contributes to the etiology and pathogenesis of AD, emerging evidence suggests that it causes "synaptic failure" before the formation of senile plaques and the occurrence of neuron death [1, 2]. Soluble oligomeric Aβ forms, rather than amyloid plaques, correlate with the severity of cognitive impairment in AD [3, 4]. Application of naturally secreted Aβ oligomers adversely affects glutamatergic synaptic transmission and plasticity [5, 6]. Neurons from transgenic mice overexpressing AD-linked mutant APP also show deficits in long-term potentiation (LTP) of glutamatergic transmission, a synaptic basis of learning and memory [7–10]. In this chapter, we summarize some of our findings on the influence of Aβ on the regulation of glutamate receptor channels and signaling molecules in cortical neurons.

## 9.2
### AMPAR-Mediated Synaptic Transmission and Ionic Current are Impaired by Aβ

Decreased expression of synaptic proteins and loss of synapses have been found with elevated Aβ levels [11–15]. This Aβ-induced synaptic dysfunction is partly attributable to the synaptic removal of AMPA (α-amino-3-hydroxy-5-methyl-4-isoxazolepropionic acid)-type glutamate receptors [16–18]. However, it is unclear how Aβ induces the loss of AMPA receptors (AMPARs) at the synapses.

Our biochemical assays [19] (Figure 9.1a) showed that the level of AMPAR subunit GluR1 at the cell surface was selectively reduced in APP transgenic mice and in Aβ-treated cortical cultures. Immunocytochemical experiments with an antibody against the extracellular domain of GluR1 in non-permeabilized cultures were also performed to test the impact of Aβ on GluR1 located at the synaptic membrane. As shown in Figure 9.1b, in untreated control cultures, most of surface GluR1 clusters (red) were co-localized with *N-methyl-D-aspartic* acid receptor (NMDAR) subunit NR1

*Lipids and Cellular Membranes in Amyloid Diseases*, First Edition. Edited by Raz Jelinek.
© 2011 Wiley-VCH Verlag GmbH & Co. KGaA. Published 2011 by Wiley-VCH Verlag GmbH & Co. KGaA.

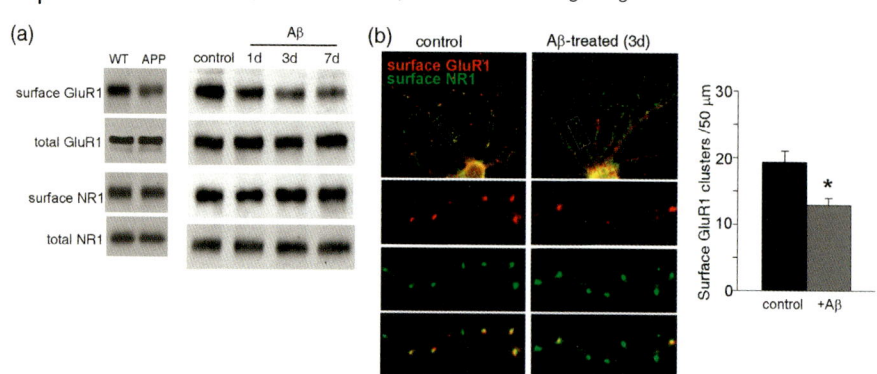

**Figure 9.1** Reduction of AMPAR surface expression by Aβ. (a) Western blots showing the surface and total GluR1 and NR1 levels in the cortex of WT versus APP mice, or in rat cortical cultures after being exposed to Aβ oligomer (1 μM, 1, 3, and 7 days). (b) Immunocytochemical images showing the co-staining of surface GluR1 and surface NR1 in cortical cultures (DIV 28) without or with Aβ (1 μM) exposure for 3 days. Shown on the right are the cumulative data (mean ± SEM) of surface GluR1 cluster density in control versus Aβ-treated cultures. $^*p < 0.01$, ANOVA (analysis of variance).

clusters at the surface (green) on dendritic spines, as indicated by numerous yellow puncta. Aβ exposure (3 days) significantly decreased surface GluR1 clusters, but not surface NR1 clusters, which led to more green puncta (NR1 only) along dendrites. These data show that Aβ decreases the number of AMPARs at the synaptic membrane.

Given the decreased level of surface GluR1 in APP mice, we investigated whether the AMPAR-mediated excitatory postsynaptic current (EPSC) is impaired in APP mice. As shown in Figure 9.2a, the amplitude of the AMPAR-EPSC evoked in frontal cortical slices was significantly decreased in APP mice compared with age-matched wild-type (WT) mice. In contrast, the amplitude of NMDAR-EPSC was not significantly changed. We further examined the direct effect of Aβ exposure on AMPAR-mediated ionic currents in cultured cortical neurons. As shown in Figure 9.2b, Aβ exposure (1–7 days) significantly decreased the AMPAR current density, but not the NMDAR current density. A short (10 min–1 h) application of Aβ failed to alter AMPAR or NMDAR current density. Taken together, these data suggest that AMPAR function is selectively impaired by Aβ, which is consistent with our biochemical and immunocytochemical results.

## 9.3
### CaMKII is Causally Involved in Aβ Impairment of AMPAR Trafficking and Function

Since CaMKII plays a key role in regulating AMPAR trafficking and function [20, 21], we wanted to establish whether the Aβ-induced decrease in AMPAR surface expression and channel currents is caused by Aβ-induced reduction of CaMKII synaptic distribution. Biochemical fractionation experiments (Figure 9.3a) showed

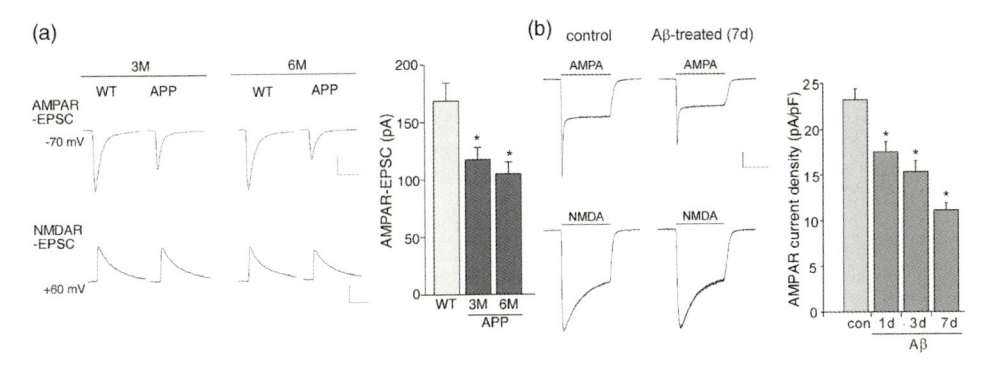

**Figure 9.2** Decrease in AMPAR synaptic responses and ionic currents by Aβ. (a) Representative AMPAR-EPSC and NMDAR-EPSC traces evoked by electrical stimulation of cortical slices from WT versus APP mice (3 months old). Shown on the right are the cumulative data on AMPAR-EPSC and NMDAR-EPSC amplitudes in WT versus APP mice (3 or 6 months old). *$p < 0.01$, ANOVA. (b) Representative whole-cell AMPAR- or NMDAR-mediated ionic current traces in cultured cortical neurons without or with Aβ (1 μM) exposure for 7 days. Shown on the right are cumulative data on AMPAR and NMDAR current densities in control versus Aβ-treated cultures. *$p < 0.01$, ANOVA.

that the synaptic pool of CaMKII is significantly decreased, whereas the cytosolic pool of CaMKII is increased, in cortical neurons from APP transgenic mice or in cortical cultures treated with Aβ oligomers. Immunocytochemical experiments (Figure 9.3b) further demonstrated that the density of CaMKII clusters at synapses is reduced by Aβ treatment.

Next, we manipulated CaMKII expression in cultured cortical neurons and examined the impact of Aβ on AMPARs. CaMKII was down-regulated by transfecting with a CaMKII siRNA or up-regulated by overexpressing WT CaMKII. As shown in Figure 9.3c, in neurons transfected with a scrambled siRNA, Aβ exposure (3 days) significantly decreased the surface GluR1 cluster density. In CaMKII siRNA-transfected neurons, the surface GluR1 cluster density was significantly reduced, and Aβ treatment failed to induce a further reduction. On the other hand, in neurons overexpressing CaMKII, the surface GluR1 cluster density was slightly increased, and the reducing effect of Aβ was blocked. Moreover, knocking down CaMKII significantly decreased the AMPAR current density, and prevented Aβ from reducing the AMPAR current density further. CaMKII overexpression slightly increased the AMPAR current density and blocked the reducing effect of Aβ on the AMPAR current density. These data suggest that Aβ decreases the number of surface AMPARs via reducing CaMKII at the synapses.

## 9.4
### PIP2 Regulation of NMDAR Currents is Lost by Aβ

The membrane phospholipid phosphatidylinositol 4,5-bisphosphate ($PIP_2$) has been implicated in the regulation of several ion channels and transporters. Recently, we

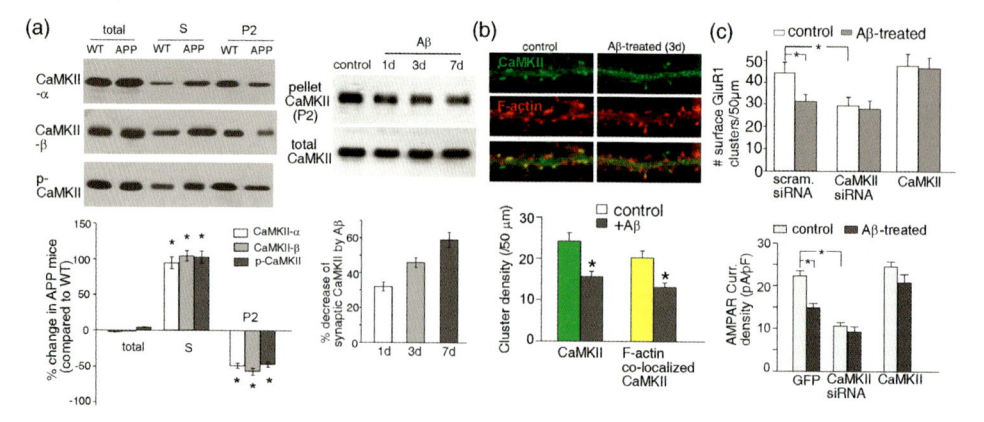

**Figure 9.3** Aberrant CaMKII subcellular distribution in AD, which is linked to Aβ regulation of AMPARs. (a) Western blots showing the expression of CaMKII (α subunit, β subunit, and auto-phosphorylated) in different subcellular fractions in the cortex from WT versus APP mice. S, the cytosolic fraction; P2, Triton-insoluble fraction in the crude synaptosome fraction, which mainly includes membrane-associated proteins in synapses. Shown below are the cumulative data on the percentage change of CaMKII in different fractions from APP mice. $^*p < 0.01$, ANOVA. Western blots and statistics on the synaptic membrane-associated CaMKII α subunit (in P2 fraction) in rat cortical cultures after being exposed to Aβ oligomer (1 μM, 1, 3, and 7 days) are also shown. (b) Immunocytochemical images showing the co-staining of CaMKII α subunit with F-actin in cortical cultures (DIV 28) without or with Aβ (1 μM, 3 days) exposure. Shown below are the cumulative data on the density of total CaMKII clusters and synaptic CaMKII clusters (co-localized with F-actin) in control versus Aβ-treated cultures. $^*p < 0.01$, ANOVA. (c) Cumulative data for immunocytochemical and electrophysiological studies showing the surface GluR1 cluster densities (left) or AMPAR current densities (right) in control or Aβ-treated neurons transfected with a scrambled siRNA, CaMKII siRNA, or WT CaMKII without (control) or with Aβ (1 μM, 3 days) exposure. $^*p < 0.01$, ANOVA.

discovered the potential mechanism by which $PIP_2$ influences the number of surface NMDARs in native neurons [22] (Figure 9.4a). Under basal conditions, $PIP_2$ synthesis is unperturbed, and cofilin remains bound to $PIP_2$ preferentially over actin, and thus is unable to depolymerize F-actin. With the actin cytoskeleton intact, NMDAR channels are stabilized at the synaptic membrane by binding to adapter proteins such as α-actinin. Activation of phospholipase C (PLC) causes the hydrolysis of $PIP_2$, leading to the release of cofilin, which now becomes available to bind to F-actin and depolymerize it. With the actin cytoskeleton disintegrated, NMDARs are internalized via clathrin-coated pits, causing the reduction of NMDAR responses.

Reduced levels of $PIP_2$ have been found in the frontal cortex of AD brains [23, 24]. Moreover, Aβ peptide is known to disrupt $PIP_2$ metabolism in a $Ca^{2+}$-dependent manner [23, 24]. Therefore, we examined whether the $PIP_2$ regulation of NMDAR channels is altered in AD-related conditions. We used the specific inhibitor of PI-4 kinase, phenylarsine oxide (PAO), which inhibits the synthesis of $PIP_2$ from phosphatidylinositol (PI), thus lowering the membrane concentration of $PIP_2$. As shown in

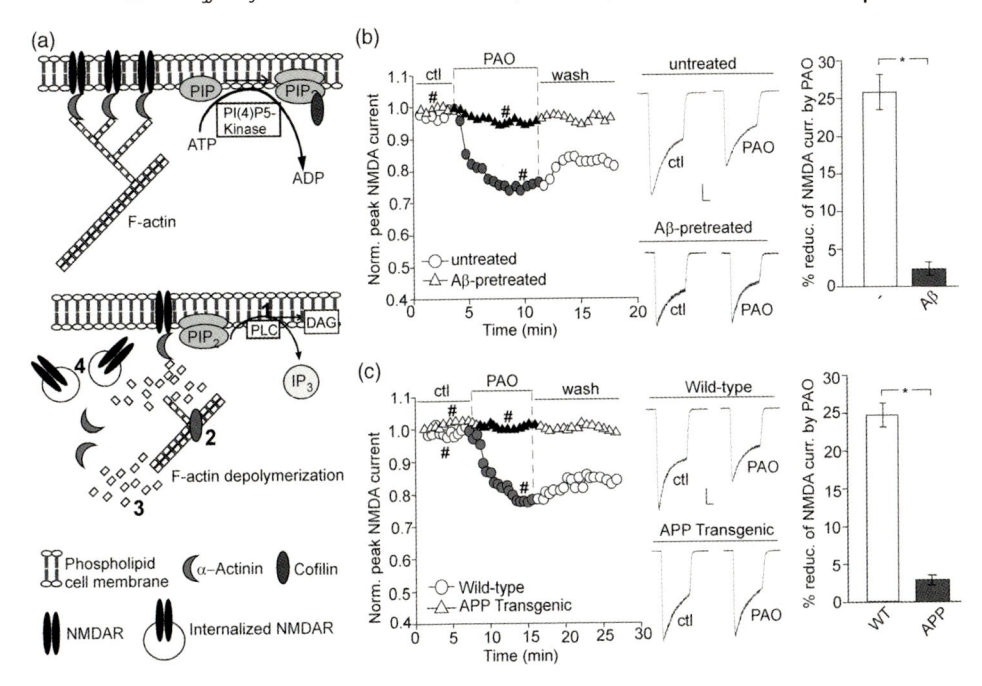

**Figure 9.4** PIP$_2$ regulation of NMDAR channels is abolished by Aβ and in APP transgenic mice. (a) A schematic model demonstrating the potential mechanism for PIP$_2$ regulation of NMDAR channels. (b, c) Plot of normalized peak NMDAR currents showing the effect of PAO (10 μM) in cultured cortical neurons pretreated with or without Aβ oligomer (1 μM, 1 h), or in neurons isolated from APP transgenic versus WT mice. Shown ion the right are the cumulative data on the percentage reduction of NMDAR currents by PAO in neurons treated with Aβ or from APP transgenic mice. *$p < 0.001$, $t$-test.

Figure 9.4b, in neurons treated with Aβ (1 h), PAO failed to reduce NMDAR currents, in contrast to the reducing effect of PAO in untreated neurons. These results are in agreement with our expectation that Aβ pretreatment disrupts basal levels of cellular PIP$_2$ [24], which might explain why no further modulation by PAO is observed.

We further validated these *in vitro* findings in the animal model of AD. As shown in Figure 9.4c, PAO failed to cause a reduction in NMDAR currents in cortical neurons from APP transgenic mice, which was significantly different from the effect of PAO in neurons from WT mice. These results suggest that the PIP$_2$ regulation of NMDAR channels is lost in AD, probably due to the disrupted PIP$_2$ metabolism by Aβ.

## 9.5
### The Effect of AChE Inhibitor on NMDAR Response is Impaired in APP Transgenic Mice

Multiple lines of evidence suggest that the central cholinergic system plays a key role in cognitive processes [25] and deterioration of the cholinergic system contributes to memory failure and cognitive decline associated with AD [26, 27]. Drugs that potentiate central cholinergic function, such as the acetylcholinesterase (AChE) inhibitor physostigmine, have been found to enhance significantly the storage of information into long-term memory and improve retrieval of information from long-term memory [28]. Although the most commonly used therapeutic strategy in AD treatment is to enhance cholinergic transmission with AChE inhibitors [29, 30], their molecular targets and cellular mechanisms remain largely unknown. Our studies found that the NMDAR, a key player implicated in the regulation of learning and memory, is one of the targets of AChE inhibitors [31].

Since Aβ has pleiotropic actions on the cholinergic system [32], including the suppression of acetylcholine (ACh) synthesis, the inhibition of ACh release, and the disruption of muscarinic receptor–G protein coupling, we examined whether the effect of AChE inhibitors on NMDAR-EPSC is altered in the mouse model of AD. As shown in Figure 9.5a, application of physostigmine caused a significantly smaller effect on NMDAR-EPSC in cortical pyramidal neurons from an APP transgenic

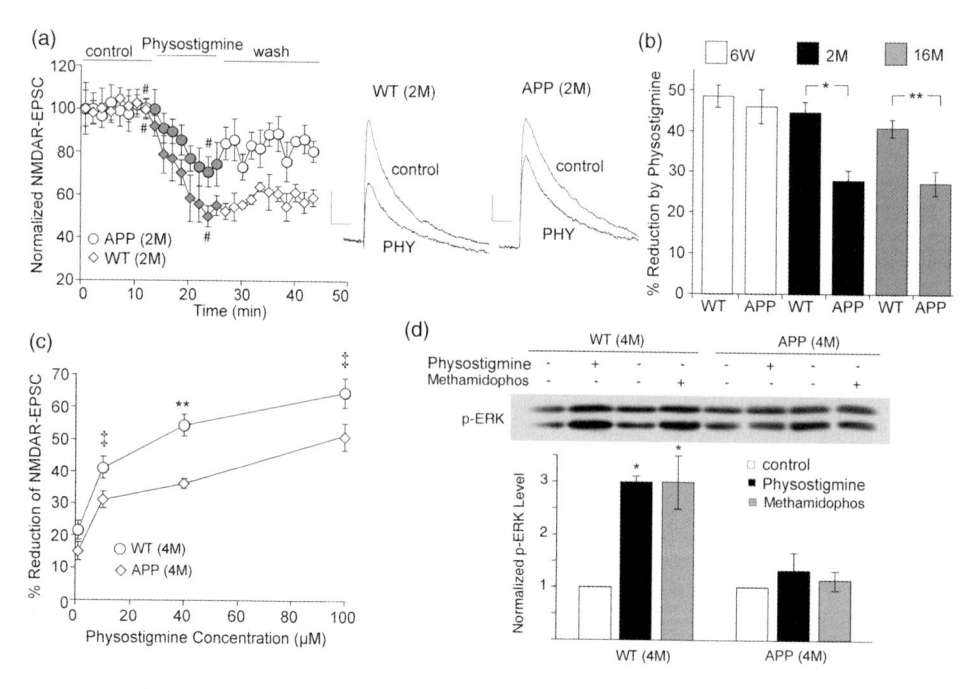

**Figure 9.5** The regulation of NMDAR-EPSC by AChE enhancers is impaired in APP transgenic mice. (a) Plot of normalized NMDAR-EPSC as a function of time and physostigmine (40 μM) application in cortical pyramidal neurons from 2 month old WT versus APP mice. Shown on the right are the representative current traces. (b) Bar plot summary of the percentage reduction of NMDAR-EPSC by physostigmine in WT and APP mice at different ages (6 weeks, 2 months, and 16 months). $^*p < 0.001$, $^{**}p < 0.005$, t-test. (c) Dose–response curves showing the effect of physostigmine on NMDAR-EPSC in 4 month old WT versus APP mice. $\ddagger p < 0.01$, $^{**}p < 0.005$, t-test. (d) Western blots and bar plot summary of phospho-ERK and total ERK in cortical slices treated without or with AChE inhibitors (physostigmine, 40 μM; methomidophos, 40 μM) from WT and APP transgenic mice (4 months old). $^*p < 0.005$, t-test.

mouse (2 months old) than an age-matched WT mouse. The percentage reduction in NMDAR-EPSC by physostigmine in three age-matched groups is summarized in Figure 9.5b. The effect of physostigmine was unchanged in the 6-week group. However, it was significantly smaller in the 2- and 16-month groups. We further compared the dose responses of physostigmine-induced reduction of NMDAR-EPSC in WT and APP transgenic mice in the 4-month group. As shown in Figure 9.5c, physostigmine produced significantly smaller effects at different doses in APP transgenic mice. The impairment of endogenous ACh effects on NMDAR-EPSC occurs at 2 months of age, supporting the notion that synaptic dysfunction is an early event in AD [1]. These results suggest that under pathological conditions, Aβ acts as a cholinergic neuromodulator to disturb the cholinergic enhancer-mediated inhibition of synaptic NMDAR responses, which could contribute to the perturbed cellular calcium homeostasis that plays an important role in the progressive neuronal loss underlying the evolving dementia [33].

We next sought to determine mechanisms underlying the impaired regulation of NMDAR responses by AChE inhibitors in the AD model. Given the role of ERK (extracellular signal-regulated kinase) in this regulatory event [31], we examined the activation of ERK by AChE inhibitors in APP transgenic mice. As shown in Figure 9.5d, both physostigmine and methomidophos (10 min treatment) significantly induced the activation of ERK (as indicated by phosphorylated ERK) in cortical slices from WT mice; however, this effect was significantly attenuated in APP mice. This suggests that Aβ impairs the regulation of NMDAR responses by AChE inhibitors via the interference of ERK activation.

## 9.6
## Aβ Impairs PKC-Dependent Signaling and Functions

Our previous study found that treatment with Aβ diminished two PKC-dependent functions of the metabotropic glutamate receptor (mGluR): the group I mGluR-induced enhancement of spontaneous IPSC (inhibitory postsynaptic current) amplitude, and the group II mGluR-induced increase of NMDAR currents [34]. In Aβ-treated slices, both group I and II mGluR agonists failed to activate protein kinase C (PKC). These results suggest that the Aβ impairment of mGluR functions is probably due to the interference of mGluR activation of PKC. Consistently, an earlier study of ours also demonstrated that the muscarinic regulation of GABAergic transmission in cortical pyramidal neurons is impaired in APP transgenic mice, which is due to the loss of PKC activation by mAChRs [35].

To find out how PKC signaling is impaired by Aβ, we turned to RACK1 (receptor for activated C-kinase 1), an anchoring protein that shuttles activated PKC to cellular membranes, which plays an important role in PKC-mediated signal transduction pathways [36, 37]. Several reports have shown that RACK1 is decreased by ∼50% in membrane fractions of aging rat brains [38–40]. Reduced RACK1 has also been found in postmortem brains of AD patients [41]. This suggests that loss of RACK1 may contribute to decreased PKC activity in the aging brain or AD.

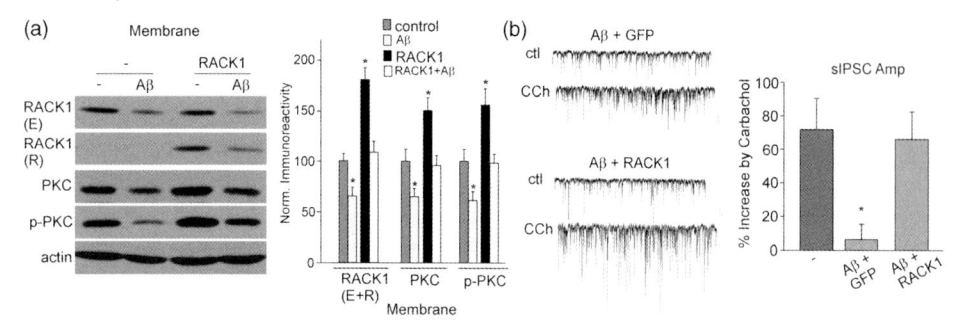

**Figure 9.6** Aβ decreases RACK1 and PKC in membrane fractions, and RACK1 expression restores the Aβ-induced loss of PKC functions *in vivo*. (a) Western blots showing endogenous (E) and recombinant (R) RACK1, PKC, p-PKC, and actin in cytosolic and membrane fractions from cortical cultures (uninfected or infected with GFP-RACK1 virus) treated without or with Aβ oligomer (1 μM, 48 h). Shown on the right are the quantification data. $^*p < 0.01$, ANOVA. (b) Traces of sIPSC showing the effect of carbachol (CCh, 20 μM) in cortical neurons from Aβ (1 μM)-injected rats that were infected with GFP or GFP-RACK1 Sindbis viruses. Shown on the right are the cumulative data. $^*p < 0.01$, ANOVA.

Our biochemical assays [42] found that Aβ treatment (48 h) induced a marked reduction in RACK1 in the membrane fraction of cortical cultures. The distribution of PKC and phospho-protein kinase C (p-PKC) (activated) in the membrane fraction was also decreased by Aβ treatment. Overexpression of RACK1 restored the Aβ-induced loss of the membrane level of PKC and p-PKC (Figure 9.6a).

To examine the impact of Aβ and RACK1 *in vivo*, we delivered oligomeric Aβ to rat frontal cortex via a stereotaxic injection. Transient RACK1 expression *in vivo* was achieved by injecting the green fluorescent protein (GFP)-RACK1 Sindbis virus. After 2–3 days of infection, the GFP or GFP-RACK1 virus was efficiently expressed in neurons at the proximity of injected sites, and the GFP + neurons showed normal morphology. In cortical pyramidal neurons without Aβ injection, application of the muscarinic receptor agonist carbachol significantly increased the sIPSC (spontaneous inhibitory postsynaptic current) amplitude, an effect dependent on PKC activation [35]. However, in cortical neurons from Aβ-injected animals, the enhancing effect of carbachol on sIPSC amplitudes was largely abolished, suggesting the loss of PKC activation. Interestingly, in RACK1-infected cortical neurons from Aβ-injected animals, the enhancing effect of carbachol on sIPSC amplitudes was restored (Figure 9.6b). These data indicate that *in vivo* RACK1 expression could rescue the Aβ-induced impairment of mAChR/PKC signaling in rat cortical neurons.

## 9.7
### Conclusion

AD is a multi-factorial disorder that involves complex changes at the molecular and cellular level. The current "Aβ hypothesis" proposes that the gradual accumulation of soluble Aβ oligomers initiates a slow but deadly cascade that adversely affects synaptic function [43]. This chapter has summarized several of the Aβ-induced impairments in glutamatergic transmission and regulation, which provides a potential mechanism for the synaptic failure that may underlie cognitive deficits at the early stage of AD. Key molecules involved in this process, such as CaMKII, $PIP_2$, ERK, PKC and RACK1, could be new therapeutic targets for AD treatment.

## References

1 Selkoe, D.J. (2002) Alzheimer's disease is a synaptic failure. *Science*, 298, 789–791.
2 Selkoe, D.J. and Schenk, D. (2003) Alzheimer's disease: molecular understanding predicts amyloid-based therapeutics. *Annu. Rev. Pharmacol. Toxicol.*, 43, 545–584.
3 Lue, L.F., Kuo, Y.M., Roher, A.E., Brachova, L., Shen, Y., Sue, L., Beach, T., Kurth, J.H., Rydel, R.E., and Rogers, J. (1999) Soluble amyloid beta peptide concentration as a predictor of synaptic change in Alzheimer's disease. *Am. J. Pathol.*, 155, 853–862.
4 McLean, C.A., Cherny, R.A., Fraser, F.W., Fuller, S.J., Smith, M.J., Beyreuther, K., Bush, A.I., and Masters, C.L. (1999) Soluble pool of Abeta amyloid as a determinant of severity of neurodegeneration in Alzheimer's disease. *Ann. Neurol.*, 46, 860–866.
5 Walsh, D.M., Klyubin, I., Fadeeva, J.V., Cullen, W.K., Anwyl, R., Wolfe, M.S., Rowan, M.J., and Selkoe, D.J. (2002) Naturally secreted oligomers of amyloid beta protein potently inhibit hippocampal long-term potentiation *in vivo*. *Nature*, 416, 535–539.
6 Cleary, J.P., Walsh, D.M., Hofmeister, J.J., Shankar, G.M., Kuskowski, M.A., Selkoe, D.J., and Ashe, K.H. (2005) Natural oligomers of the amyloid-beta protein specifically disrupt cognitive function. *Nat. Neurosci.*, 8, 79–84.
7 Chapman, P.F., White, G.L., Jones, M.W., Cooper-Blacketer, D., Marshall, V.J., Irizarry, M., Younkin, L., Good, M.A., Bliss, T.V., Hyman, B.T. *et al.* (1999) Impaired synaptic plasticity and learning in aged amyloid precursor protein transgenic mice. *Nat. Neurosci.*, 2, 271–276.
8 Oddo, S., Caccamo, A., Shepherd, J.D., Murphy, M.P., Golde, T.E., Kayed, R., Metherate, R., Mattson, M.P., Akbari, Y., and LaFerla, F.M. (2003) Triple-transgenic model of Alzheimer's disease with plaques and tangles: intracellular Abeta and synaptic dysfunction. *Neuron*, 39, 409–421.
9 Stern, E.A., Bacskai, B.J., Hickey, G.A., Attenello, F.J., Lombardo, J.A., and Hyman, B.T. (2004) Cortical synaptic integration *in vivo* is disrupted by amyloid-beta plaques. *J. Neurosci.*, 24, 4535–4540.
10 Billings, L.M., Oddo, S., Green, K.N., McGaugh, J.L., and Laferla, F.M. (2005) Intraneuronal Abeta causes the onset of early Alzheimer's disease-related cognitive deficits in transgenic mice. *Neuron*, 45, 675–688.
11 Hsia, A., Masliah, E., McConlogue, L., Yu, G., Tatsuno, G., Hu, K., Kholodenko, D., Malenka, R., Ricoll, R., and Mucke, L. (1999) Plaque-independent disruption of neuroal circuits in Alzheimer's disease mouse models. *Proc. Natl. Acad. Sci. USA*, 96, 3228–3233.
12 Lanz, T.A., Carter, D.B., and Merchant, K.M. (2003) Dendritic spine loss in the hippocampus of young PDAPP and Tg2576 mice and its prevention by the ApoE2 genotype. *Neurobiol. Dis.*, 13, 246–253.
13 Spires, T.L., Meyer-Luehmann, M., Stern, E.A., McLean, P.J., Skoch, J., Nguyen, P.T., Bacskai, B.J., and Hyman, B.T. (2005) Dendritic spine abnormalities in amyloid precursor protein transgenic mice demonstrated by gene transfer and intravital multiphoton microscopy. *J. Neurosci.*, 25, 7278–7287.
14 Snyder, E.M., Nong, Y., Almeida, C.G., Paul, S., Moran, T., Choi, E.Y., Nairn, A.C., Salter, M.W., Lombroso, P.J., Gouras, G.K., and Greengard, P. (2005) Regulation of NMDA receptor trafficking by amyloidbeta. *Nat. Neurosci.*, 8, 1051–1058.

15 Roselli, F., Tirard, M., Lu, J., Hutzler, P., Lamberti, P., Livrea, P., Morabito, M., and Almeida, O.F. (2005) Soluble beta-amyloid1–40 induces NMDA-dependent degradation of postsynaptic density-95 at glutamatergic synapses. *J. Neurosci.*, **25**, 11061–11070.

16 Kamenetz, F., Tomita, T., Hsieh, H., Seabrook, G., Borchelt, D., Iwatsubo, T., Sisodia, S., and Malinow, R. (2003) APP processing and synaptic function. *Neuron*, **37**, 925–937.

17 Almeida, C.G., Tampellini, D., Takahashi, R.H., Greengard, P., Lin, M.T., Snyder, E.M., and Gouras, G.K. (2005) Beta-amyloid accumulation in APP mutant neurons reduces PSD-95 and GluR1 in synapses. *Neurobiol. Dis.*, **20**, 187–198.

18 Hsieh, H., Boehm, J., Sato, C., Iwatsubo, T., Tomita, T., Sisodia, S., and Malinow, R. (2006) AMPAR removal underlies Abeta-induced synaptic depression and dendritic spine loss. *Neuron*, **52**, 831–843.

19 Gu, Z., Liu, W., and Yan, Z. (2009) β-Amyloid impairs AMPA receptor trafficking and function by reducing CaMKII synaptic distribution. *J. Biol. Chem.*, **284**, 10639–10649.

20 Hayashi, Y., Shi, S.H., Esteban, J.A., Piccini, A., Poncer, J.C., and Malinow, R. (2000) Driving AMPA receptors into synapses by LTP and CaMKII: requirement for GluR1 and PDZ domain interaction. *Science*, **287**, 2262–2267.

21 Poncer, J.C., Esteban, J.A., and Malinow, R. (2002) Multiple mechanisms for the potentiation of AMPA receptor-mediated transmission by alpha-$Ca^{2+}$/calmodulin-dependent protein kinase II. *J. Neurosci.*, **22**, 4406–4411.

22 Mandal, M. and Yan, Z. (2009) $PIP_2$ regulation of NMDA receptor channels in cortical neurons. *Mol. Pharmacol.*, **76**, 1349–1359.

23 Stokes, C.E. and Hawthorne, J.N. (1987) Reduced phosphoinositide concentrations in anterior temporal cortex of Alzheimer-diseased brains. *J. Neurochem.*, **48** (4), 1018–1021.

24 Berman, D.E., Dall'Armi, C., Voronov, S.V., McIntire, L.B., Zhang, H., Moore, A.Z., Staniszewski, A., Arancio, O., Kim, T.W., and Di Paolo, G. (2008) Oligomeric amyloid-beta peptide disrupts phosphatidylinositol-4,5-bisphosphate metabolism. *Nat. Neurosci.*, **11** (5), 547–554.

25 Winkler, J., Suhr, S.T., Gage, F.H., Thal, L.J., and Fisher, L.J. (1995) Essential role of neocortical acetylcholine in spatial memory. *Nature*, **375** (6531), 484–487.

26 Whitehouse, P.J., Price, D.L., Struble, R.G., Clark, A.W., Coyle, J.T., and Delon, M.R. (1982) Alzheimer's disease and senile dementia: loss of neurons in the basal forebrain. *Science*, **215** (4537), 1237–1239.

27 Coyle, JT., Price, DL., and DeLong, MR. (1983) Alzheimer's disease: a disorder of cortical cholinergic innervation. *Science*, **219** (4589), 1184–1190.

28 Davis, K.L., Mohs, R.C., Tinklenberg, J.R., Pfefferbaum, A., Hollister, L.E., and Kopell, B.S. (1978) Physostigmine: improvement of long-term memory processes in normal humans. *Science*, **201** (4352), 272–274.

29 Sitaram, N., Weingartner, H., and Gillin, JC. (1978) Human serial learning: enhancement with arecholine and choline impairment with scopolamine. *Science*, **201** (4352), 274–276.

30 Weinstock, M. (1995) The pharmacotherapy of Alzheimer's disease based on the cholinergic hypothesis: an update. *Neurodegeneration*, **4** (4), 349–356.

31 Chen, G., Chen, P., Tan, H., Ma, D., Dou, F., Feng, J., and Yan, Z. (2008) Regulation of the NMDA receptor-mediated synaptic response by acetylcholinesterase inhibitors and its impairment in an animal model of Alzheimer's disease. *Neurobiol. Aging*, **29**, 1795–1804.

32 Auld, D.S., Kar, S., and Quirion, R. (1998) Beta-amyloid peptides as direct cholinergic neuromodulators: a missing link? *Trends Neurosci.*, **21** (1), 43–49.

33 Hynd, M.R., Scott, H.L., and Dodd, P.R. (2004) Glutamate-mediated excitotoxicity and neurodegeneration in Alzheimer's disease. *Neurochem. Int.*, **45** (5), 583–595.

34 Tyszkiewicz, J.P. and Yan, Z. (2005) β-Amyloid peptides impair PKC-dependent functions of metabotropic

glutamate receptors in prefrontal cortical neurons. *J. Neurophysiol.*, **93**, 3102–3111.

35 Zhong, P., Gu, Z., Wang, X., Jiang, H., Feng, J., and Yan, Z. (2003) Impaired modulation of GABAergic transmission by muscarinic receptors in a mouse transgenic model of Alzheimer's disease. *J. Biol. Chem.*, **278**, 26888–26896.

36 Mochly-Rosen, D., Khaner, H., and Lopez, J. (1991) Identification of intracellular receptor proteins for activated protein kinase C. *Proc. Natl. Acad. Sci. USA*, **88**, 3997–4000.

37 Ron, D., Chen, C.H., Caldwell, J., Jamieson, L., Orr, E., and Mochly-Rosen, D. (1994) Cloning of an intracellular receptor for protein kinase C: a homolog of the beta subunit of G proteins. *Proc. Natl. Acad. Sci. USA*, **91**, 839–843.

38 Pascale, A., Fortino, I., Govoni, S., Trabucchi, M., Wetsel, W.C., and Battaini, F. (1996) Functional impairment in protein kinase C by RACK1 (receptor for activated C kinase 1) deficiency in aged rat brain cortex. *J. Neurochem.*, **67**, 2471–2477.

39 Battaini, F., Pascale, A., Paoletti, R., and Govoni, S. (1997) The role of anchoring protein RACK1 in PKC activation in the ageing rat brain. *Trends Neurosci.*, **20**, 410–415.

40 McCahill, A., Warwicker, J., Bolger, G.B., Houslay, M.D., and Yarwood, S.J. (2002) The RACK1 scaffold protein: a dynamic cog in cell response mechanisms. *Mol. Pharmacol.*, **62**, 1261–1273.

41 Battaini, F., Pascale, A., Lucchi, L., Pasinetti, G.M., and Govoni, S. (1999) Protein kinase C anchoring deficit in postmortem brains of Alzheimer's disease patients. *Exp. Neurol.*, **159**, 559–564.

42 Liu, W., Dou, F., Feng, J., and Yan, Z. (2009) RACK1 is involved in β-amyloid impairment of muscarinic regulation of GABAergic transmission. *Neurobiol. Aging*, in press, Epub ahead of print.

43 Haass, C. and Selkoe, D.J. (2007) Soluble protein oligomers in neurodegeneration: lessons from the Alzheimer's amyloid beta-peptide. *Nat. Rev. Mol. Cell Biol.*, **8**, 101–112.

# 10
# Membrane Changes in BSE and Scrapie

*Cecilie Ersdal, Gillian McGovern, and Martin Jeffrey*

## 10.1
## Prion Diseases

Prion diseases or transmissible spongiform encephalopathies (TSEs) are fatal neurodegenerative diseases where vacuolation in the central nervous system (CNS) is a histopathological hallmark. TSEs include among others Creutzfeldt–Jakob disease (CJD) and variant CJD (vCJD) in humans, bovine spongiform encephalopathy (BSE) in cattle, scrapie in sheep, chronic wasting disease (CWD) in cervids, and feline spongiform encephalopathy in felids (reviewed in [1]). These diseases are transmissible within or, in some cases, between mammalian species. In humans they can occur in sporadic, infectious, or familial forms. There is a genetic predisposition for many of the prion diseases, including infectious forms. BSE appeared for the first time in the UK in 1986 [2] and it is likely [3] that the disease in cattle led to the emergence of vCJD primarily in young people [4]. Scrapie is an old disease described 250 years ago. The UK sheep population has most likely been heavily exposed to BSE. Although there are no reports of natural BSE transmission, sheep are highly susceptible to experimental cattle BSE when infected by the oral route [5]. The distribution of infectivity in different tissues varies between the different prion diseases. As in scrapie, but unlike BSE in cattle, sheep experimentally infected with BSE have significant infectivity and evidence of disease in many visceral tissues.

## 10.2
## The Cellular Prion Protein (PrP$^c$) and Conversion to Disease-Associated Prion Protein (PrP$^d$)

Prion diseases are associated with conversion of the normal cellular prion protein (PrP$^c$), which is essential for transmission and appearance of disease [6], into disease-associated isoforms that may accumulate in the CNS, lymphoid tissues, and a range of viscera [7, 8]. These disease-associated accumulations of the prion protein can be detected in tissue sections by immunohistochemistry, or in tissue homogenates by

*Lipids and Cellular Membranes in Amyloid Diseases*, First Edition. Edited by Raz Jelinek.
© 2011 Wiley-VCH Verlag GmbH & Co. KGaA. Published 2011 by Wiley-VCH Verlag GmbH & Co. KGaA.

immunoblotting after partial protease digestion, and are often referred to as disease-specific prion protein ($PrP^d$) and protease-resistant prion protein ($PrP^{res}$), respectively. The "protein only hypothesis" is currently the most accepted model for the nature of the TSE agent [9, 10]. This hypothesis suggests that the infectious agent or prion is composed mainly of infectious PrP molecules or aggregates and is devoid of nucleic acid. However, it has not yet been conclusively shown that the infectious prions do not contain other molecules, and under appropriate conditions both $PrP^d$ and $PrP^{res}$ may be separated from infectivity. $PrP^d$ and $PrP^{res}$ are therefore only surrogate markers of infectivity [11].

$PrP^c$ is a cell-surface sialoglycoprotein that is ubiquitously expressed, but is particularly abundant in neurons [12]. Although three different membrane topologies of $PrP^c$ have been reported, the majority of $PrP^c$ follows a secretory pathway during synthesis that involves removal of the N-terminal signal peptide and addition of a C-terminal glycophosphatidylinositol (GPI) anchor as it transits through the endoplasmic reticulum (ER). $PrP^c$ is then transported via the Golgi to the cell membrane, where it is inserted into the outer leaflet of the plasma membrane by the GPI anchor. In common with many other GPI-anchored proteins, $PrP^c$ is thought to reside predominantly in membrane rafts rich in sphingolipids and cholesterol. Cytosolic $PrP^c$ has been reported, but it is not clear whether this form has any physiological significance (for a detailed review of $PrP^c$ trafficking, see [13]).

Membrane rafts appears to be important in distributing $PrP^c$ to distinct regions of the plasma membrane and controlling its internalization and recycling. When cultured cells are depleted of sphingolipids and cholesterol, $PrP^c$ is redistributed across the whole of the plasma membrane [14]. In developing hippocampal neurons, $PrP^c$ is equally distributed along all processes. However, in fully mature neurons, $PrP^c$ is restricted to axons, apparently because of increased partitioning of $PrP^c$ into membrane rafts [14]. $PrP^c$ appears to leave rafts *en route* to endocytosis. Although $PrP^c$ situated on the exterior of the plasma membrane lacks a transmembrane signaling domain, a number of studies have shown that in neuronal cell lines $PrP^c$ translocates to coated pits where endocytosis is mediated by clathrin mechanisms probably involving an as yet unidentified transmembrane signaling partner (Figure 10.1a) [15, 16]. In other non-neuronal cell lines, $PrP^c$ is internalized by non-clathrin, or caveolin-mediated pathways [17, 18]. Caveolin-mediated endocytosis is probably not relevant to brain as most neuronal cells do not express caveolin-1 [19]. Following internalization, $PrP^c$ is found in both the Golgi and the recycling endosomal compartment of SN56 cells derived from septum neurons. In these cells, internalized $PrP^c$ accumulates in Rab5-positive early endosomes, probably via a dynamin-dependent clathrin-mediated endocytic pathway, and is targeted to the recycling endosomal compartment. These observations indicate that trafficking of $PrP^c$ is not determined by the GPI anchor and that in contrast to other GPI-anchored proteins, $PrP^c$ is delivered to classic endosomes after internalization [20].

Several morphological studies of the localization of $PrP^c$ have been performed on rodent, simian, and human nervous tissues. By light microscopy, $PrP^c$ is readily detected within the cytoplasm of neurons of the enteric nervous system [21, 22],

## 10.2 The Cellular Prion Protein (PrP$^c$) and Conversion to Disease-Associated Prion Protein (PrP$^d$)

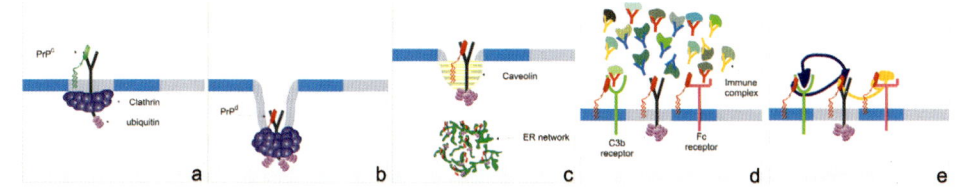

**Figure 10.1** Diagrams illustrating membrane–molecular interactions of PrP$^d$ in different cell types. (a) Normal PrP$^c$ is internalized from non-raft membrane domains. PrP$^c$ requires a transmembrane ligand to signal to clathrin and ubiquitin for internalization. (b) PrP$^d$ is internalized from neuronal membranes by a clathrin-mediated endocytic mechanism. In contrast to normal PrP$^c$, the transmembrane ligand–clathrin complex is excessively ubiquitinated and inefficiently excised from the membrane. (c) PrP$^d$ is internalized from TBM membranes with excess ubiquitin via a non-clathrin mechanism, possibly caveolin. Although efficiently excised from plasma membranes, the PrP$^d$–ligand–endocytotic protein–ubiquitin complex elicits fusion of endoplasmic reticular membranes. (d) PrP$^d$ is not internalized from the plasma membrane of FDCs, although it is ubiquitinated putatively via a non-internalized transmembrane ligand. PrP$^d$ also facilitates Fc- or C3b-mediated immune complex retention. (e) Diagram of proposed PrP$^d$–membrane interactions, suggesting, as previously proposed for PrP$^c$ [13], that it does not undergo interaction with a single protein, but with many molecular partners.

but in the CNS it is found only within a minor subpopulation of neuronal cell bodies [23, 24] with most of the expression being detected in the neuropil [25]. Results of subcellular morphological studies fall into two distinct groups. The findings of one body of research largely replicates cell biology findings and demonstrates PrP$^c$ in the Golgi and endosomes and on plasma membranes of neurons and glial cells [24, 26–28]. In contrast, other studies have described PrP$^c$ localization in synaptic boutons without any plasma membrane labeling [29–31]. The reasons for these discrepancies are not easily resolved, as two well-conducted studies using very similar pre-embedding immuno-electron microscopy methods produced conflicting results [28, 30]. Different monoclonal antibodies reportedly have different affinities for or capabilities of detecting different glycosylated forms of PrP$^c$ [32], and some PrP$^c$ glycoforms may be selectively transported along axons [33]. It is therefore possible that different antibodies used in different studies may be detecting different glycosylation variants of PrP$^c$ in different subcellular localizations. In addition, different tissue fixation methods may influence whether PrP$^c$ remains at the cell surface or diffuses away from it. Hence apparently minor technical differences between studies may determine the outcome of subcellular localization investigations.

Considerably less is known about the trafficking or localization of PrP$^c$ in non-neuronal tissues. Although PrP$^c$ expression is generally of significantly lower magnitude outside the CNS, abundant PrP$^c$ is detected in association with follicular dendritic cells (FDCs) in secondary germinal centers [34, 35]. Detailed morphological studies of the subcellular localizations of PrP$^c$ in non-nervous tissues are virtually absent, although confocal studies of human lymphoid cells have confirmed its expected location with respect to the plasma membrane [36].

The conversion from $PrP^c$ to $PrP^{res}$ involves a major conformational change from an α-helical rich structure to a β-sheet-rich isoform that results in amorphous aggregates of oligomers and/or fibrillar amyloid deposits of $PrP^{res}$ [37]. $PrP^c$ and possible cofactors such as sulfated glycans [38, 39] and the laminin receptor [40] may induce $PrP^c$ conversion via a seeded polymerization mechanism which results in $PrP^c$ joining the oligomeric or polymeric $PrP^{res}$ forms and acquiring protease resistance [41]. In scrapie-infected cells, pulse chase experiments suggest that the conversion from $PrP^c$ occurs at the plasma membrane [42], probably in rafts [43]. Both $PrP^c$ and $PrP^{res}$ are localized to rafts [44], but the membrane association is probably different. $PrP^c$ is readily released from the cell surface by bacterial phosphatidylinositol-specific phospholipase, whereas $PrP^{res}$ is not, suggesting that either conformational changes or the aggregation state of $PrP^{res}$ have blocked the enzyme cleavage site [45]. The GPI anchor seems to influence the conversion of $PrP^c$ to $PrP^{res}$, although results are conflicting. In model raft membranes, $GPI^+$ $PrP^c$ resists conversion by exogenous $PrP^{res}$, whereas $GPI^-$ $PrP^c$ is converted to $PrP^{res}$ [46]. In this cell-free system, conversion of $PrP^c$ requires that $PrP^{res}$ is inserted into the model membrane [46, 47]. *In vitro* studies have shown that the GPI anchor of $PrP^c$ is important in both the establishment and persistence of prion infection, but in an acute infection $PrP^{res}$ formation is possible with anchorless $PrP^c$ [48]. However, transgenic mice expressing anchorless $PrP^c$ are susceptible to scrapie and accumulate $PrP^d$ [49]. Disruption of rafts by squalestatin results in redistribution of $PrP^c$ to other areas of the cell membrane and subsequent $PrP^{res}$ formation is dose-dependently reduced, indicating that cellular components in rafts are required for $PrP^{res}$ conversion [50]. Using recombinant PrP in model membranes, it was shown that recombinant PrP rich in α-helical content were structurally protected in rafts. When recombinant PrP rich in β-sheets was introduced, α-PrP was converted to amyloid fibrils, demonstrating that rafts are a favorable environment for the conversion of $PrP^c$ into abnormal isoforms [51].

## 10.3
### $PrP^d$ Accumulation in the Central Nervous System and Lymphatic Tissues

Some of the pathological features of BSE and scrapie and associated immunohistochemistry are summarized in Table 10.1. In the CNS of animals with scrapie and classical BSE, $PrP^d$ is found as diffuse neuropil deposits that are not aggregated in amyloid plaques, or it is located within the neuronal perikaryonal cytoplasm (Figure 10.2a). $PrP^d$ may also accumulate around neuronal perikarya, dendrites, and glial cells (Figure 10.2a) [7, 52]. At the subcellular level, $PrP^d$ is localized to the plasma membranes of neurons and glial cells, and within lysosomes in the perikaryonal cytoplasm in BSE [53], ovine [54], and murine [55, 56] scrapie. In classical BSE and sheep scrapie, most of the $PrP^d$ in the CNS is attached to the cell membrane and is not released to the extracellular space to form fibrils. As a consequence, naturally occurring disease in domestic ruminants, deer, and cats does not result in classical amyloid plaques, although a vascular amyloid in the CNS is found in a minority of

Table 10.1 Selected pathological features and immunohistochemistry of BSE and scrapie (data from [53, 54, 61–64])[a].

| Pathological changes | BSE | | | Scrapie | | |
|---|---|---|---|---|---|---|
| | Presence | Immunohistochemistry | | Presence | Immunohistochemistry | |
| | | PrP$^d$ | Ubiquitin | | PrP$^d$ | Ubiquitin |
| *CNS* | | | | | | |
| Vacuolation of neuropil and neuronal perikarya | + | − | NE | + | − | − |
| Neuronal membrane invaginations | + | + | NE | + | + | + |
| Membrane microfolding of astrocytes and dendrites | + | + | NE | + | + | − |
| Increase in neuronal endo-lysosomes | + | + | NE | + | + | − |
| Amyloid fibrils | Few | + | NE | Few | ++ | − |
| Amyloid plaques | − | − | NE | Few | − | − |
| *Lymphatic tissues* | | | | | | |
| Cytoplasmic tubular network in TBMs | NE | NE | NE | + | + | + |
| Hyperplasia of FDC dendrites | NE | NE | NE | + | + | + |

a) Pathological changes and immunohistochemistry are ordered as present, +; not present, −; or not examined, NE. TBM, tingible body macrophage; FDC, follicular dendritic cell.

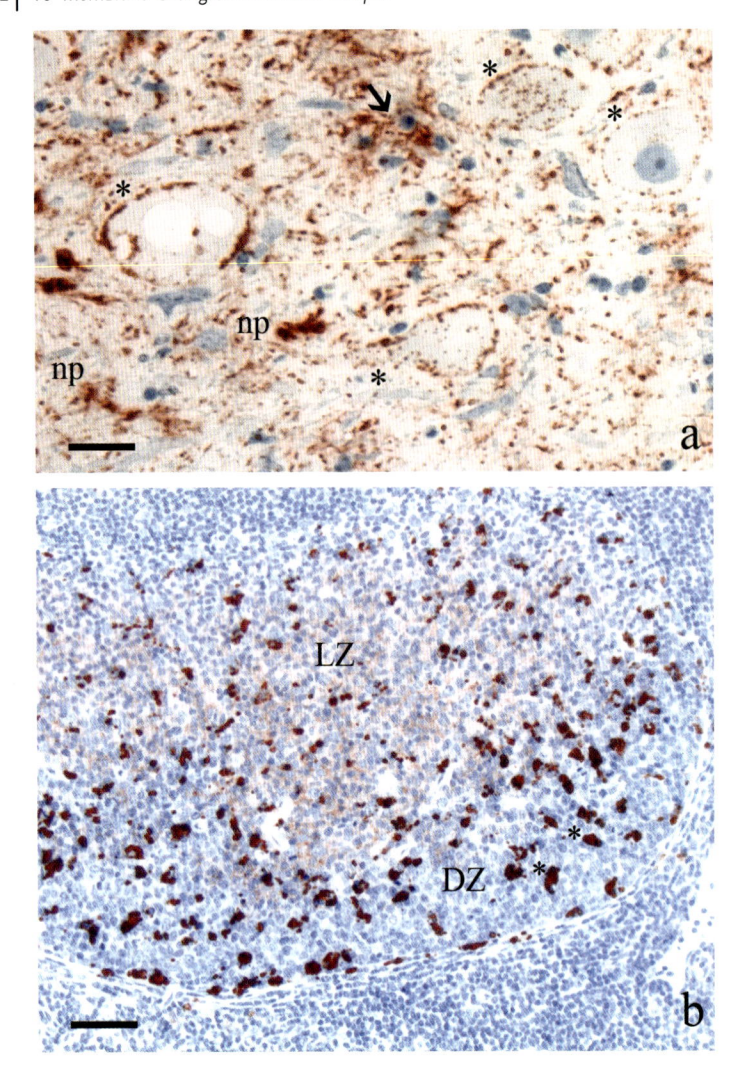

**Figure 10.2** Immunohistochemical labeling for PrP$^d$ in the CNS and lymphatic tissue. (a) Dorsal motor nucleus of the vagus nerve, sheep scrapie. Distinct perineuronal labeling is present on neuronal perikarya (asterisks); there is less intracytoplasmic labeling in the same neurons. An example of glia-associated labeling is marked with an arrow, and examples of neuropil labeling are marked np. PrP antibody F89/160.1.5. Bar: 28 μm. (b) Retropharyngeal lymph node, sheep scrapie. In the dark zone (DZ) of the lymphoid follicle there is mainly labeling of TBMs (asterisks). In the light zone (LZ) of the lymphoid follicle there is a reticular labeling pattern indicative of FDC labeling. PrP antibody F89/160.1.5. Bar: 47 μm.

sheep [57, 58] and deer TSEs [59]. Classical plaques which may be stained with tinctorial dyes such as Congo Red and are composed of mature bundles of amyloid are conspicuous in some experimental murine strains such as 87V [60] and 111a scrapie [58].

With the exception of a transgenic murine model [65], PrP$^d$ accumulation in animal TSEs is found on dendritic (Figure 10.3a) and neuronal perikaryonal plasma membranes rather than on axons [53, 54, 66]. The transgenic PG14 (TgPG14) mouse is a non-transmissible Gerstmann–Sträussler–Scheinker disease homolog with a mutant PrP molecule containing extra-octapeptide repeat sequences [65]. In the TgPG14 mouse, mutant PrP accumulates on both axonal and dendritic plasma membranes [67]. Therefore, in most classical animal TSEs, we can infer from the sites of accumulation and presumptive conversion of PrP$^d$ that normal trafficking of PrP$^c$ to mature axons is perturbed and redistributed in favor of dendrites. We would anticipate that a parallel altered composition or distribution of rafts in axons and dendrites will also be present. However, not all PrP$^d$ molecules may follow this pattern as at least one mutant PrP molecule appears to accumulate abnormally on axonal plasma membranes.

In most scrapie-affected sheep, PrP$^d$ accumulates in large amounts in different lymphoid tissues (Figure 10.2b) [8, 68]. In cattle BSE, there is a limited deposition of PrP$^d$ in the same tissues, but this has been reported only in experimentally infected cattle and not in naturally diseased animals [69]. In lymphoid tissues of scrapie-affected sheep, PrP$^d$ is localized intracellularly in endosomes, lysosomes, and cytoplasmic tubular networks of tingible body macrophages (TBMs) and at the plasma membrane of FDCs, TBMs, and B-cells in secondary germinal centers [63, 70].

## 10.4
## Aberrant Endocytosis and Trafficking of PrP$^d$ in Neurons and Tingible Body Macrophages

We have shown previously that the initial accumulation of PrP$^d$ in murine [71] and ovine [54] scrapie occurs at the plasma membrane of neurons in the absence of pathological changes (Figure 10.3a). Plasma membrane-associated PrP$^d$ increases with disease progression, but PrP$^d$ is also found within endosomes and lysosomes. Neuronal plasma membrane changes and endo-lysosomal PrP$^d$ are often found in the same cells. PrP$^d$-labeled coated pits and vesicles are concentrated at regular intervals and not uniformly distributed along membranes of shafts of dendrites (Figure 10.3b and c). Many coated pits have a bizarre structure that includes long and extended necks (Figure 10.3d and e) that are invaginations of the surface plasma membrane which may also be branched or twisted [53, 54, 61, 62]. A similar spiral membrane invagination is also found in lower numbers on axon terminals, although the membrane coating is lucent rather than electron dense (Figure 10.3c) [53, 54]. In addition, PrP$^d$-labeled coated membranes can be found adjacent to PrP$^d$-positive endosomes or lysosomes of the same neuron. We can infer from this that PrP$^d$ on neuronal cell membranes is endocytosed via clathrin-coated pits in TSE-affected cattle [53], sheep [61], cats (M. Jeffrey, unpublished observations), and mice [66].

The molecular changes which result in this morphological evidence of perturbed clathrin-mediated endocytosis are not known. In the course of normal endocytosis, binding of clathrin to the cytoplasmic face of the plasma membrane initiates the formation of the endocytotic vesicle, and amphiphysin and dynamin complete formation and excision of the coated vesicle from the plasma membrane [72, 73]. Once the vesicle has been excised from the plasma membrane, clathrin is released into the cytosol for re-utilization. Immunohistochemical studies of scrapie-affected tissues have not shown any altered distribution of these three proteins, but ubiquitin, also located on the cytoplasmic face of the membrane, is conspicuous within the plasma membrane invaginations and less so in the lysosomes [54]. These observations suggest that $PrP^d$ may signal to cytoplasmic clathrin and ubiquitin via a membrane-spanning ligand. Furthermore, as clathrin is abnormally retained on coated membrane invaginations, this suggests that the $PrP^d$–ligand–clathrin–ubiquitin array forms a stable molecular complex within the membrane that is not efficiently excised into vesicles by dynamin and related molecules (Figure 10.1b).

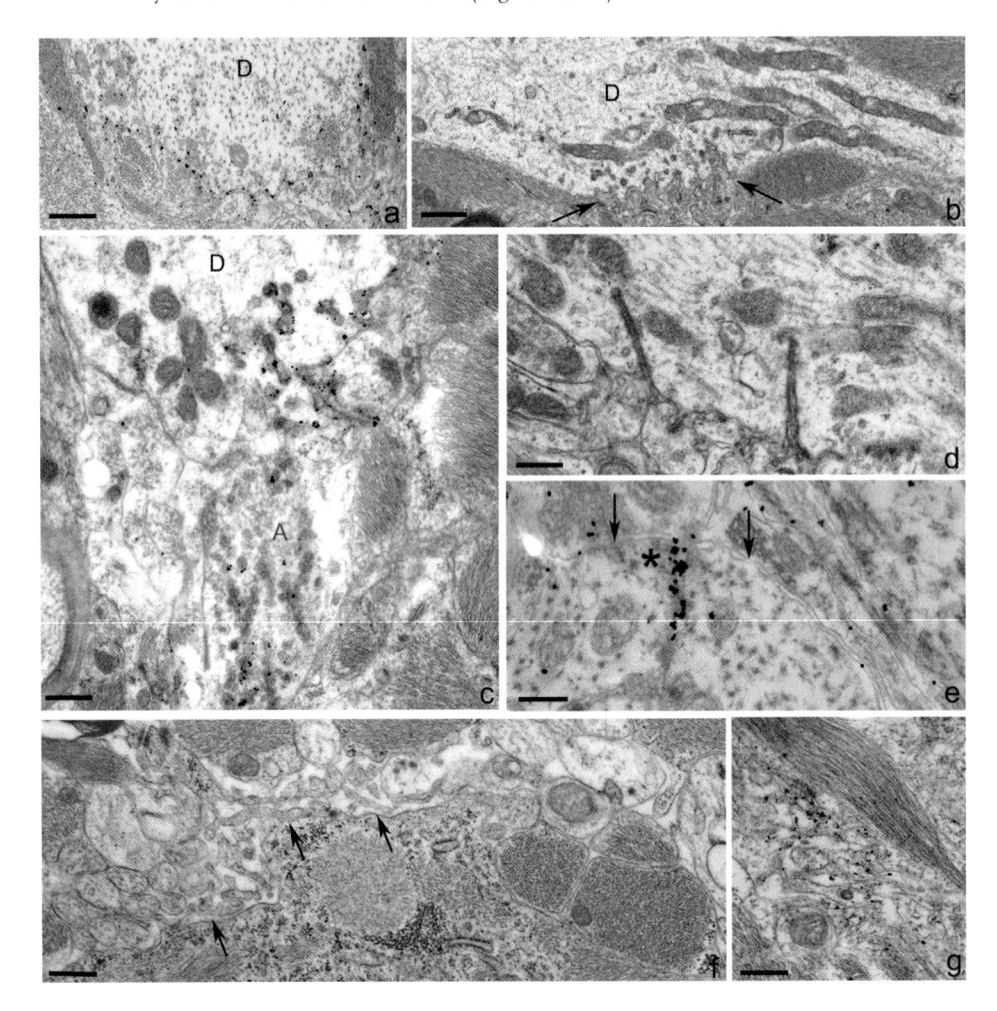

Ubiquitin is covalently attached to the target protein via an isopeptide bond between glycine in ubiquitin and lysine in the target protein. Monoubiquitination involves the addition of single ubiquitin molecules to the target protein, while several lysine residues in the protein can be tagged, giving rise to multiple monoubiquitation. Polyubiquitination of a target protein results in the attachment of several ubiquitin molecules to a single lysine residue. Monoubiquitination is implicated in clathrin-mediated endocytosis of plasma membrane proteins and sorting of proteins to multivesicular bodies (MVBs), while polyubiquitination is the first step in targeting proteins to the ubiquitin proteasome system for degradation. Multiple monoubiquitination of target proteins at the plasma membrane leads to internalization and degradation, directing proteins to the endosomal–lysosomal system while preventing recycling to the cell membrane [74]. Abnormal ubiquitin distribution was first demonstrated by immunohistochemistry in the brains of two murine scrapie strains [75]. Subsequent studies described co-localization of ubiquitin and PrP$^d$ in lysosomes and these data were used to hypothesize that the conversion of PrP$^c$ to infectious prions may take place in lysosomes [76]. However, these data are probably the result of technical artifacts. The primary PrP antibody was used at ~2 log excess concentration, resulting in artifactual false clustering of immunogold particles. When labeled with antibodies of known sequence specificity, intralysosomal PrP$^d$ accumulations are truncated and lack epitopes to the N-terminus [77], whereas plasma membrane-associated PrP$^d$ accumulations and PrP$^d$ in extracellular classical amyloid plaques are full length and labeled by antibodies which recognize epitopes throughout the N- and C-terminal domains of the PrP molecule [77, 78]. The lack of

**Figure 10.3** Plasma membrane changes and immunogold PrP$^d$ labeling in the CNS.
(a) Sheep scrapie. PrP$^d$ localized to the plasma membrane of a morphologically normal dendrite (D). Immunogold for PrP$^d$ using 523.7 antibody. Bar: 660 nm. (b) Sheep scrapie. A longitudinal section of a dendrite (D) showing localized pathology of the plasma membrane. There are numerous pits and membrane invaginations in a small segment (delimited by arrows) of the dendrite. Uranyl acetate and lead citrate counterstain. Bar: 670 nm. (c) Sheep scrapie. An area similar to that shown in (b) is shown in cross-section of dendrite (D). PrP$^d$ labeling is associated with increased numbers of coated pits which lie beneath the plasma membrane and on membrane invaginations. An axon terminal adjacent to this dendrite (A) contains numerous inclusions comprising coated tubular structure. The coats of the axonal structures are more electron lucent than the coats of the vesicles and membrane invaginations present on the dendrite. Immunogold for PrP$^d$ using 523.7 antibody. Bar: 320 nm. (d) BSE. A dendrite with two coated plasma membrane invaginations. The plasma membranes in between the invagination and the nearby processes are irregular. Uranyl acetate and lead citrate counterstain. Bar: 410 nm. (e) BSE. There is marked PrP$^d$ labeling of a plasma membrane invagination (asterisk) in a dendrite. The plasma membrane is marked by arrows. Immunogold for PrP$^d$ using 523.7 antibody. Bar: 170 nm. (f) Sheep scrapie. Astrocytic processes showing polyp-like extensions or outward folding of the cell membrane (arrows). Uranyl acetate and lead citrate counterstain. Bar: 420 nm. (g) PrP$^d$ labeling of the membrane folds on astrocytes. Immunogold for PrP$^d$ using 523.7 antibody. Bar: 420 nm. (b) Reproduced with kind permission from Springer Science and Business Media: [61], Figure 2b. (d and e) Reproduced with kind permission from John Wiley and Sons Inc.: [53], Figure 3b and d.

N-terminal epitopes of lysosomal PrP$^d$ suggests that it is degraded within the lysosomal compartment rather than created in these organelles. The increased number of normal and the presence of abnormal MVBs, and in some cattle autophagic vacuoles, in neurons are indications of an upregulated catabolic activity. Increased numbers of MVBs and lysosomes in neurons have also been reported in CJD [79]. As full-length PrP$^d$ is not located within cells, its accumulation on plasma membranes suggests that *in vivo* conversion of PrP$^c$ to PrP$^d$ also occurs at the cell membrane.

PrP$^d$ at the cell surface may also elicit polyp-like micro-folding of the plasma membrane, which can be seen on dendrites and glia (Figure 10.3f and g) [53, 54]. Whereas abnormal clathrin-mediated endocytosis is a common feature of neuronal perikaryonal and dendritic plasma membranes which accumulate PrP$^d$, it does not occur at the cell membrane of PrP$^d$-accumulating astrocytes. Conversely, membrane micro-folding occurs much more commonly and with greater magnitude on astrocyte membranes than dendritic membranes [54]. It is likely that the presence of PrP$^d$ at the cell surface can generate a proliferative response or folding of both astrocytic and dendritic plasma membranes. Immunogold electron microscopy limits analysis of membranes to a subjective appraisal of the width and curvature of membranes and does not provide any data on physical structural properties or composition. Studies of recombinant PrP folded into different structural forms have shown that altered ratios of α- and β-sheet may be associated with altered biophysical properties [51, 80]. In particular, PrP rich in β-sheet can cause destabilization of the membrane. It is perhaps reasonable to infer that membranes with morphological changes may also have altered physiochemical properties.

In both the light and dark zones of lymphoid follicles, TBMs in ovine [63] and murine [64, 81] scrapie accumulate PrP$^d$ mainly intracellularly, where it is located within lysosomes and at the membrane of fused tubular networks (Figure 10.4a and b). It may also accumulate, albeit sparsely, at the plasma membrane in association with non-coated structures morphologically similar to caveolae [82]. The primary function of TBMs in secondary lymphoid follicles is to phagocytose B cells that fail to produce viable immunoglobulins by affinity selection. TBMs can also ingest material from the extracellular space via clathrin- or caveolin-mediated endocytosis [63]. In sheep, intracytoplasmic tubular networks are ubiquitin positive (Figure 10.4c) [63], whereas in mice intracytoplasmic tubular networks are not present and endosomes and to a lesser extent lysosomes are ubiquitin positive [64]. Clathrin-coated pits on the cell membrane of TBMs are not labeled with either PrP$^d$ or ubiquitin.

Fused membrane networks formed from tubular extensions of the ER are formed following infection with simian virus 40 [83] or internalization of autocrine motility factor [84]. Alteration to the normal polygonal network of the ER occurs following extension of the ER membrane and the subsequent absence of concomitant organizing membrane proteins, leading to the collapse of the rigidly organized structure into the random arrangement [63, 85]. We hypothesize that in lymph nodes of scrapie-affected sheep PrP$^d$ at the cell surface of TBMs may complex with other proteins, including ubiquitin within the endocytic pathway, and interfere with

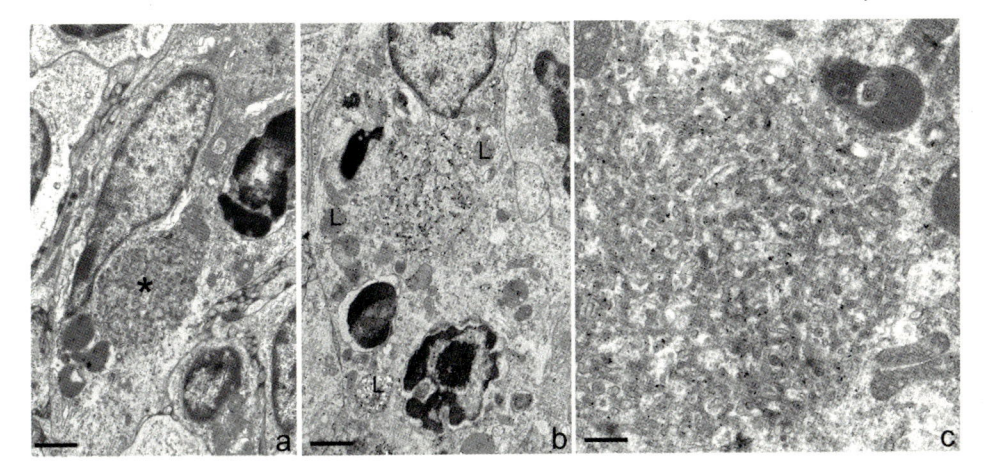

**Figure 10.4** Subcellular changes and ubiquitin and PrP$^d$ labeling of tingible body macrophages. (a) Tonsil, sheep scrapie. TBM showing a discrete intracytoplasmic structure composed of fused membranes forming a tubulo-reticular network (asterisk). Uranyl acetate and lead citrate counterstain. Bar: 1110 nm. (b) Tonsil, sheep scrapie. TBM showing PrP$^d$ labeling of mature lysosomes (L) and also of a discrete structure composed of fused tubular membranes similar to that shown in (a). Imunogold for PrP$^d$ using antibody 523.7. Bar: 1110 nm. (c) Tonsil, sheep scrapie. Ubiquitin labeling of a fused membrane network in a TBM similar to that shown in (a). Immunogold for ubiquitin (Dako). Bar: 420 nm.

organizing membrane proteins of the ER. These PrP$^d$–ligand complexes break down the normal organization of the ER, resulting in PrP$^d$ co-localization with the randomly fused tubular network (Figure 10.1c). The presence of both ubiquitin and PrP$^d$ in endosomes, but primarily PrP$^d$ alone in lysosomes in mice, is probably due to enzymatic dissociation of PrP$^d$ and ubiquitin within late endosomes prior to their fusion with lysosomes. This is not evident in ovine scrapie. Thus the mechanism of cell membrane endocytosis of PrP$^d$ and its subsequent intracellular trafficking in lymphoid tissue is different from the CNS and influenced by cell type. Some species or strain differences may also occur.

There is no convincing evidence that TBMs may actively amplify infectivity. Most likely PrP$^d$ is transferred from the infected FDCs to TBMs and then internalized. Alternatively, or additionally, PrP$^d$ could accumulate incidentally in the process of phagocytosis of effete B cells.

## 10.5
## Abnormal Maturation Cycle and Immune Complex Trapping of Follicular Dendritic Cells in Lymphoid Germinal Centers

FDC-associated PrP$^d$ is not present within the cytoplasm and is limited to the plasma membrane of FDC dendrites (processes) [63, 64]. FDCs are stromal antigen-presenting cells located in germinal centers. They are able to hold antigens for a

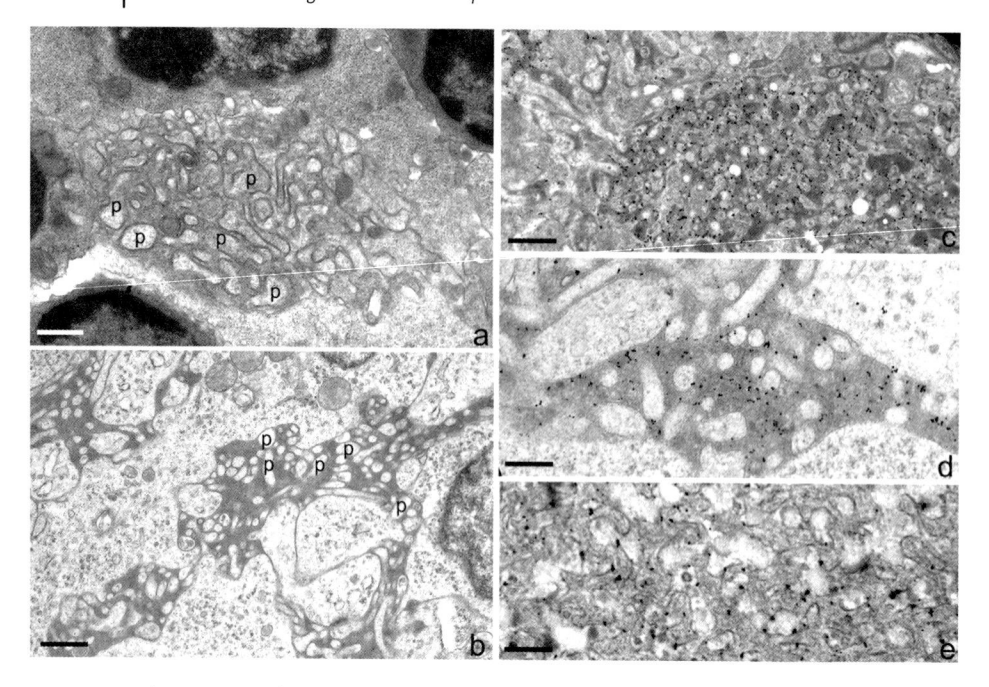

**Figure 10.5** Subcellular changes and ubiquitin, IgG, and PrP$^d$ labeling of follicular dendritic cells. (a) Mesenteric lymph node, normal mouse. Normal FDC from a secondary follicle. The FDC processes are regular and the electron-dense material between processes (p) is of uniform thickness. Uranyl acetate and lead citrate counterstain. Bar: 870 nm. (b) Tonsil, sheep scrapie. There is marked accumulation of abundant electron-dense material between irregular and enlarged FDC processes (p). Uranyl acetate and lead citrate counterstain. Bar: 780 nm. (c) Spleen, murine scrapie. PrP$^d$ is present on the cell membranes of hypertrophic FDC processes, but is largely absent from the electron-dense extracellular material. Immunogold labeling for PrP$^d$ using 1A8 antibody. Bar: 850 nm. (d) Tonsil, sheep scrapie. The extracellular electron-dense material in between FDC dendrites shows a strong signal for globulins, consistent with excess trapping of immune complexes. Immunogold labeling for IgG (Zymed). Bar: 380 nm. (e) Mesenteric lymph node, murine scrapie. The immunogold-labeled ubiquitin molecules are situated mainly on cell membranes of FDC dendrites. Immunogold labeling for ubiquitin (Dako). Bar: 370 nm.

long time as intact antigen–antibody complexes on the cell surface, and are important in B-cell differentiation, proliferation, and survival [86]. Normal FDCs may mature and regress in a continuous cycle [87]. However, in TSE-infected animals there are different morphological forms of FDCs that are not recognized in controls. These may be indicative of stages in FDC maturation and regression. Normal immature FDCs have relatively small and simple processes that only become activated when circulating immune complexes attach to C3b and Fc immunoglobulin receptors on their cell surfaces. The phenotypic expression of FDC cell surface receptors and ligands changes markedly upon activation and in parallel there is a marked increase in the length and complexity of branching of normal mature FDC dendritic processes [88]. In normal mature FDCs, the trapped immune complexes lie equidistant from adjacent dendrites, creating a space of uniform thickness between adjacent FDC processes (Figure 10.5a) [64].

In scrapie-affected mice at early stages of disease, plasma membrane PrP$^d$ accumulations may be found on immature FDCs with no discernible morphological effect when viewed under the electron microscope [64]. However, most PrP$^d$ is associated with mature, activated FDCs which show a marked hyper-proliferation of their processes (Figure 10.5b and c). This hyperplasia of processes is accompanied by an excess accumulation of electron dense material that is PrP$^d$ negative in the adjacent extracellular space (Figure 10.5c). This material contains immunoglobulins IgG (Figure 10.5d) and IgM presumptively derived from excess trapping of circulating immune complexes (Figure 10.1d). Regressing FDCs have less PrP$^d$ labeling and this may be due to decreased production of PrP$^c$ and endocytosis of existing PrP$^d$ by TBMs. The fate of infected FDCs is not known, but regressing cells may revert to a normal resting phenotype stripped of PrP$^d$ or they may die. These features suggest that the presence of PrP$^d$ on the cell membrane, although not altering the morphological structure of the membrane, nevertheless increases the retention of immune complexes and perturbs the maturation cycle of FDCs and secondary follicle maturation and regression [63, 64]. B lymphocyte alterations in clonal selection, proliferation, and maturation may occur as a consequence to the described changes and thus affect immune system function.

As also observed in neurons, increased ubiquitin labeling is found in association with PrP$^d$ at the plasma membrane of FDCs (Figure 10.5e) [64]. In contrast with neurons, immunogold electron microscopy does not reveal that FDCs recycle PrP$^d$ through endosomes, suggesting that a PrP$^d$–ligand–ubiquitin complex is retained at the plasma membrane (Figure 10.1d.).

## 10.6
### Molecular Changes of Plasma Membranes Associated with PrP$^d$ Accumulation

There are a large number of reports describing ligands of the prion protein and a wide array of molecules with which PrP$^c$ has been shown to interact *in vitro* and *in vivo* (reviewed in [13]). The differences in membrane pathology observed by electron microscopy for different sources and cell types may be due to differences in PrP$^d$ membrane–protein interactions in different cells as described above or, in the case of inherited forms of disease and related transgenic models, due to primary changes in the sequence of PrP$^c$ and its subsequent alteration to PrP$^d$. Different mechanisms of endocytosis in the CNS and lymphoid tissues and altered immune complex trapping in lymphoid tissues suggest that interactions between PrP$^d$ and different cell membrane molecules may modulate cell trafficking and implement cell membrane distortions. The common association between different inferred PrP$^d$–molecular interactions and ubiquitin suggest that these PrP$^d$–membrane molecular complexes are stable and inefficiently disposed of by cellular quality control mechanisms. Thus, both PrP$^c$ and PrP$^d$ can be implicated in a wide range of membrane molecular interactions that arise via inferred multimolecular complexes. Because of the range of molecules now linked to possible $PrP^c$ function, it has been suggested that $PrP^c$ may not have just a single function, but rather may be a scaffolding molecule used for numerous molecular interactions at the cell membrane [13]. We further suggest that the transformation of $PrP^c$ to $PrP^d$ results in stabilization of these complexes within the membrane (Figure 10.1e) and that subsequent membrane pathology is influenced by the primary $PrP^c$ sequence, the cell type, and strains of infection.

## 10.7
### Transfer of $PrP^d$ Between Cells

In the CNS of both scrapie- and BSE-infected animals, there is evidence of transfer of $PrP^d$ between directly opposed plasma membranes of different processes [53, 54]. Spiral membrane invaginations are found on axon terminals immediately adjacent to dendrites with features of abnormal endocytosis, and probably represent attempted internalization of plasma membrane $PrP^d$ transferred to axons from the adjacent dendrites. Evidence that $PrP^d$ transferred from one membrane can elicit pathological changes in the recipient cell membrane can be deduced from studies of the scrapie-infected TG3 mouse [66]. The TG3 mouse expresses $PrP^c$ under the control of the hamster GFAP promoter on a PrP-null mouse background [89]. When such mice are infected with 263K hamster scrapie, only astrocytes are susceptible to infection yet scrapie-specific pathology, including abnormal endocytosis, can be seen on neuronal perikarya and dendrites [66]. In order for this to occur, $PrP^d$ must accumulate at the cell surface of astrocytes expressing $PrP^c$ and be transferred to non-$PrP^c$-expressing neuronal processes.

Both $PrP^c$ [90–92] and $PrP^{res}$ [90, 91] have been shown to be excreted by exosomes in neuronal cell cultures, and could possibly be transferred between cells by these MVB-derived vesicles. Exosomes are released to the extracellular space when MVBs fuse with the cell membrane. In all our subcellular studies of animal TSEs, we have not been able to identify exosomes in the brain. Vesicles 50–200 nm in diameter are present in small numbers at the surface of normal mature FDCs. Such vesicles, or exosomes, contain MHC-II molecules and are putatively derived from B cells [93, 94]. During the course of scrapie infection, a large number of exosomes are found around FDCs, but they do not accumulate $PrP^d$. It is likely that as the FDC matures, its ability to trap immune complexes deteriorates while deposition of B cell derived exosomes continues to occur [63]. Recently, tunneling nanotubes that are membrane bridges between cells have been implicated in the cell-to-cell spread of $PrP^d$ *in vitro* [95].

GPI-anchored proteins can readily be detached from one membrane and reintegrated into another [96], and indeed this is also the case for $PrP^c$ [97]. We therefore find it more likely that intercellular spread of $PrP^d$ is achieved by GPI anchor transfer. This transfer seems to be indiscriminate and independent of cell type and occurs in both nervous and lymphoid tissues.

## 10.8
### Extracellular Amyloid Form of PrP$^d$

Not all PrP$^d$ remains fixed to membranes: some may be released into the extracellular space, where it may aggregate into individual amyloid fibrils. The pathogenetic process of fibril formation is not precisely understood. PrP$^d$-accumulating cell membranes adjacent to these individual fibrils often show an unusual linear regularity [54], suggesting either that membrane conformation is affected by adjacent fibrils or that membranes contribute to the assembly of PrP$^d$ molecules. Mature plaques are formed from parallel bundles of these ~8–10 nm diameter filaments.

In our studies of cattle BSE and sheep scrapie, there was only subtle evidence of amyloid fibrils in the extracellular space in one cow [53] and two sheep [61]. As mentioned above, plaques are a rare feature of sheep scrapie and are absent from BSE-affected cattle, but there are several murine scrapie strains [58] that produce abundant plaques, and mini-plaques are the defining feature of a recently recognized prion disease of cattle known as bovine amyloidotic spongiform encephalopathy [98]. Amyloid plaques are especially abundant in a transgenic mouse line in which PrP$^c$ is engineered without the GPI anchor [99]. Thus, species, prion strains, and primary structure of PrP$^c$ all influence plaque formation.

Several lines of investigation suggest that membrane-attached PrP$^c$ and PrP$^d$ interact with extracellular matrix components. Perineuronal nets surround some neuron cell bodies, dendrites, and initial segments of axons [100, 101], and are composed of hyaluronic acid, tenascin-R, and lecticans. In sporadic CJD there is loss of these perineuronal nets [102], although these observations could not be confirmed in scrapie-affected sheep when assessed with *Wisteria floribunda* agglutinin, a plant lectin with affinity for N-acetylgalactosamine visualizing the perineuronal nets [103]. Recently, glypican-1, a major highly sulfated GPI-anchored proteoglycan, has been shown to coprecipitate with PrP$^c$ and PrP$^d$. It has been suggested that glypican-1 facilitates the interaction of PrP$^c$ and PrP$^d$ in lipid rafts [38]. The N-terminus of PrP$^c$ has been shown to bind to heparin sulfate, another highly sulfated extracellular matrix proteoglycan [104–106]. Both preamyloid and amyloid deposits show abundant labeling with heparan sulfate [107, 108]. Hence there are a number of extracellular matrix molecules which may facilitate interactions between PrP$^c$ and PrP$^d$, or contribute to pathological changes including the formation of amyloid.

PrP$^d$ produced by the scrapie-infected Tg GPI anchorless mouse is not retained at the cell membrane, but is free to diffuse through the interstitial space away from its releasing cell. Amyloid plaques in these mice are located around blood vessels and the initial fibrillization of PrP$^d$ begins in their basement membranes [49]. In this model, the plaque pathogenesis is closely similar to cerebral amyloid angiopathy of Alzheimer's disease [109]. Widespread amyloid formation is a consequence of impaired drainage of PrP$^d$ from the brain along interstitial fluid-draining pathways [49]. The aggregation of PrP$^d$ into plaques in this model appears to be initiated by extracellular factors associated with basement membranes.

Intravascular and perivascular amyloid plaques are a feature of some sheep scrapie sources [110], CWD of deer [59], and at least one murine scrapie model [111]. It is possible that plaques in these species are formed from a minor population of PrP$^d$ molecules lacking a GPI anchor or some other modification which facilitates release and diffusion of soluble PrP$^d$ molecules with subsequent fibrillization within the vascular basement membranes.

These data suggest that extracellular matrix components may facilitate the lateral trafficking of PrP$^d$ from the membrane into the extracellular space where constraints on aggregation may be less, or extracellular matrix molecules may facilitate aggregation of PrP$^d$ within the membrane itself. Certain conformations of soluble PrP$^d$ molecules, such as those lacking the GPI anchor, may be able to utilize extracellular scaffolding molecules other than those available to membrane-attached PrP$^d$ and aggregate into fibrils at sites distant to their release.

The homozygous GPI anchorless mice have a longer incubation time than wild-type mice and eightfold less expression of PrP$^c$, but higher levels of PrP$^d$ at the terminal stage of disease [49]. This could be explained by a lower pathogenicity of the amyloid form of PrP$^d$, or that these transgenic mice are less susceptible to PrP$^d$ because GPI-anchored PrP$^c$ is required for neurotoxic membrane interactions. Prefibrillar oligomers from different amyloid diseases permeate lipid bilayers in what could be a common mechanism for amyloid-related degenerative changes [112]. Hence it is possible that smaller PrP$^{res}$ oligomers are the most neurotoxic to membranes. Packing diseased proteins into amyloid aggregates most likely is a protective mechanism [113], but large amyloid plaques also cause localized brain damage [49,109].

## 10.9
### Strain-Directed Effects of Prion Infection

In one study of sheep scrapie, based on morphological differences, the affected sheep were divided into two groups, one with predominantly perturbed endocytosis and one where membrane proliferations dominate [54]. The source of the animals in this study suggests that the cellular processing of PrP$^d$ and the quantity of cytopathological changes correspond to the infecting strain. Earlier light microscopic studies suggested that different strains may target some cell populations rather than others and that this leads to different processing of PrP$^d$ shown as different labeling patterns [110]. As opposed to scrapie, where several strains have been described, only one strain of classical BSE is recognized [114].

## 10.10
### Conclusion and Perspectives

In this chapter, we have illustrated that different cell types such as neurons, astrocytes, FDCs, and TBMs show different cytopathological changes when infected with the same strain of mouse or sheep scrapie. Diverse glial and neuronal pathology that is similar to sheep scrapie is also found in cattle BSE. This is most likely caused by biological differences between these cell types as exemplified by different mechanisms of endocytosis of $PrP^d$ between neurons and TBMs. We propose that different cellular changes are the result of the formation of different complexes of membrane proteins with $PrP^d$. The nature of the $PrP^d$ complexes directs differing trafficking and processing pathways and causes physiochemical changes to membranes. Ubiquitin molecules are commonly, but not exclusively, part of these protein complexes. Whereas $PrP^c$ is uniformly distributed on neuronal processes, $PrP^d$ is found mainly concentrated in multiple localized segments of dendritic membranes. The molecular signals that determine the distribution pattern of $PrP^d$ on membranes is unclear. It also remains to be shown how morphological changes of membranes correlate with functional or physical changes.

## 10.11
## Summary

Disease-specific prion protein ($PrP^d$) is associated with changes in membranes of lymphoid and nervous tissues in cattle bovine spongiform encephalopathy (BSE), sheep and murine scrapie, and some transgenic rodent prion diseases. Abnormal endocytosis of membrane $PrP^d$ occurs both on neurons and tingible body macrophages (TBMs). Follicular dendritic cell (FDC) changes include abnormal maturation and regression and excess immune complex trapping. The nature of the ultrastructural membrane changes differs between neurons, glia, and lymphoid cells, suggesting that $PrP^d$ interacts with different membrane molecules in these cells. Ubiquitin commonly co-localizes with subcellular membrane changes, suggesting that it may form complexes with $PrP^d$ and other membrane molecules. These multimolecular complexes may be inefficiently excised from the plasma membrane. In the natural prion diseases, the same pathological changes are found in and between species, but there is a clear difference in the degree of neuropathological changes between groups of sheep and mice that can be attributed to the strain of agent. When transgenic models of scrapie are studied, it becomes clear that different mutant forms of PrP produce different neuropathological lesions that can be correlated with differences in cellular trafficking of $PrP^d$. It is evident that fundamental aspects of prion diseases, such as conversion, accumulation, and trafficking of $PrP^d$, occur at plasma membranes and seem to perturb cellular processes of neurons, TBMs and FDCs.

## References

1 Johnson, R.T. (2005) Prion diseases. Lancet Neurol., 4, 635–642.
2 Wells, G.A.H., Scott, A.C., Johnson, C.T., Gunning, R.F., Hancock, R.D., Jeffrey, M., Dawson, M., and Bradley, R. (1987) A novel progressive spongiform encephalopathy in cattle. Vet. Rec., 121, 419–420.

3 Bruce, M.E., Will, R.G., Ironside, J.W., McConnell, I., Drummond, D., Suttie, A., McCardle, L., Chree, A., Hope, J., Birkett, C., Cousens, S., Fraser, H., and Bostock, C.J. (1997) Transmission to mice indicate that "new variant" CJD is caused by the BSE agent. *Nature*, **389**, 498–501.

4 Will, R.G., Ironside, J.W., Zeidler, M., Cousens, S.N., Estibeiro, K., Alperovitch, A., Poser, S., Pocchiari, M., Hofman, A., and Smith, P.G. (1996) A new variant of Creutzfeldt–Jakob disease in the UK. *Lancet*, **347**, 921–925.

5 Jeffrey, M., Ryder, S., Martin, S., Hawkins, S.A., Terry, L., Berthelin-Baker, C., and Bellworthy, S.J. (2001) Oral inoculation of sheep with the agent of bovine spongiform encephalopathy (BSE). 1. Onset and distribution of disease-specific PrP accumulation in brain and viscera. *J. Comp. Pathol.*, **124**, 280–289.

6 Büeler, H., Aguzzi, A., Sailer, A., Greiner, R.A., Autenried, P., Aguet, M., and Weissmann, C. (1993) Mice devoid of PrP are resistant to scrapie. *Cell*, **73**, 1339–1347.

7 Jeffrey, M. and González, L. (2007) Classical sheep transmissible spongiform encephalopathies: pathogenesis, pathological phenotypes and clinical disease. *Neuropathol. Appl. Neurobiol.*, **33**, 373–394.

8 Ersdal, C., Ulvund, M.J., Espenes, A., Benestad, S.L., Sarradin, P., and Landsverk, T. (2005) Mapping $PrP^{Sc}$ propagation in experimental and natural scrapie in sheep with different PrP genotypes. *Vet. Pathol.*, **42**, 258–274.

9 Prusiner, S.B. (1982) Novel proteinaceous infectious particles cause scrapie. *Science*, **216**, 136–144.

10 Prusiner, S.B. (1998) Prions. *Proc. Natl. Acad. Sci. USA*, **95**, 13363–13383.

11 Prusiner, S.B., Peters, P., Kaneko, K., Taraboulos, A., Lingappa, V., Cohen, F.E., and DeArmond, S.J. (1999) Cell biology of prions, in *Prion Biology and Diseases* (ed. S.B. Prusiner), Cold Spring Harbor Laboratory Press, Cold Spring Harbor, NY, pp. 349–391.

12 Bendheim P.E., Brown, H.R., Rudelli, R.D., Scala, L.J., Goller, N.L., Wen, G.Y., Kascsak, R.J., Cashman, N.R., and Bolton, D.C. (1992) Nearly ubiquitous tissue distribution of the scrapie agent precursor protein. *Neurology*, **42**, 149–156.

13 Linden, R., Martins, V.R., Prado, M.A., Cammarota, M., Izquierdo, I., and Brentani, R.R. (2008) Physiology of the prion protein. *Physiol. Rev.*, **88**, 673–728.

14 Galvan, C., Camoletto, P.G., Dotti, C.G., Aguzzi, A., and Ledesma, M.D. (2005) Proper axonal distribution of $PrP^{C}$ depends on cholesterol-sphingomyelin-enriched membrane domains and is developmentally regulated in hippocampal neurons. *Mol. Cell. Neurosci.*, **30**, 304–315.

15 Sunyach, C., Jen, A., Deng, J., Fitzgerald, K., Frobert, Y., Grassi, J., McCaffrey, M., and Morris, R. (2003) The mechanism of internalization of glycosylphosphatidylinositol-anchored prion protein. *EMBO J.*, **22**, 3591–3601.

16 Taylor, D.R., Watt, N.T., Perera, W.S., and Hooper, N.M. (2005) Assigning functions to distinct regions of the N-terminus of the prion protein that are involved in its copper-stimulated, clathrin-dependent endocytosis. *J. Cell Sci.*, **118**, 5141–5153.

17 Marella, M., Lehmann, S., Grassi, J., and Chabry, J. (2002) Filipin prevents pathological prion protein accumulation by reducing endocytosis and inducing cellular PrP release. *J. Biol. Chem.*, **277**, 25457–25464.

18 Peters, P.J., Mironov, A., Jr., Peretz, D., van Donselaar, E., Leclerc, E., Erpel, S., DeArmond, S.J., Burton, D.R., Williamson, R.A., Vey, M., and Prusiner, S.B. (2003) Trafficking of prion proteins through a caveolae-mediated endosomal pathway. *J. Cell Biol.*, **162**, 703–717.

19 Gorodinsky, A. and Harris, D.A. (1995) Glycolipid-anchored proteins in neuroblastoma cells form detergent-resistant complexes without caveolin. *J. Cell Biol.*, **129**, 619–627.

20 Magalhães, A.C., Silva, J.A., Lee, K.S., Martins, V.R., Prado, V.F., Ferguson, S.S., Gomez, M.V., Brentani, R.R., and Prado, M.A. (2002) Endocytic intermediates

involved with the intracellular trafficking of a fluorescent cellular prion protein. *J. Biol. Chem.*, **277**, 33311–33318.

21. González, L., Terry, L., and Jeffrey, M. (2005) Expression of prion protein in the gut of mice infected orally with the 301V murine strain of the bovine spongiform encephalopathy agent. *J. Comp. Pathol.*, **132**, 273–282.

22. Shmakov, A.N., McLennan, N.F., McBride, P., Farquhar, C.F., Bode, J., Rennison, K.A., and Ghosh, S. (2000) Cellular prion protein is expressed in the human enteric nervous system. *Nat. Med.*, **6**, 840–841.

23. Ford, M.J., Burton, L.J., Li, H., Graham, C.H., Frobert, Y., Grassi, J., Hall, S.M., and Morris, R.J. (2002a) A marked disparity between the expression of prion protein and its message by neurones of the CNS. *Neuroscience*, **111**, 533–551.

24. Mironov, A., Jr., Latawiec, D., Wille, H., Bouzamondo-Bernstein, E., Legname, G., Williamson, R.A., Burton, D., DeArmond, S.J., Prusiner, S.B., and Peters, P.J. (2003) Cytosolic prion protein in neurons. *J. Neurosci.*, **23**, 7183–7193.

25. Moya, K.L., Sales, N., Hassig, R., Creminon, C., Grassi, J., and Di Giamberardino, L. (2000) Immunolocalization of the cellular prion protein in normal brain. *Microsc. Res. Tech.*, **50**, 58–65.

26. Bailly, Y., Haeberlé, A.M., Blanquet-Grossard, F., Chasserot-Golaz, S., Grant, N., Schulze, T., Bombarde, G., Grassi, J., Cesbron, J.Y., and Lemaire-Vieille, C. (2004) Prion protein (PrP$^C$) immunocytochemistry and expression of the green fluorescent protein reporter gene under control of the bovine PrP gene promoter in the mouse brain. *J. Comp. Neurol.*, **473**, 244–269.

27. Godsave, S.F., Wille, H., Kujala, P., Latawiec, D., DeArmond, S.J., Serban, A., Prusiner, S.B., and Peters, P.J. (2008) Cryo-immunogold electron microscopy for prions: toward identification of a conversion site. *J. Neurosci.*, **28**, 12489–12499.

28. Laine, J., Marc, M.E., Sy, M.S., and Axelrad, H. (2001) Cellular and subcellular morphological localization of normal prion protein in rodent cerebellum. *Eur. J. Neurosci.*, **14**, 47–56.

29. Fournier, J.G., Escaig-Haye, F., Billette de Villemeur, T., and Robain, O. (1995) Ultrastructural localization of cellular prion protein (PrP$^C$) in synaptic boutons of normal hamster hippocampus. *C. R. Acad. Sci. III*, **318**, 339–344.

30. Haeberlé, A.M., Ribaut-Barassin, C., Bombarde, G., Mariani, J., Hunsmann, G., Grassi, J., and Bailly, Y. (2000) Synaptic prion protein immuno-reactivity in the rodent cerebellum. *Microsc. Res. Tech.*, **50**, 66–75.

31. Sales, N., Rodolfo, K., Hassig, R., Faucheux, B., Di Giamberardino, L., and Moya, K.L. (1998) Cellular prion protein localization in rodent and primate brain. *Eur. J. Neurosci.*, **10**, 2464–2471.

32. Beringué, V., Mallinson, G., Kaisar, M., Tayebi, M., Sattar, Z., Jackson, G., Anstee, D., Collinge, J., and Hawke, S. (2003) Regional heterogenity of cellular prion protein isoforms in the mouse brain. *Brain*, **126**, 2065–2073.

33. Rodolfo, K., Hassig, R., Moya, K.L., Frobert, Y., Grassi, J., and Di Giamberardino, L. (1999) A novel cellular prion protein isoform present in rapid anterograde axonal transport. *Neuroreport*, **10**, 3639–3644.

34. Ford, M.J., Burton, L.J., Morris, R.J., and Hall, S.M. (2002b) Selective expression of prion protein in peripheral tissues of the adult mouse. *Neuroscience*, **113**, 177–192.

35. McBride, P.A., Eikelenboom, P., Kraal, G., Fraser, H., and Bruce, M.E. (1992) PrP protein is associated with follicular dendritic cells of spleens and lymph nodes in uninfected and scrapie-infected mice. *J. Pathol.*, **168**, 413–418.

36. Mattei, V., Garofalo, T., Misasi, R., Circella, A., Manganelli, V., Lucania, G., Pavan, A., and Sorice, M. (2004) Prion protein is a component of the multimolecular signaling complex involved in T cell activation. *FEBS Lett.*, **560**, 14–18.

37 Pinheiro, T.J. (2006) The role of rafts in the fibrillization and aggregation of prions. *Chem. Phys. Lipids*, **141**, 66–71.

38 Taylor, D.R., Whitehouse, I.J., and Hooper, N.M. (2009) Glypican-1 mediates both prion protein lipid raft association and disease isoform formation. *PLoS Pathog.*, **5**, e1000666.

39 Wong, C., Xiong, L.W., Horiuchi, M., Raymond, L., Wehrly, K., Chesebro, B., and Caughey, B. (2001) Sulfated glycans and elevated temperature stimulate PrP$^{Sc}$-dependent cell-free formation of protease-resistant prion protein. *EMBO J.*, **20**, 377–386.

40 Leucht, C., Simoneau, S., Rey, C., Vana, K., Rieger, R., Lasmézas, C.I., and Weiss, S. (2003) The 37kDa/67kDa laminin receptor is required for PrP$^{Sc}$ propagation in scrapie-infected neuronal cells. *EMBO Rep.*, **4**, 290–295.

41 Caughey, B., Baron, G.S., Chesebro, B., and Jeffrey, M. (2009) Getting a grip on prions: oligomers, amyloids, and pathological membrane interactions. *Annu. Rev. Biochem.*, **78**, 177–204.

42 Caughey, B., Raymond, G.J., Ernst, D., and Race, R.E. (1991) N-terminal truncation of the scrapie-associated form of PrP by lysosomal protease(s): implications regarding the site of conversion of PrP to the protease-resistant state. *J. Virol.*, **65**, 6597–6603.

43 Taylor, D.R. and Hooper, N.M. (2007) Role of lipid rafts in the processing of the pathogenic prion and Alzheimer's amyloid-beta proteins. *Semin. Cell Dev. Biol.*, **18**, 638–648.

44 Naslavsky, N., Stein, R., Yanai, A., Friedlander, G., and Taraboulos, A. (1997) Characterization of detergent-insoluble complexes containing the cellular prion protein and its scrapie isoform. *J. Biol. Chem.*, **272**, 6324–6331.

45 Stahl, N., Borchelt, D.R., and Prusiner, S.B. (1990) Differential release of cellular and scrapie prion proteins from cellular membranes by phosphatidylinositol-specific phospholipase C. *Biochemistry*, **29**, 5405–5412.

46 Baron, G.S., Wehrly, K., Dorward, D.W., Chesebro, B., and Caughey, B. (2002) Conversion of raft associated prion protein to the protease-resistant state requires insertion of PrP-res (PrP$^{Sc}$) into contiguous membranes. *EMBO J.*, **21**, 1031–1040.

47 Baron, G.S. and Caughey, B. (2003) Effect of glycosylphosphatidylinositol anchor-dependent and -independent prion protein association with model raft membranes on conversion to the protease-resistant isoform. *J. Biol. Chem.*, **278**, 14883–14892.

48 Priola, S.A. and McNally, K.L. (2009) The role of the prion protein membrane anchor in prion infection. *Prion*, **3**, 134–138.

49 Chesebro, B., Race, B., Meade-White, K., LaCasse, R., Race, R., Klingeborn, M., Striebel, J., Dorward, D., McGovern, G., and Jeffrey, M. (2010) Fatal transmissible amyloid encephalopathy: a new type of prion disease associated with lack of prion protein membrane anchoring. *PLoS Pathog.*, **6**, e1000800.

50 Bate, C., Salmona, M., Diomede, L., and Williams, A. (2004) Squalestatin cures prion-infected neurons and protects against prion neurotoxicity. *J. Biol. Chem.*, **279**, 14983–14990.

51 Kazlauskaite, J., Sanghera, N., Sylvester, I., Venien-Bryan, C., and Pinheiro, T.J. (2003) Structural changes of the prion protein in lipid membranes leading to aggregation and fibrillization. *Biochemistry*, **42**, 3295–3304.

52 Wells, G.A.H. and Wilesmith, J.W. (1995) The neuropathology and epidemiology of bovine spongiform encephalopathy. *Brain Pathol.*, **5**, 91–103.

53 Erdal, C., Goodsir, C.M., Simmons, M.M., McGovern, G., and Jeffrey, M. (2009) Abnormal prion protein is associated with changes of plasma membranes and endocytosis in bovine spongiform encephalopathy (BSE)-affected cattle brains. *Neuropathol. Appl. Neurobiol.*, **35**, 259–271.

54 Jeffrey, M., McGovern, G., Goodsir, C.M., Sisó, S., and González, L. (2009a) Strain-associated variations in abnormal PrP trafficking of sheep scrapie. *Brain Pathol.*, **19**, 1–11.

55 Jeffrey, M., Goodsir, C.M., Bruce, M.E., McBride, P.A., Scott, J.R., and Halliday, W.G. (1992) Infection specific prion protein (PrP) accumulates on neuronal plasmalemma in scrapie infected mice. *Neurosci. Lett.*, **147**, 106–109.

56 Jeffrey, M., Goodsir, C.M., Bruce, M., McBride, P.A., Scott, J.R., and Halliday, W.G. (1994) Correlative light and electron microscopy studies of PrP localisation in 87V scrapie. *Brain Res.*, **656**, 329–343.

57 González, L., Martin, S., Begara-McGorum, I., Hunter, N., Houston, F., Simmons, M., and Jeffrey, M. (2002) Effects of agent strain and host genotype on PrP accumulation in the brain of sheep naturally and experimentally affected with scrapie. *J. Comp. Pathol.*, **126**, 17–29.

58 Jeffrey, M., Goodsir, C.M., Holliman, A., Higgins, R.J., Bruce, M.E., McBride, P.A., and Fraser, J.R. (1998) Determination of the frequency and distribution of vascular and parenchymal amyloid with polyclonal and N-terminal-specific PrP antibodies in scrapie-affected sheep and mice. *Vet. Rec.*, **142**, 534–537.

59 Liberski, P.P., Guiroy, D.C., Williams, E.S., Walis, A., and Budka, H. (2001) Deposition patterns of disease-associated prion protein in captive mule deer brains with chronic wasting disease. *Acta Neuropathol.*, **102**, 496–500.

60 Bruce, M.E., McBride, P.A., and Farquhar, C.F. (1989) Precise targeting of the pathology of the sialoglycoprotein, PrP, and vacuolar degeneration in mouse scrapie. *Neurosci. Lett.*, **102**, 1–6.

61 Ersdal, C., Simmons, M.M., Goodsir, C., Martin, S., and Jeffrey, M. (2003) Subcellular pathology of scrapie: coated pits are increased in PrP codon 136 alanine homozygous scrapie-affected sheep. *Acta Neuropathol. (Berl.)*, **106**, 17–28.

62 Ersdal, C., Simmons, M.M., González, L., Goodsir, C.M., Martin, S., and Jeffrey, M. (2004) Relationships between ultrastructural scrapie pathology and patterns of abnormal prion protein accumulation. *Acta Neuropathol. (Berl.)*, **107**, 428–438.

63 McGovern, G. and Jeffrey, M. (2007) Scrapie-specific pathology of sheep lymphoid tissues. *PLoS ONE*, **2**, e1304.

64 McGovern, G., Mabbott, N., and Jeffrey, M. (2009) Scrapie affects the maturation cycle and immune complex trapping by follicular dendritic cells in mice. *PLoS ONE*, **4**, e8186.

65 Chiesa, R., Piccardo, P., Ghetti, B., and Harris, D.A. (1998) Neurological illness in transgenic mice expressing a prion protein with an insertional mutation. *Neuron*, **21**, 1339–1351.

66 Jeffrey, M., Goodsir, C.M., Race, R.E., and Chesebro, B. (2004) Scrapie-specific neuronal lesions are independent of neuronal PrP expression. *Ann. Neurol.*, **55**, 781–792.

67 Jeffrey, M., Goodsir, C., McGovern, G., Barmada, S.J., Medrano, A.Z., and Harris, D.A. (2009b) Prion protein with an insertional mutation accumulates on axonal and dendritic plasmalemma and is associated with distinctive ultrastructural changes. *Am. J. Pathol.*, **175**, 1208–1217.

68 Heggebø, R., Press, C.M., Gunnes, G., Ulvund, M.J., Tranulis, M.A., and Landsverk, T. (2003) Detection of $PrP^{Sc}$ in lymphoid tissues of lambs experimentally exposed to the scrapie agent. *J. Comp. Pathol.*, **128**, 172–181.

69 van Keulen, L.J., Bossers, A., and van Zijderveld, F. (2008) TSE pathogenesis in cattle and sheep. *Vet. Res.*, **39**, 24.

70 Jeffrey, M., McGovern, G., Goodsir, C.M., Brown, K.L., and Bruce, M.E. (2000) Sites of prion protein accumulation in scrapie-infected mouse spleen revealed by immuno-electron microscopy. *J. Pathol.*, **191**, 323–332.

71 Jeffrey, M., Goodsir, C.M., Bruce, M.E., McBride, P.A., and Fraser, J.R. (1997) *In vivo* toxicity of prion protein in murine scrapie: ultrastructural and immunogold studies. *Neuropathol. Appl. Neurobiol.*, **23**, 93–101.

72 Hinshaw, J.E. (2000) Dynamin and its role in membrane fission. *Annu. Rev. Cell Dev. Biol.*, **16**, 483–519.

73 Takei, K., Slepnev, V.I., Haucke, V., and De Camilli, P. (1999) Functional partnership between amphiphysin and dynamin in clathrin-mediated endocytosis. *Nat. Cell Biol.*, **1**, 33–39.

74 Haglund, K., Di Fiore, P.P., and Dikic, I. (2003) Distinct monoubiquitin signals in receptor endocytosis. *Trends Biochem. Sci.*, **28**, 598–603.

75 Lowe, J., McDermott, H., Kenward, N., Landon, M., Mayer, R.J., Bruce, M., McBride, P., Somerville, R.A., and Hope, J. (1990) Ubiquitin conjugate immunoreactivity in the brains of scrapie infected mice. *J. Pathol.*, **162**, 61–66.

76 László, L., Lowe, J., Self, T., Kenward, N., Landon, M., McBride, T., Farquhar, C., McConnell, I., Brown, J., and Hope, J. (1992) Lysosomes as key organelles in the pathogenesis of prion encephalopathies. *J. Pathol.*, **166**, 333–341.

77 Jeffrey, M., Martin, S., and González, L. (2003) Cell-associated variants of disease-specific prion protein immunolabelling are found in different sources of sheep transmissible spongiform encephalopathy. *J. Gen. Virol.*, **84**, 1033–1045.

78 Jeffrey, M., Goodsir, C., Bruce, M., McBride, P., and Fraser, J. (1996) Subcellular localization and toxicity of pre-amyloid and fibrillar prion protein accumulations in murine scrapie, in *Transmissible Subacute Spongiform Encephalopathies: Prion Diseases* (eds. L. Court and B. Dodet), Elsevier, Paris, pp. 129–135.

79 Kovács, G.G., Gelpi, E., Strobel, T., Ricken, G., Nyengaard, J.R., Bernheimer, H., and Budka, H. (2007) Involvement of the endosomal–lysosomal system correlates with regional pathology in Creutzfeldt–Jakob disease. *J. Neuropathol. Exp. Neurol.*, **66**, 628–636.

80 Kazlauskaite, J., Young, A., Gardner, C.E., Macpherson, J.V., Venien-Bryan, C., and Pinheiro, T.J. (2005) An unusual soluble beta-turn-rich conformation of prion is involved in fibril formation and toxic to neuronal cells. *Biochem. Biophys. Res. Commun.*, **328**, 292–305.

81 McGovern, G., Brown, K.L., Bruce, M.E., and Jeffrey, M. (2004) Murine scrapie infection causes an abnormal germinal centre reaction in the spleen. *J. Comp. Pathol.*, **130**, 181–194.

82 Pelkmans, L. and Helenius, A. (2002) Endocytosis via caveolae. *Traffic*, **3**, 311–320.

83 Kartenbeck, J., Stukenbrok, H., and Helenius, A. (1989) Endocytosis of simian virus 40 into the endoplasmic reticulum. *J. Cell Biol.*, **109**, 2721–2729.

84 Benlimame, N., Simard, D., and Nabi, I.R. (1995) Autocrine motility factor receptor is a marker for a distinct membranous tubular organelle. *J. Cell Biol.*, **129**, 459–471.

85 Sprocati, T., Ronchi, P., Raimondi, A., Francolini, M., and Borgese, N. (2006) Dynamic and reversible restructuring of the ER induced by PDMP in cultured cells. *J. Cell Sci.*, **119**, 3249–3260.

86 Park, C.S. and Choi, Y.S. (2005) How do follicular dendritic cells interact intimately with B cells in the germinal centre? *Immunology*, **114**, 2–10.

87 Rademakers, L.H. (1992) Dark and light zones of germinal centres of the human tonsil: an ultrastructural study with emphasis on heterogeneity of follicular dendritic cells. *Cell Tissue Res.*, **269**, 359–368.

88 Heinen, E., Bosseloir, A., and Bouzahzah, F. (1995) Follicular dendritic cells: origin and function. *Curr. Top. Microbiol. Immunol.*, **201**, 15–47.

89 Raeber, A.J., Race, R.E., Brandner, S., Priola, S.A., Sailer, A., Bessen, R.A., Mucke, L., Manson, J., Aguzzi, A., Oldstone, M.B., Weissmann, C., and Chesebro, B. (1997) Astrocyte-specific expression of hamster prion protein (PrP) renders PrP knockout mice susceptible to hamster scrapie. *EMBO J.*, **16**, 6057–6065.

90 Vella, L.J., Sharples, R.A., Lawson, V.A., Masters, C.L., Cappai, R., and Hill, A.F. (2007) Packaging of prions into exosomes is associated with a novel pathway of PrP processing. *J. Pathol.*, **211**, 582–590.

91 Fevrier, B., Vilette, D., Archer, F., Loew, D., Faigle, W., Vidal, M., Laude, H., and Raposo, G. (2004) Cells release prions in association with exosomes. *Proc. Natl. Acad. Sci. USA*, **101**, 9683–9688.

92 Fauré, J., Lachenal, G., Court, M., Hirrlinger, J., Chatellard-Causse, C., Blot, B., Grange, J., Schoehn, G., Goldberg, Y., Boyer, V., Kirchhoff, F., Raposo, G., Garin, J., and Sadoul, R. (2006) Exosomes are released by cultured cortical neurones. *Mol. Cell. Neurosci.*, **31**, 642–648.

93 Denzer, K., Kleijmeer, M.J., Heijnen, H.F., Stoorvogel, W., and Geuze, H.J. (2000a) Exosome: from internal vesicle of the multivesicular body to intercellular signaling device. *J. Cell Sci.*, **113**, 3365–3374.

94 Denzer, K., van Eijk, M., Kleijmeer, M.J., Jakobson, E., de Groot, C., and Geuze, H.J. (2000b) Follicular dendritic cells carry MHC class II-expressing microvesicles at their surface. *J. Immunol.*, **165**, 1259–1265.

95 Gousset, K. and Zurzolo, C. (2009) Tunnelling nanotubes: a highway for prion spreading? *Prion*, **3**, 94–98.

96 Ilangumaran, S., Robinson, P.J., and Hoessli, D.C. (1996) Transfer of exogenous glycosylphosphatidylinositol (GPI)-linked molecules to plasma membranes. *Trends Cell Biol.*, **6**, 163–167.

97 Liu, T., Li, R., Pan, T., Liu, D., Petersen, R.B., Wong, B.S., Gambetti, P., and Sy, M.S. (2002) Intercellular transfer of the cellular prion protein. *J. Biol. Chem.*, **277**, 47671–47678.

98 Casalone, C., Zanusso, G., Acutis, P., Ferrari, S., Capucci, L., Tagliavini, F., Monaco, S., and Caramelli, M. (2004) Identification of a second bovine amyloidotic spongiform encephalopathy: molecular similarities with sporadic Creutzfeldt–Jakob disease. *Proc. Natl. Acad. Sci. USA*, **101**, 3065–3070.

99 Chesebro, B., Trifilo, M., Race, R., Meade-White, K., Teng, C., LaCasse, R., Raymond, L., Favara, C., Baron, G., Priola, S., Caughey, B., Masliah, E., and Oldstone, M. (2005) Anchorless prion protein results in infectious amyloid disease without clinical scrapie. *Science*, **308**, 1435–1439.

100 Brückner, G., Brauer, K., Härtig, W., Wolff, J.R., Rickmann, M.J., Derouiche, A., Delpech, B., Girard, N., Oertel, W.H., and Reichenbach, A. (1993) Perineuronal nets provide a polyanionic, glia-associated form of microenvironment around certain neurons in many parts of the rat brain. *Glia*, **8**, 183–200.

101 Brückner, G., Szeöke, S., Pavlica, S., Grosche, J., and Kacza, J. (2006) Axon initial segment ensheathed by extracellular matrix in perineuronal nets. *Neuroscience*, **138**, 365–375.

102 Belichenko, P.V., Miklossy, J., Belser, B., Budka, H., and Celio, M.R. (1999) Early destruction of the extracellular matrix around parvalbumin-immunoreactive interneurons in Creutzfeldt–Jakob disease. *Neurobiol. Dis.*, **6**, 269–279.

103 Vidal, E., Bolea, R., Tortosa, R., Costa, C., Domènech, A., Monleón, E., Vargas, A., Badiola, J.J., and Pumarola, M. (2006) Assessment of calcium-binding proteins (parvalbumin and calbindin D-28K) and perineuronal nets in normal and scrapie-affected adult sheep brains. *J. Virol. Methods*, **136**, 137–146.

104 Pan, T., Wong, B.S., Liu, T., Li, R., Petersen, R.B., and Sy, M.S. (2002) Cell-surface prion protein interacts with glycosaminoglycans. *Biochem. J.*, **368**, 81–90.

105 Parkin, E.T., Watt, N.T., Hussain, I., Eckman, E.A., Eckman, C.B., Manson, J.C., Baybutt, H.N., Turner, A.J., and Hooper, N.M. (2007) Cellular prion protein regulates beta-secretase cleavage of the Alzheimer's amyloid precursor protein. *Proc. Natl. Acad. Sci. USA*, **104**, 11062–11067.

106 Shyng, S.L., Lehmann, S., Moulder, K.L., and Harris, D.A. (1995) Sulfated glycans stimulate endocytosis of the cellular isoform of the prion protein, $PrP^C$, in cultured cells. *J. Biol. Chem.*, **270**, 30221–30229.

107 McBride, P.A., Wilson, M.I., Eikelenboom, P., Tunstall, A., and Bruce, M.E. (1998) Heparan sulfate proteoglycan is associated with amyloid plaques and neuroanatomically targeted

PrP pathology throughout the incubation period of scrapie-infected mice. *Exp. Neurol.*, **149**, 447–454.

108 Snow, A.D., Kisilevsky, R., Willmer, J., Prusiner, S.B., and DeArmond, S.J. (1989) Sulfated glycosaminoglycans in amyloid plaques of prion diseases. *Acta Neuropathol.*, **77**, 337–342.

109 Revesz, T., Holton, J.L., Lashley, T., Plant, G., Rostagno, A., Ghiso, J., and Frangione, B. (2002) Sporadic and familial cerebral amyloid angiopathies. *Brain Pathol.*, **12**, 343–357.

110 González, L., Martin, S., and Jeffrey, M. (2003) Distinct profiles of PrP$^d$ immunoreactivity in the brain of scrapie- and BSE-infected sheep: implications for differential cell targeting and PrP processing. *J. Gen. Virol.*, **84**, 1339–1350.

111 Bruce, M., Chree, A., Williams, E.S., and Fraser, H. (2000) Perivascular PrP amyloid in the brains of mice infected with chronic wasting disease. *Brain Pathol.*, **10**, 662–663.

112 Kayed, R., Sokolov, Y., Edmonds, B., McIntire, T.M., Milton, S.C., Hall, J.E., and Glabe, C.G. (2004) Permeabilization of lipid bilayers is a common conformation-dependent activity of soluble amyloid oligomers in protein misfolding diseases. *J. Biol. Chem.*, **279**, 46363–46366.

113 Winklhofer, K.F. and Tatzelt, J. (2006) The role of chaperones in Parkinson's disease and prion diseases. *Handb. Exp. Pharmacol.*, **172**, 221–258.

114 Bruce, M.E., Boyle, A., Cousens, S., McConnell, I., Foster, J., Goldmann, W., and Fraser, H. (2002) Strain characterization of natural sheep scrapie and comparison with BSE. *J. Gen. Virol.*, **83**, 695–704.

# 11
# Interaction of Alzheimer Amyloid Peptide with Cell Surfaces and Artificial Membranes

*David A. Bateman and Avijit Chakrabartty*

## 11.1
## Introduction

The hallmarks of Alzheimer's disease (AD), first characterized by Alois Alzheimer in 1907 [1], include extensive loss of neurons and the presence of intracellular neurofibrillary tangles and extracellular senile plaques. The neurofibrillary tangles are predominantly comprised of intraneuronal inclusions of paired helical filaments of hyperphosphorylated tau protein. The senile plaques display red–green birefringence when stained with Congo Red, the definitive diagnostic test for the presence of amyloid. Glenner and Wong in 1984 successfully isolated and purified senile plaques, and determined that the major constituent was a peptide [2]. This peptide, denoted amyloid β (Aβ), varies from 39 to 43 amino acids in length, with the most abundant forms being 40 and 42 amino acids ($A\beta_{40}$ and $A\beta_{42}$, respectively). $A\beta_{40}$ and $A\beta_{42}$ can form amyloid fibrils, but are also associated with other structural forms in the progression to the fibril state. The monomeric form of the Aβ peptide has been generally considered as not being the major neurotoxic species. *In vitro* studies have indicated the formation of fibril intermediates prior to fibril formation including Aβ dimers, trimers, and tetramers, large spherical particles, and protofibrils [3–10]. One group of fibril intermediates, referred to as amyloid-derived diffusible ligands (ADDLs), have been shown to be potent neurotoxins that kill neurons in cultured hippocampal brain slices from mice, and have the ability to abolish long-term potentiation [11]. Amyloid fibrils are unbranched fibers 6–10 nm in diameter that display red–green birefringence when stained with Congo Red and viewed between

*Lipids and Cellular Membranes in Amyloid Diseases*, First Edition. Edited by Raz Jelinek.
© 2011 Wiley-VCH Verlag GmbH & Co. KGaA. Published 2011 by Wiley-VCH Verlag GmbH & Co. KGaA.

crossed polarizers. $A\beta_{40}$ and $A\beta_{42}$ have also been shown to co-incorporate into fibrillar aggregates [12].

## 11.2
## Comparison of the Neurotoxicity of Oligomeric and Fibrillar Alzheimer Amyloid Peptides

The current focus of toxicity is with fibril intermediates, and whether they are more or less toxic than the mature fibrils. Initially, only the fibrillar form of $A\beta$ peptide and not amorphous aggregates of $A\beta$ were shown to cause toxicity in hippocampal cultures [13]. However, later studies utilizing $A\beta$ peptide solutions that were devoid of amyloid fibrils demonstrated that small, diffusible $A\beta$ oligomers (ADDLs) were neurotoxic to hippocampal cultures [11]. Additionally, $A\beta$ peptide oligomerization and formation of sodium dodecyl sulfate-stable oligomers have been shown to occur at low nanomolar levels [14, 15]. Another intermediate in the progression of $A\beta$ fibrillogenesis that has been characterized and found to be cytotoxic is the protofibril [16]. Protofibrils evolve into to amyloid fibrils and have similar secondary structure, but are in equilibrium with low molecular weight $A\beta$ peptide molecules [16]. The toxicity of monomeric, oligomeric, and fibrillar structural species of $A\beta_{40}$ and $A\beta_{42}$ was recently analyzed with Neuro-2A neuroblastoma cells [17]. From this study, low concentrations of both unaggregated $A\beta_{40}$ and $A\beta_{42}$ exhibited a neurotrophic effect at low concentrations (1–100 nM) and neurotoxicity at higher concentrations (1–15 µM) [17]. The oligomeric form of $A\beta_{42}$ inhibited neuronal viability at 10-fold lower concentrations than the fibrillar form [17]. The oligomeric form of $A\beta_{40}$ was 50-fold less toxic than $A\beta_{42}$ oligomeric form, but the $A\beta_{40}$ did not form fibrils under the same conditions as used for $A\beta_{42}$ [17]. Due to the difficulty in forming the oligomeric and fibril structural forms with $A\beta_{40}$, a direct comparison with $A\beta_{42}$ becomes difficult with the results reported. Also, the conformation of the various $A\beta$ species could have altered during the 20 h incubation period with these cells. The variability in the cytotoxic activity of $A\beta$ preparations reported in the literature has recently been proposed to be caused by the structural heterogeneity present in commercial preparations, which may contain variable amounts of oligomers, fibrils, or non-specific aggregates [18]. In an alternative approach, secreted oligomers and monomers of $A\beta$ peptide collected from cell culture media were microinjected into cerebral hippocampus of rats [19]. In that study, fibrils were removed by sedimentation, and the $A\beta$ oligomers and monomers were found to inhibit hippocampal long-term potentiation [19]. With removal of monomers from oligomer solutions, long-term potentiation was still inhibited, indicating that oligomeric $A\beta$ is the entity responsible for the inhibition [19]. This method of cell-excreted $A\beta$ peptide may offer a way to avoid issues observed with commercially purchased $A\beta$ peptides, but would require additional characterization of solutions to determine peptide lengths and species present. Also, this method offers limitations to $A\beta$ modifications, for studying peptide mutants or addition of peptide markers.

## 11.3
### Aβ Oligomerization at the Cell Surface

Aβ was not found to oligomerize within human cerebrospinal fluid or conditioned media from Chinese hamster ovaries [20]. Similarly, we also observed no significant oligomerization of Aβ when monitored by analytical ultracentrifugation with conditioned media obtained from PC12, N2A, and SH5Y neuronal cell lines. These findings indicate that Aβ oligomerization would occur only through interaction with cells. We developed a method to visualize directly the interactions of monomeric Aβ with neuronal cell lines [21–23]. We have demonstrated previously that attaching a fluorescent label to the N-terminus of Aβ via a flexible glycine linker does not alter its amyloidogenic properties [23]. We note that the N-terminus of Aβ peptides is more accessible and less involved in amyloidogenesis than other regions of the sequence, hence it is likely that fluorescent labeling of the N-terminus is likely to have the least influence on amyloidogenesis. Using this method, we have observed direct Aβ aggregation on the surface of neuronal cell lines (Figure 11.1). Interestingly, the rate of Aβ cell association correlates with its propensity to aggregate [23]. $A\beta_{42}$ shows a fast

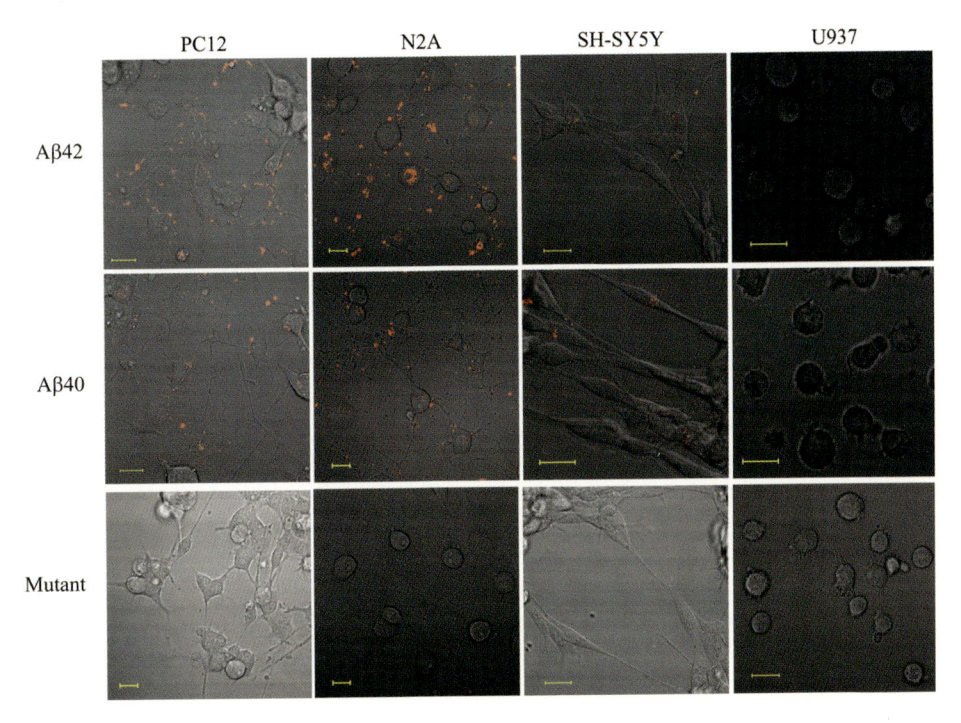

**Figure 11.1** Confocal microscopy images of various cell lines treated for 12 h with TMR-labeled peptides. All scale bars are 20 μm. The fluorescent emissions of TMR-labeled peptides are indicated in red. $A\beta_{42}$ displayed the greatest cell surface binding among the peptides to the differentiated neuronal cell lines (PC12, N2A, and SH-SY5Y). The mutant peptide was not observed to associate with any of the cell lines tested. None of the peptides tested associated with the U937 cells.

binding phase (<1 h) in which ~50% of the cells are associated with the peptide. This is followed by a slower binding phase that occurs over a 24 h period, after which the fluorescence intensity begins to decline. The fast binding phase is missing in the binding kinetics of $A\beta_{40}$; however, in all other respects, the kinetic features of $A\beta_{40}$ binding are similar to those of $A\beta_{42}$. The inability of $A\beta_{40}$ to display the fast binding phase that was exhibited by $A\beta_{42}$ may provide an explanation for why $A\beta_{42}$ is more toxic than $A\beta_{40}$. We have also found that preaggregated $A\beta_{40}$ associates rapidly to PC12 cells and is internalized, but at a slower rate than preaggregated $A\beta_{42}$.

## 11.4
### Catalysis of Aβ Oligomerization by the Cell Surface

Studies with purified Aβ have shown that the critical concentration of Aβ aggregation in cell-free systems is in the micromolar range [7]. The punctate staining observed in Figure 11.1 was obtained with 1.5 μM peptide; thus, cell surface binding appears to catalyze Aβ aggregation, allowing for aggregates to form at concentrations well below the critical concentration for aggregation in benign buffer [7]. On lowering the concentration of $A\beta_{40/42}$ to 50 nM, the lower limit of detection for the confocal microscope used for these experiments, produced a similar punctate staining pattern (Figure 11.2). This concentration is within the estimated physiological concentration of Aβ [24, 25]. Therefore, binding of Aβ to cell surface receptors appears to reduce drastically the critical concentration for Aβ aggregation.

Our investigation of the interaction of $A\beta_{40}$ and $A\beta_{42}$ with NGF differentiated PC-12 cells at concentrations approaching physiological levels has implications for the mechanism of AD. While genetic evidence from familial AD and transgenic mice studies support the amyloid cascade hypothesis [26], one shortcoming of the hypothesis is that unnaturally high concentrations of Aβ are required to demonstrate aggregation *in vitro*. When starting with Aβ preparations that are devoid of aggregation seeds, the critical concentration for aggregation at neutral pH is 10–40 μM for $A\beta_{40}$ and approximately fivefold less for $A\beta_{42}$ [7]. Most *in vitro* studies of Aβ aggregation use 5–10-fold higher concentrations [6, 27]. This raises the dilemma of how Aβ aggregates can form *in vivo*, where the Aβ concentration is considerably lower than the critical concentration [24, 25]. The presence of amyloid plaques and the identification of Aβ oligomers in AD brain [28, 29] clearly demonstrate that aggregation occurs *in vivo*; therefore, factors in the *in vivo* environment must catalyze the aggregation process. The presence of these *in vivo* catalysts is exemplified by studies utilizing brain slices, where inhibition of long-term potentiation was observed at Aβ concentrations as low as 5 nM [11].

## 11.5
### Type of Aβ Complexes that Form on the Cell Surface

To study Aβ on the surface of live cells, we synthesized and fluorescently labeled $A\beta_{42}$. We covalently attached either 6-carboxyfluorescein (FAM) or tetramethylrhodamine

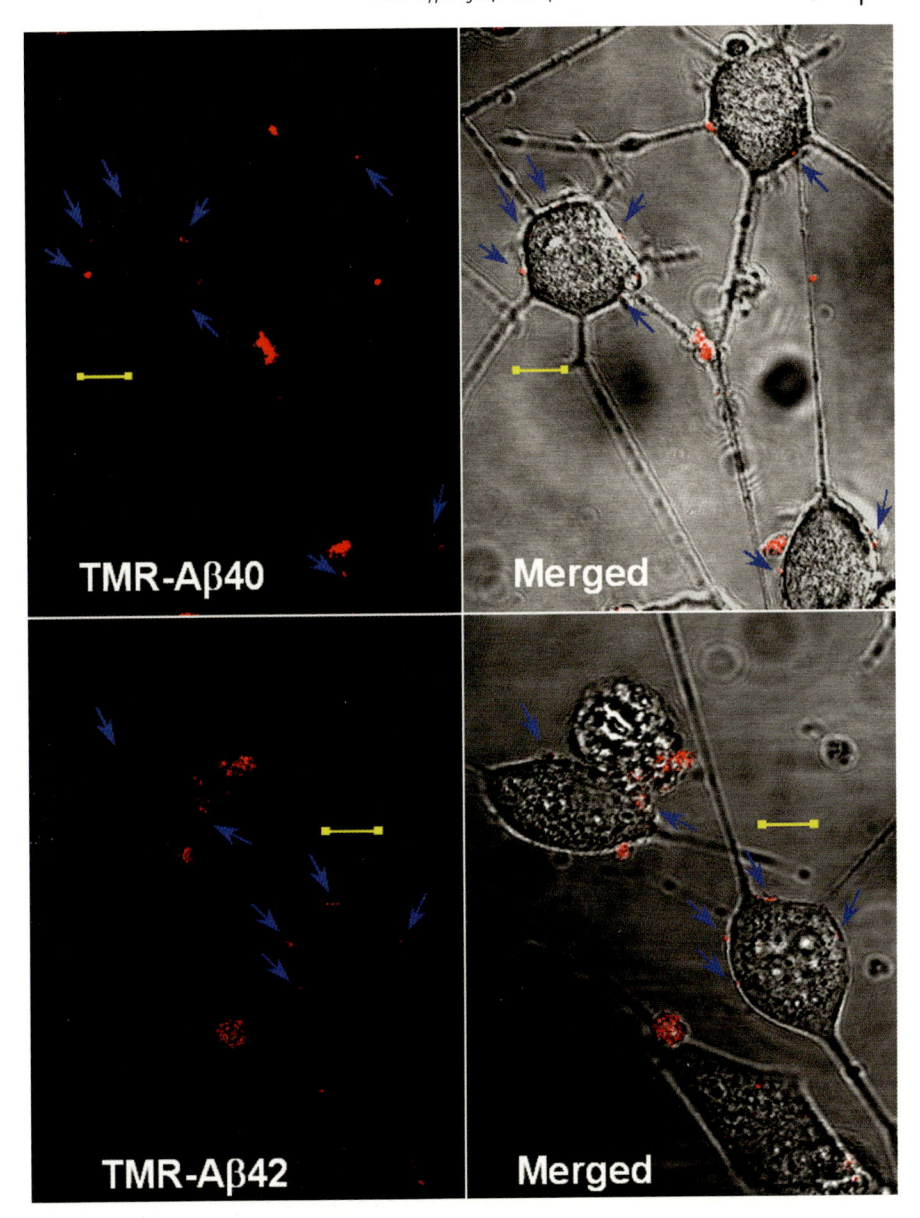

**Figure 11.2** Punctate stain of Aβ with live PC12 cells at physiological concentration. All scale bars are 10 μm. The distinct punctate staining patterns (indicated with blue arrows) are visible with treatment of 50 nM for both (a) Aβ$_{40}$ and (b) Aβ$_{42}$ after staining for 24 h. Merged phase contrast images are included to the right for (a) and (b).

(TMR) to the N-terminus of Aβ via a flexible glycine linker to generate FAM-Aβ$_{42}$ or TMR-Aβ$_{42}$, respectively. The N-terminus is highly accessible [30, 31] and attaching a fluorescent label has been shown not to alter either its amyloidogenic properties [23, 32, 33] or its solubility behavior [34, 35]. FAM and TMR fluorophores were selected because they do not selectively partition into any particular subcellular organelle or microenvironment [36, 37] and their fluorescence properties are ideal for confocal microscopy [38, 39].

We compared the effects of treating live differentiated PC12 cells (which are derived from a pheochromocytoma of the rat adrenal medulla) with FAM-Aβ$_{42}$ alone or in combination with TMR-Aβ$_{42}$ (Mix) [22]. After 24 h treatments with either FAM-Aβ$_{42}$ or Mix, confocal microscopy of a plane within individual PC12 cells was imaged. The distinct staining pattern of Aβ$_{42}$ was observed on the cell body, on neuritic extensions, and within the cell (Figure 11.3). Highlighted in yellow are regions where strong energy transfer between these fluorophores was detected (Figure 11.3f). FAM- and TMR-labeled Aβ$_{42}$ appeared to make an excellent donor–acceptor Förster or fluorescence resonance energy transfer (FRET) pair; in the Mix sample, we observed quenching of FAM emission and increased TMR fluorescence when only FAM was excited. We observed photobleaching fluorescence resonance energy transfer (pbFRET) for specific aggregates, where the excited FAM transfers its energy to TMR, reducing its photodegradation rate. Selecting two aggregates that display co-localization (Figure 11.3a and b) and photobleaching for 527 s, we observed that the first aggregate photobleached at a similar rate to FAM alone, reflected by similar time constants (Figure 11.3j). The second aggregate photobleached much more slowly with a longer time constant, indicating energy transfer (Figure 11.3k). Similar time constants were observed for the treatment of living PC12 cells with FAM alone for various treatment times [22]. For the Mix samples, we observed increasing average time constants with increasing treatment time. Monitoring aggregates on the cell body versus the cell neurite revealed that the average transfer efficiency for the aggregates located on the cell body after 24 h of treatment was similar to that for aggregates on the neurites after only 6 h of treatment [22]. However, the distribution of transfer efficiency for the various time points seemed to indicate two key populations: aggregates that have no energy transfer and aggregates that are transferring energy to varying degrees. These two aggregate populations on the cell surface indicate that there is a population of aggregates having peptide units closely packed to allow energy transfer and another aggregate population where the peptide units are spaced apart and incapable of energy transfer. This distribution is very different when comparing aggregates on the surface of living PC12 cells to internalized aggregates. The internalized aggregates were observed to have an average transfer efficiency of 0.6, with most of the observed aggregates allowing for energy transfer [22]. This finding suggests that the process of internalization causes increased compaction of the aggregates.

To investigate which regions of Aβ$_{42}$ within the aggregates were exposed, we treated live PC12 cells with unlabeled Aβ$_{42}$ for 2 h followed by treatment with one of three monoclonal mouse antibodies targeted to a specific region of Aβ$_{42}$. Antibodies were added in high excess to PC12 cells, as their detection concentration was

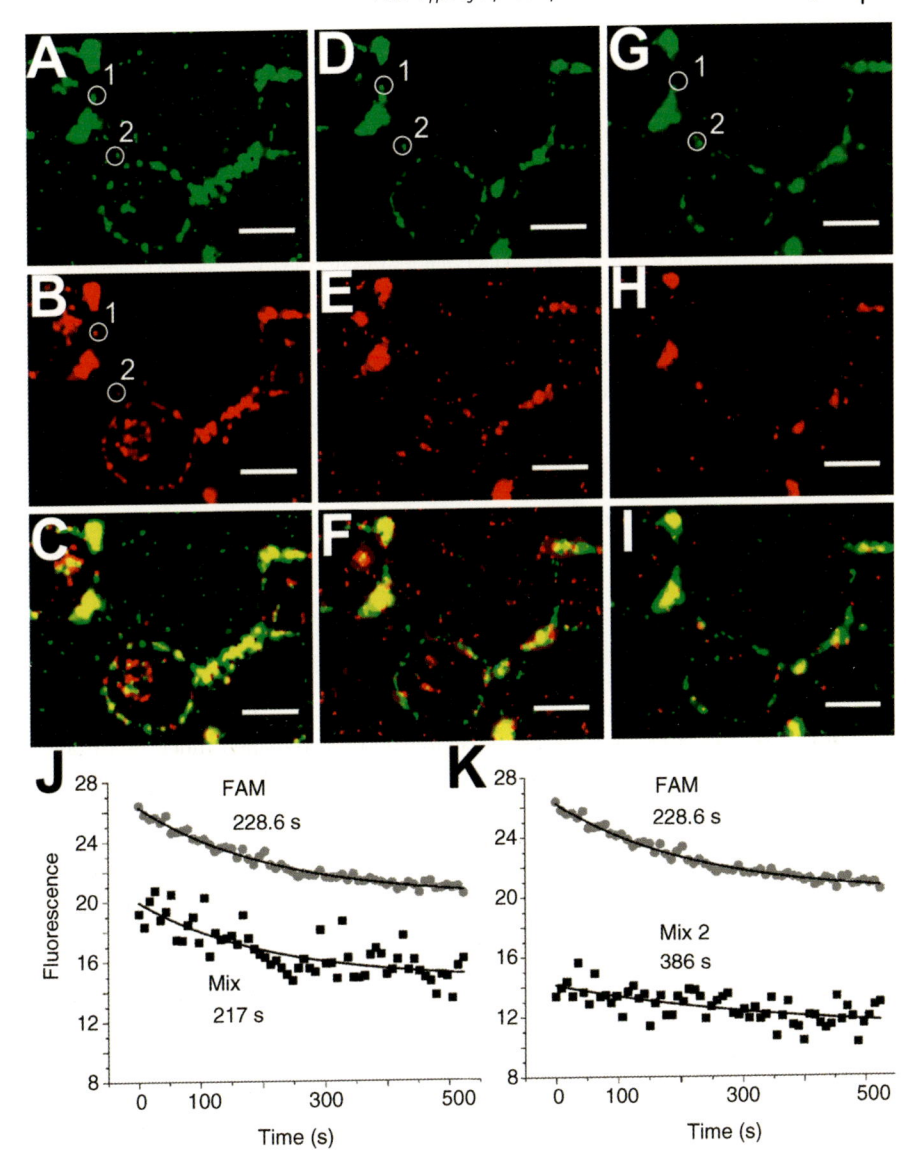

11.5 Type of Aβ Complexes that Form on the Cell Surface | 237

**Figure 11.3** Co-localization and pbFRET aggregates of Aβ$_{42}$ with live PC12 cells treated for 24 h. All images were taken through the center of the cells. All scale bars are 10 μm. (a) Mix sample monitoring with BP filter for total FAM signal. (b) Mix sample monitoring with BP filter for total TMR signal. (c) Merged imaged of FAM and TMR total signals. (d) Mix sample exciting FAM and monitoring initial FAM signal at spectral wavelength 527 nm. (e) Mix sample exciting FAM and monitoring initial TMR signal at spectral wavelength 591 nm. (f) Merge of initial FAM and TMR signals. (g) Mix sample exciting FAM and monitoring final FAM signal at spectral wavelength 527 nm. (h) Mix sample exciting FAM and monitoring final TMR signal at spectral wavelength 591 nm. (i) Merge of final FAM and TMR signals after bleaching for 527 s. (j) Fluorescent decay curve for Mix region 1 compared with FAM alone decay curve. (k) Fluorescent decay curve for Mix region 2 compared with FAM alone decay curve.

observed to vary by dot blot analysis [22]. Using flow cytometry, two key populations of aggregates were also observed: one that stained with all three antibodies and another that stained primarily with the internal sequence and N-terminus specific antibodies. The exclusion of the C-terminus antibody indicated that these aggregates may be more compact with the C-termini buried and represent a smaller fraction of the total number of observed events [22].

In order to assess further the level of aggregate compaction, we utilized potassium iodide (KI), which is a small molecule quencher of fluorescence [40, 41]. After using KI to treat aggregates that formed on the surface of PC12 cells, we found that the majority of the aggregates were quenched. With increasing exposure time of the cells to $A\beta_{42}$ there were visually more KI-resistant aggregates on the surface of living PC12 cells [22]. However, a distribution of surface aggregates was observed with KI-sensitive aggregates and KI-resistant aggregates, even for the 24 h $A\beta_{42}$ treatment.

## 11.6
### Association of Alzheimer Amyloid Peptides with Lipid Particles

Addition of $A\beta$ peptide to cultured neurons is known to cause the release of lipid particles in a dose- and time-dependent manner [42]. It was observed that oligomeric but not fibrillar or monomeric $A\beta_{40}$ caused the release of cholesterol, phospholipids, and GM1 in $A\beta$-associated lipid particles [42]. Lipid release is an active cellular process that involves protein kinase C (PKC) signaling [43], and treatment with PKC inhibitors completely inhibited lipid release induced by oligomeric $A\beta_{40}$ [42, 44]. It has also been reported previously that $A\beta$ peptide can stimulate PKC in PC12 cells [24]. The $A\beta_{40}$ lipid particles that were generated were unable to bind to cells and become internalized through regular processing routes [42]. Interactions of $A\beta$ peptides with purified plasma, endosomal, lysosomal, and Golgi membranes have also been studied, and the results indicate that plasma, endosomal, and lysosomal membranes accelerate $A\beta$ fibrillogenesis [45]. An increased surface organization was also observed with the plasma membrane [45]. $A\beta$ peptides are known to bind to GM1 ganglioside with surprising specificity [46–49]. However, a recent study also indicated the association of $A\beta_{40}$ to phosphatidylserine [50, 51]. Using annexin V, a phosphatidylserine-specific marker, competitive inhibition of $A\beta$ toxicity was observed with PC12 cells [50]. Further, PC12 cells that are more prone to $A\beta$ association have also been shown to bind more annexin V, implicating phosphatidylserine for the enhanced $A\beta$ association [51]. These studies with GM1 and phosphatidylserine indicate possible lipid receptors for $A\beta$ peptide.

## 11.7
### Future Directions

Previous studies have used dyes, radioactive $A\beta$, and antibodies to study the interactions of $A\beta$ with cell cultures [28, 52–59]. Our results reveal similar punctate staining as imaged previously with antibodies directed specifically against oligomeric structural forms of Aβ [28, 57]. However, our technique allows images to be obtained with live cells throughout the progression of Aβ aggregation, and through its multitude of intermediate states. The major drawbacks of endpoint antibody detection with fixed cells or limited detection of specific conformations of Aβ are avoided by using our approach. This approach will be especially beneficial for developing therapeutic treatments that target specific conformations of Aβ, as the global effect of Aβ on the cell can be monitored.

With hints at possible cellular interaction sites, and the multiple structures that seem to be forming on the living cell surface, further investigations need to be performed on the sequence of events governing Aβ peptide association with the surface of cells. The mechanism by which the Aβ peptides associate with the cell surface and induce neuronal loss needs to be clarified. Elucidation of the cellular active form of Aβ and identification of its specific interaction sites can pave the way to therapeutic treatments that not only prevent onset of AD but also lead to clearing of established plaques.

## Acknowledgments

This work was supported by a grant from Canadian Institutes of Health Research (to A.C.) and from the Natural Science and Engineering Research Council of Canada and a Scace Fellowship (to D.A.B.).

## References

1 Alzheimer, A. (1907) Uber eine eigenartige Erkankung der Hirnrinde. *Allg. Z Psychiatr. Psych.-Gericht. Med.,* **64**, 146–148.

2 Glenner, G.G. and Wong, C.W. (1984) Alzheimer's disease: initial report of the purification and characterization of a novel cerebrovascular amyloid protein. *Biochem. Biophys. Res. Commun.,* **120**, 885–890.

3 Huang, T.H., Yang, D.S., Plaskos, N.P., Go, S., Yip, C.M., Fraser, P.E., and Chakrabartty, A. (2000) Structural studies of soluble oligomers of the Alzheimer beta-amyloid peptide. *J. Mol. Biol.,* **297**, 73–87.

4 Nybo, M., Svehag, S.E., and Holm Nielsen, E. (1999) An ultrastructural study of amyloid intermediates in A beta1–42 fibrillogenesis. *Scand. J. Immunol.,* **49**, 219–223.

5 Blackley, H.K., Sanders, G.H., Davies, M.C., Roberts, C.J., Tendler, S.J., and Wilkinson, M.J. (2000) In-*situ* atomic force microscopy study of beta-amyloid fibrillization. *J. Mol. Biol.,* **298**, 833–840.

6 Walsh, D.M., Lomakin, A., Benedek, G.B., Condron, M.M., and Teplow, D.B. (1997) Amyloid beta-protein fibrillogenesis. Detection of a protofibrillar intermediate. *J. Biol. Chem.,* **272**, 22364–22372.

7 Harper, J.D., Wong, S.S., Lieber, C.M., and Lansbury, P.T. (1997) Observation of metastable Abeta amyloid protofibrils by atomic force microscopy. *Chem. Biol.,* **4**, 119–125.

8 Stine, W.B.J., Snyder, S.W., Ladror, U.S., Wade, W.S., Miller, M.F., Perun, T.J.,

Holzman, T.F., and Krafft, G.A. (1996) The nanometer-scale structure of amyloid-beta visualized by atomic force microscopy. *J. Protein Chem.*, **15**, 193–203.

9 Bitan, G., Lomakin, A., and Teplow, D.B. (2001) Amyloid beta-protein oligomerization: prenucleation interactions revealed by photo-induced cross-linking of unmodified proteins. *J. Biol. Chem.*, **276**, 35176–35184.

10 Gorman, P.M., Yip, C.M., Fraser, P.E., and Chakrabartty, A. (2003) Alternate aggregation pathways of the Alzheimer beta-amyloid peptide: Abeta association kinetics at endosomal pH. *J. Mol. Biol.*, **325**, 743–757.

11 Lambert, M.P., Barlow, A.K., Chromy, B.A., Edwards, C., Freed, R., Liosatos, M., Morgan, T.E., Rozovsky, I., Trommer, B., Viola, K.L., Wals, P., Zhang, C., Finch, C.E., Krafft, G.A., and Klein, W.L. (1998) Diffusible, nonfibrillar ligands derived from Abeta1–42 are potent central nervous system neurotoxins. *Proc. Natl. Acad. Sci. USA*, **95**, 6448–6453.

12 Frost, D., Gorman, P.M., Yip, C.M., and Chakrabartty, A. (2003) Co-incorporation of A beta 40 and A beta 42 to form mixed pre-fibrillar aggregates. *Eur. J. Biochem.*, **270**, 654–663.

13 Lorenzo, A. and Yankner, B.A. (1994) Beta-amyloid neurotoxicity requires fibril formation and is inhibited by Congo Red. *Proc. Natl. Acad. Sci. USA*, **91**, 12243–12247.

14 Podlisny, M.B., Ostaszewski, B.L., Squazzo, S.L., Koo, E.H., Rydell, R.E., Teplow, D.B., and Selkoe, D.J. (1995) Aggregation of secreted amyloid beta-protein into sodium dodecyl sulfate-stable oligomers in cell culture. *J. Biol. Chem.*, **270**, 9564–9570.

15 Podlisny, M.B., Walsh, D.M., Amarante, P., Ostaszewski, B.L., Stimson, E.R., Maggio, J.E., Teplow, D.B., and Selkoe, D.J. (1998) Oligomerization of endogenous and synthetic amyloid beta-protein at nanomolar levels in cell culture and stabilization of monomer by Congo Red. *Biochemistry*, **37**, 3602–3611.

16 Walsh, D.M., Hartley, D.M., Kusumoto, Y., Fezoui, Y., Condron, M.M., Lomakin, A., Benedek, G.B., Selkoe, D.J., and Teplow, D.B. (1999) Amyloid beta-protein fibrillogenesis. Structure and biological activity of protofibrillar intermediates. *J. Biol. Chem.*, **274**, 25945–25952.

17 Dahlgren, K.N., Manelli, A.M., Stine, W.B.J., Baker, L.K., Krafft, G.A., and LaDu, M.J. (2002) Oligomeric and fibrillar species of amyloid-beta peptides differentially affect neuronal viability. *J. Biol. Chem.*, **277**, 32046–32053.

18 Stine, W.B.J., Dahlgren, K.N., Krafft, G.A., and LaDu, M.J. (2003) In vitro characterization of conditions for amyloid-beta peptide oligomerization and fibrillogenesis. *J. Biol. Chem.*, **278**, 11612–11622.

19 Walsh, D.M., Klyubin, I., Fadeeva, J.V., Cullen, W.K., Anwyl, R., Wolfe, M.S., Rowan, M.J., and Selkoe, D.J. (2002) Naturally secreted oligomers of amyloid beta protein potently inhibit hippocampal long-term potentiation *in vivo*. *Nature*, **416**, 535–539.

20 Walsh, D.M., Tseng, B.P., Rydel, R.E., Podlisny, M.B., and Selkoe, D.J. (2000) The oligomerization of amyloid beta-protein begins intracellularly in cells derived from human brain. *Biochemistry*, **39**, 10831–10839.

21 Bateman, D.A. and Chakrabartty, A. (2004) Interactions of Alzheimer amyloid peptides with cultured cells and brain tissue, and their biological consequences. *Biopolymers*, **76**, 4–14.

22 Bateman, D.A. and Chakrabartty, A. (2009) Two distinct conformations of Abeta aggregates on the surface of living PC12 cells. *Biophys. J.*, **96**, 4260–4267.

23 Bateman, D.A., McLaurin, J., and Chakrabartty, A. (2007) Requirement of aggregation propensity of Alzheimer amyloid peptides for neuronal cell surface binding. *BMC Neurosci.*, **8**, 29.

24 Luo, Y., Hawver, D.B., Iwasaki, K., Sunderland, T., Roth, G.S., and Wolozin, B. (1997) Physiological levels of beta-amyloid peptide stimulate protein kinase C in PC12 cells. *Brain Res.*, **769**, 287–295.

25. Seubert, P., Vigo-Pelfrey, C., Esch, F., Lee, M., Dovey, H., Davis, D., Sinha, S., Schlossmacher, M., Whaley, J., Swindlehurst, C. et al. (1992) Isolation and quantification of soluble Alzheimer's beta-peptide from biological fluids. *Nature*, **359**, 325–327.

26. Hardy, J.A. and Higgins, G.A. (1992) Alzheimer's disease: the amyloid cascade hypothesis. *Science*, **256**, 184–185.

27. Wood, S.J., Maleeff, B., Hart, T., and Wetzel, R. (1996) Physical, morphological and functional differences between pH 5.8 and 7.4 aggregates of the Alzheimer's amyloid peptide Abeta. *J. Mol. Biol.*, **256**, 870–877.

28. Kayed, R., Head, E., Thompson, J.L., McIntire, T.M., Milton, S.C., Cotman, C.W., and Glabe, C.G. (2003) Common structure of soluble amyloid oligomers implies common mechanism of pathogenesis. *Science*, **300**, 486–489.

29. Lambert, M.P., Viola, K.L., Chromy, B.A., Chang, L., Morgan, T.E., Yu, J., Venton, D.L., Krafft, G.A., Finch, C.E., and Klein, W.L. (2001) Vaccination with soluble Abeta oligomers generates toxicity-neutralizing antibodies. *J. Neurochem.*, **79**, 595–605.

30. Kheterpal, I., Williams, A., Murphy, C., Bledsoe, B., and Wetzel, R. (2001) Structural features of the Abeta amyloid fibril elucidated by limited proteolysis. *Biochemistry*, **40**, 11757–11767.

31. Petkova, A.T., Yau, W., and Tycko, R. (2006) Experimental constraints on quaternary structure in Alzheimer's beta-amyloid fibrils. *Biochemistry*, **45**, 498–512.

32. Huang, T.H., Yang, D.S., Plaskos, N.P., Go, S., Yip, C.M., Fraser, P.E., and Chakrabartty, A. (2000) Structural studies of soluble oligomers of the Alzheimer beta-amyloid peptide. *J. Mol. Biol.*, **297**, 73–87.

33. Huang, T.H., Fraser, P.E., and Chakrabartty, A. (1997) Fibrillogenesis of Alzheimer Abeta peptides studied by fluorescence energy transfer. *J. Mol. Biol.*, **269**, 214–224.

34. Sengupta, P., Garai, K., Sahoo, B., Shi, Y., Callaway, D.J.E., and Maiti, S. (2003) The amyloid beta peptide (Abeta(1–40)) is thermodynamically soluble at physiological concentrations. *Biochemistry*, **42**, 10506–10513.

35. Tjernberg, L.O., Pramanik, A., Björling, S., Thyberg, P., Thyberg, J., Nordstedt, C., Berndt, K.D., Terenius, L., and Rigler, R. (1999) Amyloid beta-peptide polymerization studied using fluorescence correlation spectroscopy. *Chem. Biol.*, **6**, 53–62.

36. Hazum, E., Cuatrecasas, P., Marian, J., and Conn, P.M. (1980) Receptor-mediated internalization of fluorescent gonadotropin-releasing hormone by pituitary gonadotropes. *Proc. Natl. Acad. Sci. USA*, **77**, 6692–6695.

37. Smith, M.L., Carski, T.R., and Griffin, C.W. (1962) Modification of fluorescent-antibody procedures employing crystalline tetramethylrhodamine isothiocyanate. *J. Bacteriol.*, **83**, 1358–1359.

38. Young, R.M., Arnette, J.K., Roess, D.A., and Barisas, B.G. (1994) Quantitation of fluorescence energy transfer between cell surface proteins via fluorescence donor photobleaching kinetics. *Biophys. J.*, **67**, 881–888.

39. Slavik, J. (1994) *Fluorescent Probes in Cellular and Molecular Biology*, CRC Press, Boca Raton, FL, pp. 241–279.

40. Baneyx, G. and Vogel, V. (1999) Self-assembly of fibronectin into fibrillar networks underneath dipalmitoyl phosphatidylcholine monolayers: role of lipid matrix and tensile forces. *Proc. Natl. Acad. Sci. USA*, **96**, 12518–12523.

41. Umberger, J.Q. and LaMer, V.K. (1945) The kinetics of diffusion controlled molecular and ionic reactions in solution as determined by measurements of the quenching of fluorescence. *J. Am. Chem. Soc.*, **67**, 1099–1109.

42. Michikawa, M., Gong, J.S., Fan, Q.W., Sawamura, N., and Yanagisawa, K. (2001) A novel action of Alzheimer's amyloid beta-protein (Abeta): oligomeric Abeta promotes lipid release. *J. Neurosci.*, **21**, 7226–7235.

43. Mendez, A.J., Oram, J.F., and Bierman, E.L. (1991) Protein kinase C as a mediator of high density lipoprotein receptor-

dependent efflux of intracellular cholesterol. *J. Biol. Chem.*, **266**, 10104–10111.

44 Gong, J., Sawamura, N., Zou, K., Sakai, J., Yanagisawa, K., and Michikawa, M. (2002) Amyloid beta-protein affects cholesterol metabolism in cultured neurons: implications for pivotal role of cholesterol in the amyloid cascade. *J. Neurosci. Res.*, **70**, 438–446.

45 Waschuk, S.A., Elton, E.A., Darabie, A.A., Fraser, P.E., and McLaurin, J.A. (2001) Cellular membrane composition defines Abeta–lipid interactions. *J. Biol. Chem.*, **276**, 33561–33568.

46 McLaurin, J., Franklin, T., Fraser, P.E., and Chakrabartty, A. (1998) Structural transitions associated with the interaction of Alzheimer beta-amyloid peptides with gangliosides. *J. Biol. Chem.*, **273**, 4506–4515.

47 McLaurin, J. and Chakrabartty, A. (1996) Membrane disruption by Alzheimer beta-amyloid peptides mediated through specific binding to either phospholipids or gangliosides. Implications for neurotoxicity. *J. Biol. Chem.*, **271**, 26482–26489.

48 Yanagisawa, K. and Ihara, Y. (1998) GM1 ganglioside-bound amyloid beta-protein in Alzheimer's disease brain. *Neurobiol. Aging*, **19**, S65–S67.

49 Yanagisawa, K., Odaka, A., Suzuki, N., and Ihara, Y. (1995) GM1 ganglioside-bound amyloid beta-protein (Abeta): a possible form of preamyloid in Alzheimer's disease. *Nat. Med.*, **1**, 1062–1066.

50 Lee, G., Pollard, H.B., and Arispe, N. (2002) Annexin 5 and apolipoprotein E2 protect against Alzheimer's amyloid-beta-peptide cytotoxicity by competitive inhibition at a common phosphatidylserine interaction site. *Peptides*, **23**, 1249–1263.

51 Simakova, O. and Arispe, N.J. (2007) The cell-selective neurotoxicity of the Alzheimer's Abeta peptide is determined by surface phosphatidylserine and cytosolic ATP levels. Membrane binding is required for Abeta toxicity. *J. Neurosci.*, **27**, 13719–13729.

52 Klunk, W.E., Bacskai, B.J., Mathis, C.A., Kajdasz, S.T., McLellan, M.E., Frosch, M.P., Debnath, M.L., Holt, D.P., Wang, Y., and Hyman, B.T. (2002) Imaging Abeta plaques in living transgenic mice with multiphoton microscopy and methoxy-X04, a systemically administered Congo Red derivative. *J. Neuropathol. Exp. Neurol.*, **61**, 797–805.

53 Kung, M., Hou, C., Zhuang, Z., Zhang, B., Skovronsky, D., Trojanowski, J.Q., Lee, V.M., and Kung, H.F. (2002) IMPY: an improved thioflavin-T derivative for *in vivo* labeling of beta-amyloid plaques. *Brain Res.*, **956**, 202–210.

54 Maggio, J.E., Stimson, E.R., Ghilardi, J.R., Allen, C.J., Dahl, C.E., Whitcomb, D.C., Vigna, S.R., Vinters, H.V., Labenski, M.E., and Mantyh, P.W. (1992) Reversible *in vitro* growth of Alzheimer disease beta-amyloid plaques by deposition of labeled amyloid peptide. *Proc. Natl. Acad. Sci. USA*, **89**, 5462–5466.

55 Burdick, D., Kosmoski, J., Knauer, M.F., and Glabe, C.G. (1997) Preferential adsorption, internalization and resistance to degradation of the major isoform of the Alzheimer's amyloid peptide, Abeta 1–42, in differentiated PC12 cells. *Brain Res.*, **746**, 275–284.

56 Lambert, M.P., Viola, K.L., Chromy, B.A., Chang, L., Morgan, T.E., Yu, J., Venton, D.L., Krafft, G.A., Finch, C.E., and Klein, W.L. (2001) Vaccination with soluble Abeta oligomers generates toxicity-neutralizing antibodies. *J. Neurochem.*, **79**, 595–605.

57 Lacor, P.N., Buniel, M.C., Chang, L., Fernandez, S.J., Gong, Y., Viola, K.L., Lambert, M.P., Velasco, P.T., Bigio, E.H., Finch, C.E., Krafft, G.A., and Klein, W.L. (2004) Synaptic targeting by Alzheimer's-related amyloid beta oligomers. *J. Neurosci.*, **24**, 10191–10200.

58 Kokubo, H., Kayed, R., Glabe, C.G., and Yamaguchi, H. (2005) Soluble Abeta oligomers ultrastructurally localize to cell processes and might be related to synaptic dysfunction in Alzheimer's disease brain. *Brain Res.*, **1031**, 222–228.

59 Lee, E.B., Leng, L.Z., Zhang, B., Kwong, L., Trojanowski, J.Q., Abel, T., and Lee, V.M. (2006) Targeting amyloid-beta peptide (Abeta) oligomers by passive immunization with a conformation-selective monoclonal antibody improves learning and memory in Abeta precursor protein (APP) transgenic mice. *J. Biol. Chem.*, **281**, 4292–4299.

# 12
# Experimental Approaches and Technical Challenges for Studying Amyloid–Membrane Interactions

*Raz Jelinek and Tania Sheynis*

## 12.1
### Introduction

Membrane interactions of amyloidogenic proteins and peptides are increasingly considered a primary factor of toxicity in amyloid diseases. Accordingly, heightened interest in this field has led to a proliferation of studies and experimental work aimed at characterizing membrane association of amyloid peptides and their aggregate species and, conversely, the effect of lipids and lipid bilayers on the misfolding and fibrillation processes involving amyloid peptides.

This chapter aims to summarize experimental techniques and molecular models employed in recent years for the study of amyloid peptide–membrane interactions. The dramatic proliferation of literature focusing on amyloid–membrane interactions and relationships naturally allows only a brief survey of the exciting developments in this field, and we have tried to present here representative important studies. Specifically, we discuss various experimental platforms and methods, primarily employing model membrane assemblies such as liposomes, planar lipid bilayers, and others. A critical theme apparent in our discussion is the significant impact of the choice of model systems employed for studying peptide–membrane interactions in general, and amyloid peptides in particular. Indeed, diverging interpretations and conclusions pertaining to mechanisms of membrane interactions of amyloid proteins and their biological significance can be traced in many cases to the use of different membrane models. Accordingly, an important observation underscored in this chapter concerns the importance of parameter selection and molecular characterization of the experimental platforms and model systems when designing experiments aimed at elucidating amyloid–membrane interactions.

Thematically, this chapter comprises two intertwined foci – *molecular systems*, primarily biomimetic membrane models (such as vesicles, micelles, monolayers, supported bilayers, etc.) employed for studying amyloid protein interactions with the cell membrane, and the *experimental techniques* applied in order to elucidate amyloid protein interactions with the model membranes constructed.

*Lipids and Cellular Membranes in Amyloid Diseases*, First Edition. Edited by Raz Jelinek.
© 2011 Wiley-VCH Verlag GmbH & Co. KGaA. Published 2011 by Wiley-VCH Verlag GmbH & Co. KGaA.

An important objective of this chapter is to highlight the discrepancies in the interpretation of the experiments, particularly in relation to the physiological aspects of the amyloid protein systems studied. It should be noted that despite the consensus that amyloid peptides exhibit lipid interactions, there are still ongoing controversies as to the precise nature of the membrane binding and the actual biological significance of these interactions. Indeed, the wealth of studies in this exciting field still underlines the quest for direct links between results obtained through various biophysical studies and the actual scenarios in real cell systems and tissues.

## 12.2
## Unilamellar Vesicles and Micelles

Vesicles of different sizes (small unilamellar vesicles, SUVs; large unilamellar vesicles, LUVs; giant unilamellar vesicles, GUVs), and to some extent micelles, generally constitute the bedrock model membrane systems for studying interactions of membrane-active species, including amyloid peptides and proteins. Numerous studies have employed vesicles for the analysis of amyloid–membrane interactions, and a comprehensive summary of all such studies in beyond the scope of this chapter. We present below studies that point to particularly important applications of vesicle and micelle systems to the field, and promising current avenues of research.

### 12.2.1
### Fluorescence spectroscopy

*Fluorescence spectroscopy* and its plethora of diverse applications have been among the most useful and widely used tools for probing lipid-associated fibrillogenesis. Various fluorescence spectroscopic experiments have been carried out to probe membrane interactions of the Alzheimer amyloid-β (Aβ) peptide. Qiu *et al.* recently reported the application of *fluorescence anisotropy* and *fluorescence resonance energy transfer* (*FRET*) for studying the interactions of Aβ with lipid bilayers [1]. The experiments were designed to elucidate the contribution of certain bilayer components, particularly cholesterol, towards modulating membrane interactions of the peptide. Specifically, the authors monitored the fluorescence anisotropy, a measure of relative molecular mobility within lipid bilayers [2–4], of dehydroergosterol (DHE), a fluorescent cholesterol-mimic compound. The results indicated clear modulation of DHE fluorescence anisotropy and FRET from Tyr10 of Aβ to DHE following incubation of liposomes with the peptide. Particularly intriguing was the observation of distinct "kinks" in the fluorescence anisotropy and FRET graphs (as a function of sterol content in the membrane), which were interpreted as critical cholesterol concentrations required for formation of membrane domains. Only Aβ oligomers, but not monomeric species, were shown to alter lateral organization of the lipids [1].

Binding of Aβ to membrane domains, particularly regions denoted as "lipid rafts" containing sphingomyelin, cholesterol, and gangliosides, was observed through application of the 7-diethylaminocoumarin-3-carbonyl (DAC) fluorophore covalently attached to the peptide [5]. The fluorescence intensity and emission maximum of DAC are sensitive to the hydrophobicity of the microenvironment, making the probe useful for testing Aβ interactions with membranes of different lipid compositions. Through recording the boron-dipyrromethene (BODIPY) excimer formation, the authors showed that gangliosides assemble into clusters within cholesterol-containing membranes. The putative lipid rafts were hypothesized to constitute "seeding platforms" for amyloid growth through specific recognition of the ganglioside assemblies by the Aβ peptide [5]. Pyrene excimer/monomer ratios, calculated following fluorescence energy transfer from tryptophan, indicated that Aβ increases the fluidity of membranes with high cholesterol content, but has no effect on intrinsically fluid lipid bilayers [6].

More complex fluorescence experiments have utilized labeling of *both* lipid bilayers and the amyloid peptides. A representative report by Narayanan and Scarlata described the application of FRET of doubly labeled α-synuclein (Figure 12.1) and lipid labeling with laurdan to study the association of the peptide with membrane and the effect of the lipids upon its self-assembly properties [7]. The authors also made use of the *intrinsic fluorescence* of the four tyrosine residues in the α-synuclein sequence. Both FRET and tyrosine fluorescence experiments demonstrated that the aggregation process of the protein is significantly inhibited by its membrane association. Tyrosine fluorescence was also used for the systematic examination of α-synuclein insertion into vesicles comprised of various acidic and zwitterionic phospholipids [8].

Replacement of tyrosine with tryptophan, which exhibits more favorable intrinsic fluorescence properties, was carried out in several studies. Detailed time-resolved tryptophan fluorescence experiments have recently pointed to the participation of specific residues within the α-synuclein sequence in lipid binding [9]. The same

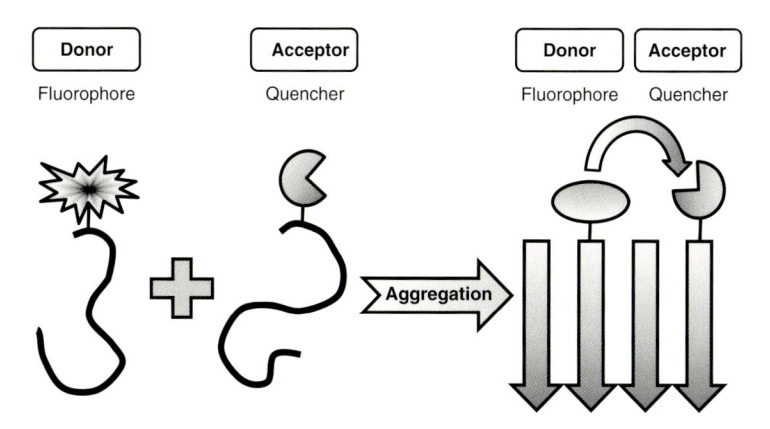

**Figure 12.1** Schematic depiction of the FRET experiment described by Narayanan and Scarlata [7]. Amyloidogenic protein (α-synuclein) was separately labeled by two distinct amino-reactive probes: fluorescent dye (donor) and quencher (acceptor). Following mixing of the two protein populations, aggregation occurred, bringing the donor and the acceptor into close proximity. As a result, fluorescence from the donor was absorbed by the quencher, and the course of fibrillation was monitored through the continuous decrease of the fluorophore emission intensity.

replacement was also applied to explore the effect of bilayer charge and fluidity upon membrane binding and insertion of the Aβ peptide [10].

*Leakage of vesicle-encapsulated fluorescent dyes* has been routinely used for the detection of membrane interactions and disruption by amyloid protein aggregates. Calcein release experiments were carried out to characterize membrane disintegration by prion isoform exhibiting a high content of β-sheet secondary structure [11]. Carboxyfluorescein leakage experiments aided in showing that two distinct regions within the islet amyloid polypeptide (IAPP) peptide are responsible for membrane interactions and fibrillation [12]. Fluorescent dye release analysis employed in a recent elegant study of fragmented fibrillar assemblies revealed an intriguing relationship between membrane disruption and the size of fibrillar aggregates otherwise bearing the same structure and morphology [13]. Through careful protocols designed to separate fibrillar species of several amyloidogenic proteins according to size, Xue *et al.* demonstrated that dye release could be directly correlated with the bilayer disruption by fibrils of specific lengths. It should be emphasized, however, that applications of dye release experiments are highly prone to local concentration effects and even simple mixing procedures occasionally yield highly divergent results in seemingly identical experimental settings. Furthermore, variations in sample preparation protocols affecting vesicle integrity and stability might significantly alter the recorded data.

*Fluorescence anisotropy* has been a widely used parameter for evaluating bilayer dynamics and fluidity. While many studies have analyzed peptide-induced modulation of the fluorescence anisotropy of lipid-embedded dyes as a probe for membrane binding and insertion, fluorescence anisotropy of dyes that are covalently attached to the amyloidogenic peptides themselves has also been used for providing information pertaining amyloid–lipid interactions. Knight and Miranker, for example, described a kinetic analysis of lipid-modulated IAPP fibrillogenesis through monitoring the fluorescence anisotropy of rhodamine in a synthetic rhodamine–IAPP peptide [14]. Specifically, rhodamine anisotropy was shown to increase dramatically following addition of negatively charged phospholipid vesicles to the peptide. Advanced fluorescence techniques were applied for analyses of lipid-induced aggregation of proteins for which amyloid fibril formation has not been widely investigated, such as endostatin [15] and lysozyme [16].

## 12.2.2
### Fluorescence microscopy

*Fluorescence microscopy* techniques have been applied to study the effect of amyloid peptide upon model membrane assemblies, particularly lipid vesicles. An obvious advantage of fluorescence microscopy is the capability in many instances to image the impact of the proteins upon lipid bilayers. Giant vesicles are particularly useful in this respect because they allow direct visualization of the effect of interacting species on the lipid membrane on a macroscopic scale. Sparr *et al.* presented a confocal fluorescence microscopy study in which the activity of IAPP on fluorescently labeled giant vesicles was investigated [17]. Time-dependent membrane disruption by the peptide was visualized using fluorescently labeled lipids reconstituted in the bilayer of the vesicles, complemented by recording an influx of the water-soluble dye carboxyfluorescein into the inner compartment of the vesicular assemblies. Furthermore, confocal microscopy was used in that study for deciphering co-localization of IAPP aggregates and membrane lipids, in which the aggregates were dyed with Congo Red while the lipids were labeled with probes exhibiting different excitation/emission wavelengths. Hamada et al. recently reported a microscopy analysis of fluorescently labeled $A\beta_{1-40}$ segregated into fluid lipid domains formed within heterogeneous lipid layers in oil–water microdroplets [18]. An interesting experiment in which fluorescence microscopy was applied to study penetratin-treated fluorescently labeled giant vesicles revealed the formation of visible fluorescent aggregates at the vesicle surface, ascribed to peptide–lipid assemblies [19]. Morita et al. demonstrated that different Aβ aggregates, such as oligomers, protofibrils, and full-length fibrils, induce dramatic modulation of the overall shapes of giant lipid vesicles [20].

## 12.2.3
### Nuclear magnetic resonance

*Nuclear magnetic resonance (NMR)* spectroscopy has been widely applied for investigating the effects of amyloid peptides on lipid membranes and the complementary structural effects of lipid bilayers upon the membrane-attached peptides. $^2$H and $^{31}$P NMR spectroscopy experiments pointed to reorganization of lipid bilayers following addition of the 29–42 fragment of Aβ peptide [21]. Both nuclei facilitate probing the *bilayer* environment within the model membranes (i.e., the lipid vesicles). Lineshape analyses of the peaks of both nuclei were employed for investigating the effects of the peptide on the lipid bilayer (either multilamellar vesicles in the case of the $^{31}$P experiments and oriented bicelles for the $^2$H analysis). A recent solution NMR study was focused on the analysis of phospholipid binding by specific residues of several α-synuclein variants [22]. The pronounced sensitivity of solution NMR to the degree of structural modulations of the protein allowed the authors to identify a distinct distribution of lipid-associated conformations in the disease-linked mutants. The study suggests that changes in the membrane-binding properties of the protein lead to exposure of a hydrophobic unordered domain within the sequence which most likely promotes aggregation.

Bokvist et al. used $^{31}$P *magic angle spinning (MAS)* NMR combined with circular dichroism (CD) spectroscopy to study adsorption on and fibrillation of $A\beta_{1-40}$ induced by lipid vesicles [23]. Through analysis of linewidths and spectral shifts of the $^{31}$P peak corresponding to the bilayer-forming phospholipids, the authors showed that interactions of Aβ species with negatively charged membranes differed markedly depending on whether the peptide was incorporated into the bilayer or added following membrane formation. The study suggested that insertion of Aβ molecules into lipid bilayers protects the peptide from fibrillation through binding to negatively charged lipids. Whereas $^{31}$P NMR spectroscopy generally provided information regarding structural modifications of lipid bilayers upon peptide interactions, $^{13}$C NMR spectroscopy (specifically through application of *cross-polarization magic angle spinning, CPMAS*) has been a useful tool for elucidating binding and reorganization of the amyloid peptides themselves. Gehman *et al.* studied lipid bilayer interactions of A$\beta_{1-42}$ and A$\beta_{25-35}$ fragment extracted from the C-terminus of the peptide [24]. They showed through careful analysis of the positions and intensities of the $^{13}$C signals that both the parent peptide and the shorter fragment undergo structural transitions to a more extended β-sheet conformation upon binding to model membranes. The solid-state $^{31}$P NMR experiments further exposed the consequences of inclusion of metal ions, such as $Cu^{2+}$, indicating that the peptide–metal complex perturbs the headgroup region of the negatively charged lipid bilayers.

Detergent micelles designed to mimic membrane environments have been used in several instances for NMR structural studies of amyloid peptides. Motta *et al.* presented a high-resolution NMR investigation of human calcitonin (hCT) incorporated within sodium dodecyl sulfate (SDS) micelles [25]. The analysis revealed that clear traces of an amphipathic helical structure within the central region of the micelle-associated molecule is followed by an extended hydrophobic domain at the C-terminus of the peptide. No indication of fibrillation events was reported; in fact, aggregation phenomena would be inherently "invisible" in these types of high-resolution micelle experiments. Hence the utilization of complementary techniques in combination with NMR spectroscopy is required for studying lipid-induced fibrillation.

A recent paper reported the three-dimensional structure of non-fibrillogenic rat IAPP in a membrane-mimic environment through application of high-resolution solution NMR spectroscopy [26]. The N-terminal loop of the peptide was found to be positioned differently to that of human IAPP (hIAPP), providing clues for distinct aggregation behavior and toxicity of the two molecules.

### 12.2.4
**Electron paramagnetic resonance**

*Electron paramagnetic resonance* (*EPR*) spectroscopy has been particularly useful for studying metal ion interactions with amyloid peptides, and also the association of fibrillar species with lipid bilayers labeled with doxyl spin probes. An important study by Curtain *et al.* showed that divalent cations, such as $Cu^{2+}$, can bridge Aβ monomers through association with the histidine side chains [27]. Using spectral line analysis of the copper X band spectra, the authors observed competitive binding between $Cu^{2+}$ and $Zn^{2+}$. The EPR experiments employing vesicles supplemented by the spin probe 1-palmitoyl-2-(16-doxylstearoyl)phosphatidylserine further validated that copper ions promote insertion of Aβ species into lipid bilayers comprising zwitterionic and negatively charged phospholipids; no membrane incorporation was detected in pure zwitterionic vesicles. Penetration of Aβ into the membrane was accompanied by subsequent transition from a β-sheet to an α-helical structure. A follow-up EPR study, in which an important spectral parameter examined was the increase in the motionally restricted lipid component (MRLC), evaluated metal ion induced bilayer interactions of Aβ in more detail.

The authors confirmed that $Cu^{2+}$- and $Zn^{2+}$-induced penetration of the peptide into the lipid bilayer depends on multiple factors, including pH, peptide length, and cholesterol content of the membrane [28].

An elegant EPR study of membrane interactions of spin-labeled hIAPP was reported recently [29]. The authors prepared a series of IAPP mutants substituted with cysteine at different positions in the peptide sequence. Following labeling of the cysteine residues with an EPR-active spin probe, EPR spectroscopy was applied with and without lipid vesicles to probe the effect of the membrane on peptide conformation. Interestingly, the authors detected a clear helical structure in the central region of the peptide, which probably accounts for the membrane binding of the molecule. Furthermore, the data pointed to disordered structure in the amyloidogenic region of the vesicle-bound peptide, unraveling a possible mechanism for lipid-induced fibrillation of hIAPP [29]. An EPR technique denoted *double-electron electron resonance* (*DEER*), in which a two-frequency pulsed EPR allows probing of both *intra-* and *inter*molecular distances, has been very useful in revealing the formation of stable, dimer-based α-synuclein aggregates upon interaction with lipid vesicles [30].

## 12.2.5
### Other experimental techniques

Other experimental techniques have been applied for studying amyloid–membrane interactions in lipid vesicle environments. *Small-angle neutron scattering* (*SANS*) has been successfully applied for evaluating amyloid peptide interactions with model membranes. SANS is a well-established technique for investigating unilamellar vesicle properties, and the effect of membrane-active peptides upon such properties [31, 32]. In the context of amyloid–membrane interactions, several reports have described the application of SANS for studying vesicle association of Aβ peptide and its shorter toxic fragment Aβ$_{25-35}$ [33–35]. Dante *et al.* evaluated the effect of Aβ$_{1-42}$ on vesicle parameters, including the size of the vesicular assemblies, size distribution, and bilayer thickness as extracted from the SANS experiments [33]. Association of the amyloidogenic peptide with the membrane was shown to induce thinning of the lipid bilayer and an increase in the vesicular radius, interpreted as vesicle fusion. *Turbidity measurements* are another sensitive technique for evaluating the binding of amyloid peptides to vesicles. A recent study employed turbidity analysis to reveal the kinetic profile of the binding of IAPP *fibrils* to vesicles comprising different lipid contents [36].

*Isothermal titration calorimetry* (*ITC*) has been applied to evaluate the effect of small-molecule inhibitors upon membrane binding of Aβ$_{1-40}$ [37]. The technique is based on monitoring the thermal changes, for example, *heat of reaction*, when lipid vesicle solutions are mixed with the peptide (with or without the presence of the inhibitors). The intrinsic high sensitivity of isothermal calorimetry to minute transformations in the organization and cooperative properties of lipid assemblies induced by membrane-active species provides a powerful means for identifying bilayer insertion on the one hand and screening inhibitors of membrane binding on the other.

## 12.3
## Black Lipid Membranes

Black lipid membranes (BLMs) have been an important tool for elucidating peptide-induced channel formation in lipid bilayers [38]. The experimental chamber in the BLM apparatus consists of two compartments separated by a thin Teflon film with a small circular hole (50–200 μm) at the center; planar lipid bilayers are formed by spreading lipid mixtures on the hole. Membrane-associated peptides are then incorporated in the thin planar bilayer, and their channel-forming properties are monitored.

BLM experiments provided some of the first experimental evidence that amyloidogenic peptides, particularly Aβ, gives rise to ion channel activity when added to lipid bilayers [39]. Other studies have confirmed those observations [40], and furthermore correlated channel formation with sterol presence in lipid bilayers [41]. Channel assembly induced by other fibrillar proteins, such as prion [42] and calcitonin [43], were detected through application of the BLM methodology.

BLMs have also been instrumental in supporting the hypothesis that amyloid *oligomers*, rather than the mature fibrils, constitute the toxic factor in amyloid diseases [44–49]. The critical distinction between the biological activity of oligomers versus microscopic amyloid fibrils has emerged as an underlying factor pointing to the putative significance of membrane interactions of amyloid proteins. The report of Kayed *et al.* has been a significant contribution, demonstrating that soluble oligomers of several amyloid protein families gave rise to increased conductance in the BLM measurements [50]. Importantly, the authors did not find evidence for discrete channel or pore formation. One of the experimentally critical facets of that work was the careful construction of defined oligomeric species which maintained stability through the experiment. However, more recent findings re-established the role of amyloid fibrils as toxic, membrane-active species with a propensity to disintegrate both model and cellular membranes [13, 36, 51]. Important factors that contribute to fibril–lipid interactions include fibril length, fibrillation in the presence of preformed lipid bilayers, and protein-to-lipid ratio [13, 17, 51–53]. An additional path that affects the toxic properties of mature fibrils involves lipid-catalyzed fragmentation of larger aggregates to spherical oligomers and protofibrils [54–56].

Whereas Kayed *et al.* did not find evidence for channels formed by amyloid protein assemblies, other reports have claimed that such structures were in fact abundant in certain amyloid systems. Arispe *et al.*, for example, have shown that $Aβ_{1-40}$ peptide induced the formation of cation channels using BLMs [39, 57]. One of the primary experimental challenges in BLM experiments is to ascertain that the peptide tested indeed associates with the bilayer placed between the two aqueous chambers, which together exhibit a much larger surface area and high solution volume. The authors verified the occurrence of direct incorporation of the Aβ peptide from the solution into the artificial 1-palmitoyl 2-oleoylphosphatidylethanolamine (POPE) membrane by using the double-dip method to form a bilayer membrane at the tip of a patch pipette [39]. A few minutes after dissolution of the Aβ peptide, dramatic channel activity was recorded, and the authors demonstrated transport of positive ions through the putative amyloid channels, including $K^+$, $Cs^+$, and $Ca^{2+}$. Based on these observations, it was concluded that synthetic $A\beta_{1-40}$ can assume a conformation that enables the molecule to enter the bilayer membranes and form cation-selective pores. The observation that soluble $A\beta_{1-40}$ peptide can associate with planar lipid bilayers and form active channels is significant, since this peptide fragment is released into the cerebrospinal fluid [48, 58]. Accordingly, changes in intracellular and extracellular $Ca^{2+}$ concentrations, induced by possible channel activity of membrane-associated $A\beta_{1-40}$, could be a major factor in affecting neurotoxicity [59, 60].

The putative formation of *ion membrane channels* by amyloid peptides has been put forward as a likely toxic feature of diverse peptide families [61, 62]. Indeed, pore formation in the cellular membrane is believed to be a common phenomenon underlying the cytolytic action of numerous membrane-active peptides and proteins [63]. There is growing experimental evidence that different amyloid peptides indeed induce channel formation in lipid bilayers [40–43, 61, 64, 65]. Alzheimer $A\beta$ peptide, in particular, has been shown to form channels in various model lipid systems [40, 65].

An interesting application of BLM experiments coupled with single-molecule fluorescence microscopy for studying membrane interactions and aggregation of amyloid peptides has recently been reported [66]. The experiment utilized a BLM device conjugated to an inverted confocal microscope, allowing evaluation of the correlation between conductivity events (e.g., channel formation within the lipid bilayer) and binding/aggregation of fluorescently labeled $A\beta_{1-40}$. This sensitive approach demonstrated a direct link between oligomer assembly and channel formation, specifically the *absence* of pore formation when $A\beta_{1-40}$ comprised monomers or dimers.

## 12.4
## Langmuir Monolayers

Langmuir monolayers constitute a convenient platform for studying peptide adsorption and insertion into lipid layers [67]. The primary advantage of the Langmuir monolayer methodology is that it allows sensitive measurements of the affinity and insertion kinetics of amphiphilic molecules or aggregates into lipid monolayers. Application of Langmuir monolayer techniques can be carried out through thermodynamic analysis (e.g., compression isotherms of *mixed* monolayers of lipids and the amyloid peptide tested [68], or alternatively through injection of the peptide *underneath* a lipid monolayer and recording peptide adsorption on the monolayer [69]. Recent Langmuir monolayer experiments demonstrated that cholesterol regulates the binding affinity of $A\beta_{1-40}$ to different glycolipids by modulating their conformation [70]. Intriguingly, the study clearly indicated that cholesterol is able to promote and inhibit membrane interactions of the peptide, depending on the glycolipids present, thus shedding light on the nature of controversial reports on this subject.

Langmuir monolayers have also been attractive as means for studying amyloid–membrane interactions because of the availability of diverse surface characterization techniques. *Infrared (IR) spectroscopy* [71, 72] has been widely and primarily employed in the context of Langmuir monolayers for identifying β-sheet structures within fibril assemblies – a common defining structure of amyloid protein aggregates [73]. Maltseva *et al.* reported on a study of Aβ$_{1-40}$ adsorption on Langmuir monolayers utilizing negatively charged and zwitterionic lipids [74]. They applied *infrared reflection adsorption spectroscopy (IRRAS)* and *grazing incidence X-ray diffraction (GIXD)*, which allowed the determination of monolayer-associated structural changes of the peptide assemblies. Specifically, the peptide was found to adsorb to monolayers in a β-sheet conformation with the main axes of β-strands lying parallel to the membrane surface. The significance of these analyses lies in the capability to evaluate *in situ* both lipid-induced conformational changes undergone by the peptide and also modulation of lipid packing affected by the interacting amyloidogenic molecule.

The application of *IRRAS* [75, 76] to analyze structural and kinetic parameters pertaining to lipid interactions of IAPP has also been described [77]. IRRAS relies upon comparison of the amide I band spectra taken with p-polarized light with varying angle of incidence. This essentially provides information on the orientation of the peptide, as p-polarized light probes the transition dipole moment components parallel and perpendicular to the surface. Through application of IRRAS, Lopes *et al.* [77] compared the binding, orientation, and secondary structure of human and rat IAPP within monolayers comprised of zwitterionic and negatively charged lipids. They particularly showed that both peptides associate with charged lipid monolayers through their N-termini possessing an α-helical conformation. However, whereas the non-amyloidogenic rat IAPP preserves its helical property during the entire course of the lipid binding, human protein is transformed from a mainly α-helical to a parallel β-sheet structure upon contact with lipids. The experiments confirmed that this IR technique constitutes a powerful tool that enables one to discriminate the conformational dynamics of IAPP underlying its amyloidogenic properties.

Advanced IR methods have been developed to elucidate peptide accumulation and structural features of lipid-associated amyloid peptide aggregates, in the context of Langmuir monolayer environments. Koppaka and Axelsen introduced a methodology denoted *polarized attenuated total internal reflection Fourier transform infrared (PATIR-FTIR) spectroscopy* [78]. This technique allows *in situ* analysis of peptide adsorption on Langmuir monolayers of lipids, the conformation adopted by the peptide, and its orientation within the lipid monolayer. One of the main advantages was the feasibility of comparison between different lipid environments; specifically, it was demonstrated that oxidized lipids clearly accelerated adsorption and β-sheet formation by the Aβ peptide, a result with obvious physiological implications. PATIR-FTIR spectroscopy has been recently applied to the study of Aβ misfolding and fibrillogenesis induced by a lipid oxidation product [79].

Recent experiments have utilized the air–water boundary for studying amyloid protein behavior at interfaces. Fu *et al.* reported elegant *sum frequency generation (SFG)* experiments in which misfolding of IAPP at the water surface and on lipid–water interfaces was investigated [80]. SFG allowed monitoring of the vibrations of the amide I region of the peptide, with and without the presence of negatively charged phospholipids, and the pronounced effect of the membrane environment upon peptide misfolding from an α-helical/random coil conformation to a parallel β-sheet structure. That work particularly emphasized the advantages of SFG for studying amyloid peptides at *interface* regions, in which the peptide concentration is inherently low (as compared with bulk), providing *in situ* kinetic and structural information.

Other studies have demonstrated dramatic lipid-modulated aggregation of amyloidogenic peptides in Langmuir monolayer environments. Dorosz *et al.* recently reported that the amyloidogenic determinant of prion protein (PrP(106–126)) formed highly ordered fibrils upon adsorption on lipid monolayers at the air–water interface [69]. Significantly, PrP(106–126) fibrillation, studied by various *in situ* and *ex situ* techniques, was dependent to a large extent upon the presence and abundance of negatively charged phospholipids in the monolayers.

## 12.5
## Solid-Supported Bilayers

Lipid bilayers deposited on solid substrates have been used as convenient platforms for studying amyloid–membrane interactions, primarily through application of optical techniques such as *surface plasmon resonance (SPR)* [81, 82]. SPR allows the detection of binding events through monitoring changes in light reflectivity due to association of membrane-active species with lipid bilayer immobilized on thin (~50 nm) metal film (Figure 12.2).

SPR was used for studying the membrane binding of Aβ peptide, particularly the dependence of binding on the presence of cholesterol [83]. Smith *et al.* examined the interaction of α-synuclein with surface-supported bilayers [84]. Binding was monitored over a period of several days and it was found that the highest affinity to the lipid layer occurred when *oligomeric* species of α-synuclein formed. Furthermore, significantly less membrane binding was observed when the oligomers were further incubated, forming larger fibrils.

*Plasmon waveguide resonance (PWR) spectroscopy* has contributed to the determination of amyloid peptide binding to lipid bilayers [85]. The technique relies on measuring the modulation of a plasmon waveguide upon binding of membrane-active species to a lipid bilayer (thus leading to mass change). Using PWR spectroscopy, Devanathan *et al.* found that two principal components of lipid rafts – sphingomyelin and cholesterol – complement each other in their impact on the activity of the amyloid peptide: while sphingomyelin initiates Aβ aggregation, cholesterol enhances membrane affinity of the peptide and facilitates Aβ insertion into the lipid bilayer [85]. Similarly to SPR, PWR constitutes a highly sensitive tool for determining biomolecule affinity to lipid surfaces. The technique allows the evaluation of the effect of lipids upon the fibrillation process of the peptides; however, it cannot elucidate the exact nature of the aggregated species involved (i.e., oligomers,

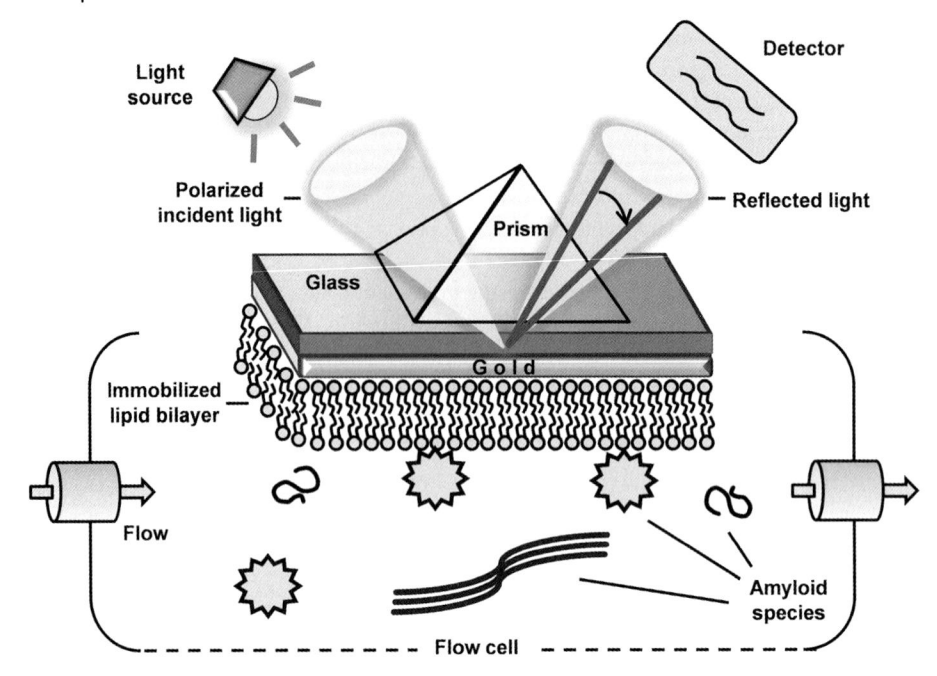

**Figure 12.2** Schematic depiction of amyloid–lipid binding assayed by the SPR technique. The sensor chip consists of a glass slide coated with a thin metal layer (typically gold). The incident light beam is set to hit the glass surface at a critical angle at which no light is refracted through the specimen, that is, total internal reflection occurs. If p-polarized and monochromatic light is used under these conditions, it generates an evanescent wave which penetrates a very short distance (tens of nanometers) into the gold layer, exciting free electrons at the metal surface. This energy transfer results in a sharp decrease in intensity at a certain angle within the reflected light (shown by a dark line on the scheme) and is denoted surface plasmon resonance. Membrane-active ligands are introduced into the system by a continuous flow of aqueous phase (buffer) containing amyloidogenic species. Binding of amyloid aggregates to the lipid bilayer immobilized on the gold film results in an increase in the SPR angle (shown by a black arrow), which is correlated with the total mass of the associated material. SPR is a highly sensitive tool for the evaluation of the membrane affinity of analytes, enabling kinetic and thermodynamic binding constants to be determined.

protofibrils, fibrils). *Dual polarization interferometry* (*DPI*) is another optical technique that makes use of solid-supported lipid bilayers for studying lipid-induced protein aggregation. DPI is based on laser waveguide interference [86, 87], which allows *in situ* monitoring of the layer thickness and density, specifically as they relate to lipids and membrane-active amyloidogenic proteins. Sanghera *et al.* recently reported a DPI investigation of the adsorption of full-length prion protein on lipid layers comprising zwitterionic and negative phospholipids [88]. The DPI results indicated pronounced self-association of the peptide adsorbed on negatively charged membranes, which corroborate various other data linking prion toxicity and its interactions with negative membrane surfaces.

Different types of supported lipid bilayers have been used as platforms for studying amyloid peptide–membrane interactions. Valincius *et al.* presented a comprehensive study of interactions by Aβ oligomers with free-standing solid-supported lipid bilayers and tethered bilayers [89]. Using primarily *electrochemical impedance spectroscopy* (*EIS*) and complemented by neutron reflection, they demonstrated that dramatic modulation of bilayer dielectric properties, increased ionic conductivity, and bilayer thinning were induced upon addition of soluble Aβ oligomers. Notably, both bilayer leaflets were affected by interaction with the oligomers, indicating substantial insertion of the aggregated species into the membrane. Solid-supported bilayers prepared through liposome deposition have been employed as a platform for following lipid-associated amyloid peptide oligomerization through application of a *quartz crystal microbalance* (*QCM*) [90]. The study found preferred binding of the aggregates to lipids containing saturated hydrocarbon chains.

*Atomic force microscopy* (*AFM*) experiments, conducted on samples comprising solid-supported lipid bilayers, have been an important factor in promoting the notion that membrane interactions of amyloid peptides, and particularly the putative formation of amyloid pores/channels, constitute important factors in toxicity and pathogenesis in amyloid diseases. Lin *et al.* were among the first to report AFM images of Aβ$_{1-42}$ peptide assemblies on planar lipid bilayers, which were ascribed by the authors to ion channels [40]. Aided primarily by the AFM analysis and complemented by electrophysiological measurements, this study demonstrated that *nonfibrillar* annular species of Aβ were responsible for bilayer adsorption and formation of the putative ion channels. Whereas no Aβ fibrillation was detected by Lin *et al.*, later *in situ* AFM studies by Yip *et al.* (operating in the "tapping mode") showed growth of Aβ fibrils in an environment of a bilayer comprising a physiological lipid mixture (total lipid extract) [91]. The main difference between these two studies is primarily the preparation and likely stabilization of oligomeric Aβ assemblies in the presence of lipid molecules. Specifically, Lin *et al.* prepared Aβ–lipid mixed vesicles which were deposited on mica in order to create planar layers, whereas Yip *et al.* injected the peptide solution over a preformed lipid bilayer.

Attribution of membrane-active properties to *specific* amyloidogenic oligomers is a complex task since these aggregates are transient intermediates within the self-assembly route. A recent study by Kayed *et al.* provided evidence that small spherical, but not annular, Aβ oligomers contribute significantly to the current detected across lipid bilayers [92]. Such a conclusion became possible due to carefully designed protocols that allowed preparation of relatively homogeneous spherical and annular Aβ species. The authors confirmed that the lipid membrane catalyzes an assembly of spherical aggregates into larger ring-shaped structures; however, the latter failed to permeate lipid bilayers. The significance of these findings is emphasized by the fact that torus-like, membrane-associated structures that visually resemble pores are widely observed in different amyloid systems and proteins [40, 61, 65, 93]. The report by Kayed *et al.* provides a very important example demonstrating that exceptional caution should be exercised during data interpretation.

AFM has been applied in investigations aimed at assessing lipid bilayer interactions of amyloid peptides other than Aβ, and at probing the effect of the lipid assemblies upon fibrillation processes of the proteins studied. Jo et al. used *in situ* AFM to visualize membrane disruption by soluble α-synuclein entities and particularly lipid-induced aggregation of the peptide [94]. That study also revealed slower membrane disruption and aggregation of a lipid-associated pathogenic A53T mutant of α-synuclein, suggesting a possible link between α-synuclein toxicity and membrane association. An AFM and transmission electron microscopy (TEM) study depicted pore-like structures formed upon binding of calcitonin to lipid bilayers, specifically to lipid rafts [93]. These last two studies took advantage of the inherent capability of AFM to facilitate real-time kinetic analysis of both membrane disruption and peptide fibrillation.

Interesting time-lapse AFM experiments investigating amylin (also known as IAPP) interactions and oligomerization upon binding to lipid bilayers, particularly lipid rafts, have been reported [95]. The investigation was made possible through *in situ* kinetic AFM analysis of lipid domains to which the human IAPP peptide was added. Taking into account the inherent difficulties with the application and interpretation of AFM images of lipid environments, tapping-mode AFM allowed the observation of amylin oligomers, protofibrils, and fibrils, accompanied by bilayer disruption [95].

Solid-supported bilayers have been employed for studying lipid-induced aggregation by using fluorescence microscopy techniques. Recently reported *epi-fluorescence microscopy* experiments revealed a pH dependence of α-synuclein binding and aggregation on supported lipid films comprising zwitterionic (phosphatidylcholine) and negatively charged (phosphatidic acid) lipids [96]. Using dual fluorescent labeling of α-synuclein and phosphatidylcholine, the microscopy experiments helped to shed light on the conditions affecting aggregation of the protein at the bilayer surface.

*Two-dimensional IR correlation (2D-IR) spectroscopy* was used for studying protein aggregation in the presence of lipids [97]. This technique increases the spectral resolution in the amide I region corresponding to the secondary structure elements of macromolecules, and also facilitates better kinetic analysis of folding/misfolding processes [98, 99]. 2D-IR spectroscopy could be particularly suited to the determination of lipid-induced aggregation because it could "freeze" the structures formed during the fibrillation process and assess the features of the aggregates formed. Specifically, Paquet et al. analyzed the fibrillation process of cytochrome c on the surface of negatively charged vesicles, and succeeded in resolving distinct folding/unfolding events occurring throughout the process [97]. Particularly interesting are observations indicating that the aggregation of the protein initiates from nearly the native state, followed by further unfolding.

*Total internal reflection fluorescence (TIRF) microscopy* has recently been used by Choucair *et al.* to investigate Aβ binding to solid-supported lipid bilayers [100]. TIRF microscopy is an interface-sensitive method that probes fluorescence within <100 nm of the surface and therefore readily detects peptide bound to the membrane with no background signal due to residual labeled peptides in solution. In particular, TIRF microscopy is useful for the identification of peptide binding to lipid surfaces *prior* to the appearance of peptide aggregates. The experiments reported by Choucair *et al.* employed solid-supported bilayers bearing saturated and unsaturated lipids, designed to illuminate the formation of membrane domains. The lipids and Aβ peptide were fluorescently labeled, utilizing dyes exhibiting different excitation and emission wavelengths. Accumulation and aggregation of the peptide on the lipid layer was detected 15 min after initiating incubation. The TIRF microscopy experiments further revealed that Aβ was preferentially associated with *lipid gel* domains, rather than with fluidic bilayer regions; the authors excluded the possibility of deeper penetration of the peptide into the bilayer. These findings are consistent with other published data indicating specific interactions of Aβ with rigid membrane domains, such as lipid rafts [101]. It should be noted, however, that the fluorescent residue attached to the membrane-active peptide examined in TIRF microscopy analysis (Aβ peptide in the work of Choucair *et al.* discussed here) might interfere with the assembly properties and fibril formation of the peptides.

## 12.6
## Other Techniques

Newly developed chromatic biomimetic membranes comprising phospholipids and *polydiacetylene* (*PDA*) have been employed for studying lipid interactions of amyloidogenic peptides [102, 103]. PDA is a lipid-like polymer which undergoes dramatic colorimetric and fluorescent transformations induced by external stimuli. Diacetylene molecules accommodate various types of physiological lipids and form vesicular bilayer structures in which the polymerization process and the chromatic transitions are preserved. The polymerized units and also the lipids exhibit distinct microscopic phases and form separate domains within the PDA-based sensor [104]. The lipid domains serve as biomimetic recognition elements, whereas the polymer moieties represent a signal transducer which translates molecular interactions with the lipid bilayer into a measurable blue → red colorimetric shift and increase in fluorescence intensity. Biosensing applications of the system include membrane-associated biological molecules such as peptides, hormones, and ionophores. In the context of amyloid peptide–membrane interactions, PDA-based vesicle assays have shown that transient oligomeric aggregates of IAPP are the primary membrane-disruptive species, not the monomers or mature fibrils formed by this peptide [102] (Figure 12.3).

Another study employing lipid–PDA vesicle systems elucidated the membrane-active properties of casein, the main protein constituent of bovine milk, which was shown to form amyloid aggregates in certain pathologies [103]. The advantage of lipid–PDA vesicle assays for studying amyloid–membrane interactions stems from the simple and robust biosensing mechanism inherent in the system, that is, easily recorded colorimetric/fluorescent changes of the vesicles directly induced by membrane-bound peptide species [105, 106].

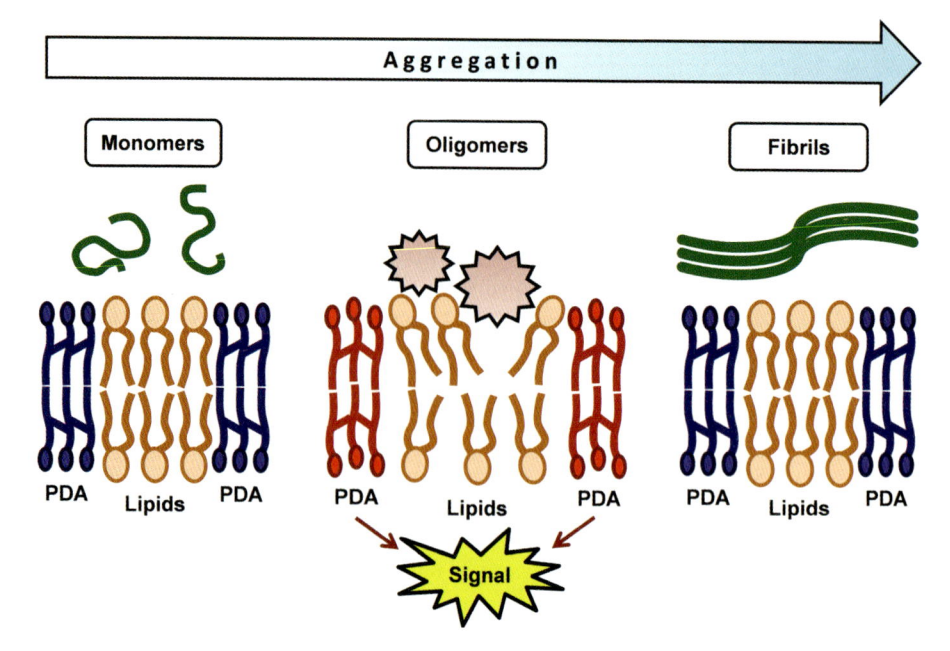

**Figure 12.3** Schematic description of the PDA–IAPP experiment [102]. Aliquots were withdrawn from the protein solution in buffer at different time points during the fibrillation course and incubated with the PDA–lipids vesicles. Only transient oligomers, and not monomers or full-length fibrils, interacted with the lipid domains of the polymer–lipid mixed vesicles, resulting in chromatic transition from blue to red and an increase in fluorescence of the PDA sensor. The existence of specific aggregated species (oligomers, fibrils) was confirmed by complementary techniques.

*Mass spectrometry* (MS) has been applied for deciphering lipid-associated oligomeric aggregates of amyloid peptides. Liu *et al.* utilized *matrix-assisted laser desorption/ionization time-of-flight mass spectrometry* (*MALDI-TOF-MS*) to identify modifications of Aβ histidine residues induced by a lipid oxidation product [79]. Giannakis *et al.* recently introduced a lipid binding assay using *surface-enhanced laser desorption/ionization time-of-flight mass spectrometry* (*SELDI-TOF-MS*) designed to characterize the oligomeric profile of α-synuclein aggregates formed upon lipid binding [107]. To prepare samples for the MS analysis, the authors developed a new solid-phase binding system, in which lipid monolayers dried on a protein chip surface are used for targeting the α-synuclein oligomers. Based on these experimental constructs, they carefully analyzed the extent of lipid binding of the wild type sequence and also single-residue mutants of the peptide. The experimental data revealed preferential binding of the α-synuclein *dimers* to the lipid-coated substrates, and also differences in lipid affinity among the mutants examined. The advantage of the MS approach presented by Giannakis *et al.* is the capability to identify precisely the mass (and thus composition) of membrane-bound species and quantify the relative abundance of the aggregates. A limitation of the technique is the poor structural and dynamic similarities between the chip-deposited lipid layer used in the experiments and actual cellular membranes.

## 12.7
## Challenges and Future Work

The accumulated body of work obtained through application of the various biophysical techniques described in this chapter clearly demonstrates that membrane binding of amyloid species is prevalent and most likely exhibits important biological functions. Although this general conclusion is backed up by numerous experimental results, determination of the exact mechanisms of lipid bilayer insertion, and the relationships between bilayer interaction and amyloid pathogenesis, are still under debate.

The issue of membrane interactions of amyloid peptides is inherently complex because of the occurrence and interrelation between two main parallel processes: membrane binding and peptide aggregation. For this reason, special care should be exercised when applying and interpreting commonly used experimental techniques. For example, chromatography and ultracentrifugation, which have been routinely used for resolving vesicle–peptide complexes, require several hours; during this period, proteins can continue to aggregate, complicating data interpretation. CD spectroscopy reveals conformational changes of amyloid proteins which are induced upon interactions with lipids; however, quantitative analysis of aggregate formation using this technique can be difficult. SPR, which provides highly sensitive readings upon lipid affinity of amyloid peptide species, generally does not provide information upon the specific oligomers which interact with the lipid layers.

Some of the controversies and divergence of opinions in the field can be traced to the question of whether the model systems being used actually mimic the real physiological scenarios. For example, the inclusion of few lipid species in vesicle models, such as sphingomyelin and cholesterol in "lipid raft" models, might indeed highlight a putative contribution of raft domains, although such an approach does not take into account the interplay between various molecular species in the complex cellular membrane. Similarly, supported lipid bilayers constitute a useful platform for application of powerful bioanalytical techniques; however, it is obvious that the macroscopic rearrangement of the lipid bilayer in such film systems is very different from the cellular membrane, which is curved and is not anchored to a solid support.

Other published reports pose challenges in terms of the accurate interpretation of the experimental data. Microscopy results depicting, for example, pore or channel formation are a case in point. The hypothesis that "amyloid pores" constitute a primary toxic determinant in amyloid diseases has been among the most important driving forces in the quest for understanding the significance of amyloid–membrane interactions. However, over the years there have been many published TEM and AFM images claiming to demonstrate pore formation or other membrane modulation events induced by amyloid peptides, which appear ambiguous and/or inconsistent.

One of the most important conclusions arising from the plethora of reports and ongoing studies pertaining to membrane interactions of amyloid proteins is that precise control of the experimental conditions is pivotal for the interpretation of the experimental data and the veracity of the conclusions drawn. In particular, the characteristics of the amyloid aggregates in terms of size appear as primary factors affecting membrane activity. Hence the observations of channel or pore formation induced by amyloid proteins are most likely closely dependent on the state of the amyloid species – monomers, oligomers, protofibrillar aggregates, or mature amyloid fibrils. In each experimental technique described in this chapter – such as BLM and AFM – the state of the protein assemblies studied had a profound effect on the experimental outcome and interpretation.

Sample preparation is a highly important element affecting the amyloid–membrane interactions recorded and their presumed biological significance. In particular, the choice of the *lipid composition* of the various artificial membranes prepared (vesicles, planar bilayers, Langmuir monolayers, etc.) was shown to be a major factor contributing to variations among reported data. It seems that a fundamental prerequisite for reliable biological interpretation of data in the field would be the assessment of the optimal lipid compositions that would mimic the scenario in actual cellular membranes and physiological membrane tissues. In addition, pretreatment and proper care of amyloidogenic protein samples are required in order to eliminate the majority of pre-existing aggregates, which otherwise could affect the route and kinetics of the self-assembly process.

Another cautionary aspect of studies of amyloid peptide interactions is the inherent deficiencies and limitations associated with each of the techniques employed. Specifically, *BLM measurements* rely on interactions of soluble macromolecules with a *planar* lipid bilayer with limited surface area – an unlikely scenario for the highly curved and high surface area membrane tissue environments for an amyloid peptide. *Fluorescence techniques* generally exploit site-specific labeling of peptides and/or membranes which might affect the structural features of the molecules examined and the interactions between them. Furthermore, the pronounced sensitivity of fluorescence dyes to their environments often lead to variations between different sample batches. *AFM* and *SPR* investigate solid-supported bilayers, and this configuration might affect the bilayer properties and/or the bimolecular interactions probed. *NMR* experimentation often requires the use of lipid assemblies and model membranes, such as multilamellar vesicles or detergent micelles, which only remotely mimic actual biological environments. *Langmuir monolayer* techniques focus on interactions of peptides and fibrillar species with *monolayers* at the air–water interface, which obviously only partly resemble true membrane bilayer environments.

Overall, it is clear that no single technique could be regarded as the "holy grail" for studying amyloid–membrane interactions. Indeed, the overwhelming majority of published reports include several experimental techniques that complement and corroborate each other. New experimental approaches and advances made in existing techniques promise to push the field forwards, and further promote the realization that membrane interactions of amyloidogenic peptides are critical factors both in shaping the aggregation properties of the peptides, and also their toxicity and adverse physiological effects.

## References

1 Qiu, L., Lewis, A., Como, J., Vaughn, M.W., Huang, J., Somerharju, P., Virtanen, J., and Cheng, K.H. (2009) Cholesterol modulates the interaction of beta-amyloid peptide with lipid bilayers. *Biophys. J.*, **96**, 4299–4307.

2 Gidwani, A., Holowka, D., and Baird, B. (2001) Fluorescence anisotropy measurements of lipid order in plasma membranes and lipid rafts from RBL-2H3 mast cells. *Biochemistry*, **40**, 12422–12429.

3 Jahnig, F. (1979) Structural order of lipids and proteins in membranes: evaluation of fluorescence anisotropy data. *Proc. Natl. Acad. Sci. USA*, **76**, 6361–6365.

4 Lakowicz, J.R. (1999) *Principles of Fluorescence Spectroscopy*, 2nd edn., Kluwer Academic/Plenum Publishers, New York.

5 Kakio, A., Nishimoto, S., Yanagisawa, K., Kozutsumi, Y., and Matsuzaki, K. (2002) Interactions of amyloid beta-protein with various gangliosides in raft-like membranes: importance of GM1 ganglioside-bound form as an endogenous seed for Alzheimer amyloid. *Biochemistry*, **41**, 7385–7390.

6 Chochina, S.V., Avdulov, N.A., Igbavboa, U., Cleary, J.P., O'Hare, E.O., and Wood, W.G. (2001) Amyloid beta-peptide1–40 increases neuronal membrane fluidity: role of cholesterol and brain region. *J. Lipid Res.*, **42**, 1292–1297.

7 Narayanan, V. and Scarlata, S. (2001) Membrane binding and self-association of alpha-synucleins. *Biochemistry*, **40**, 9927–9934.

8 Zhu, M., Li, J., and Fink, A.L. (2003) The association of alpha-synuclein with membranes affects bilayer structure, stability, and fibril formation. *J. Biol. Chem.*, **278**, 40186–40197.

9 Pfefferkorn, C.M. and Lee, J.C. (2010) Tryptophan probes at the alpha-synuclein and membrane interface. *J. Phys. Chem. B*, **114**, 4615–4622.

10 Wong, P.T., Schauerte, J.A., Wisser, K.C., Ding, H., Lee, E.L., Steel, D.G., and Gafni, A. (2009) Amyloid-beta membrane binding and permeabilization are distinct processes influenced separately by membrane charge and fluidity. *J. Mol. Biol.*, **386**, 81–96.

11 Kazlauskaite, J., Sanghera, N., Sylvester, I., Venien-Bryan, C., and Pinheiro, T.J. (2003) Structural changes of the prion protein in lipid membranes leading to aggregation and fibrillization. *Biochemistry*, **42**, 3295–3304.

12 Brender, J.R., Lee, E.L., Cavitt, M.A., Gafni, A., Steel, D.G., and Ramamoorthy, A. (2008) Amyloid fiber formation and membrane disruption are separate processes localized in two distinct regions of IAPP, the type-2-diabetes-related peptide. *J. Am. Chem. Soc.*, **130**, 6424–6429.

13 Xue, W.F., Hellewell, A.L., Gosal, W.S., Homans, S.W., Hewitt, E.W., and Radford, S.E. (2009) Fibril fragmentation enhances amyloid cytotoxicity. *J. Biol. Chem.*, **284**, 34272–34282.

14 Knight, J.D. and Miranker, A.D. (2004) Phospholipid catalysis of diabetic amyloid assembly. *J. Mol. Biol.*, **341**, 1175–1187.

15 Zhao, H., Jutila, A., Nurminen, T., Wickstrom, S.A., Keski-Oja, J., and Kinnunen, P.K. (2005) Binding of endostatin to phosphatidylserine-containing membranes and formation of amyloid-like fibers. *Biochemistry*, **44**, 2857–2863.

16 Gorbenko, G.P., Ioffe, V.M., and Kinnunen, P.K. (2007) Binding of lysozyme to phospholipid bilayers: evidence for protein aggregation upon membrane association. *Biophys. J.*, **93**, 140–153.

17 Sparr, E., Engel, M.F., Sakharov, D.V., Sprong, M., Jacobs, J., de Kruijff, B., Hoppener, J.W., and Killian, J.A. (2004) Islet amyloid polypeptide-induced membrane leakage involves uptake of lipids by forming amyloid fibers. *FEBS Lett.*, **577**, 117–120.

18 Hamada, T., Morita, M., Kishimoto, Y., Komatsu, Y., Vestergaard, M., and Takagi, M. (2010) Biomimetic microdroplet membrane interface: detection of the lateral localization of amyloid beta peptides. *J. Phys. Chem. Lett.*, **1**, 170–173.

19 Lee, C.C., Sun, Y., and Huang, H.W. (2010) Membrane-mediated peptide conformation change from alpha-monomers to beta-aggregates. *Biophys. J.*, **98**, 2236–2245.

20 Morita, M., Vestergaard, M., Hamada, T., and Takagi, M. (2010) Real-time observation of model membrane dynamics induced by Alzheimer's amyloid beta. *Biophys. Chem.*, **147**, 81–86.

21 Mingeot-Leclercq, M.P., Lins, L., Bensliman, M., Van Bambeke, F., Van Der Smissen, P., Peuvot, J., Schanck, A., and Brasseur, R. (2002) Membrane destabilization induced by beta-amyloid peptide 29–42: importance of the amino-terminus. *Chem. Phys. Lipids*, **120**, 57–74.

22 Bodner, C.R., Maltsev, A.S., Dobson, C.M., and Bax, A. (2010) Differential phospholipid binding of alpha-synuclein variants implicated in Parkinson's disease revealed by solution NMR spectroscopy. *Biochemistry*, **49**, 862–871.

23 Bokvist, M., Lindstrom, F., Watts, A., and Grobner, G. (2004) Two types of Alzheimer's beta-amyloid (1–40) peptide membrane interactions: aggregation preventing transmembrane anchoring versus accelerated surface fibril formation. *J. Mol. Biol.*, **335**, 1039–1049.

24 Gehman, J.D., O'Brien, C.C., Shabanpoor, F., Wade, J.D., and Separovic, F. (2008) Metal effects on the membrane interactions of amyloid-beta peptides. *Eur. Biophys. J.*, **37**, 333–344.

25 Motta, A., Andreotti, G., Amodeo, P., Strazzullo, G., and Castiglione Morelli, M.A. (1998) Solution structure of human calcitonin in membrane-mimetic environment: the role of the amphipathic helix. *Proteins*, **32**, 314–323.

26 Nanga, R.P., Brender, J.R., Xu, J., Hartman, K., Subramanian, V., and Ramamoorthy, A. (2009) Three-dimensional structure and orientation of rat islet amyloid polypeptide protein in a membrane environment by solution NMR spectroscopy. *J. Am. Chem. Soc.*, **131**, 8252–8261.

27 Curtain, C.C., Ali, F., Volitakis, I., Cherny, R.A., Norton, R.S., Beyreuther, K., Barrow, C.J., Masters, C.L., Bush, A.I., and Barnham, K.J. (2001) Alzheimer's disease amyloid-beta binds copper and zinc to generate an allosterically ordered membrane-penetrating structure containing superoxide dismutase-like subunits. *J. Biol. Chem.*, **276**, 20466–20473.

28 Curtain, C.C., Ali, F.E., Smith, D.G., Bush, A.I., Masters, C.L., and Barnham, K.J. (2003) Metal ions, pH, and cholesterol regulate the interactions of Alzheimer's disease amyloid-beta peptide with membrane lipid. *J. Biol. Chem.*, **278**, 2977–2982.

29 Apostolidou, M., Jayasinghe, S.A., and Langen, R. (2008) Structure of alpha-helical membrane-bound human islet amyloid polypeptide and its implications for membrane-mediated misfolding. *J. Biol. Chem.*, **283**, 17205–17210.

30 Drescher, M., van Rooijen, B.D., Veldhuis, G., Subramaniam, V., and Huber, M. (2010) A stable lipid-induced aggregate of alpha-synuclein. *J. Am. Chem. Soc.*, **132**, 4080–4082.

31 Pencer, J., Nieh, M.P., Harroun, T.A., Krueger, S., Adams, C., and Katsaras, J. (2005) Bilayer thickness and thermal response of dimyristoylphosphatidyl-choline unilamellar vesicles containing cholesterol, ergosterol and lanosterol: a small-angle neutron scattering study. *Biochim. Biophys. Acta*, **1720**, 84–91.

32 Schmiedel, H., Almasy, L., and Klose, G. (2006) Multilamellarity, structure and hydration of extruded POPC vesicles by SANS. *Eur. Biophys. J.*, **35**, 181–189.

33 Dante, S., Hauss, T., Brandt, A., and Dencher, N.A. (2008) Membrane fusogenic activity of the Alzheimer's peptide A beta(1–42) demonstrated by small-angle neutron scattering. *J. Mol. Biol.*, **376**, 393–404.

34 Dante, S., Hauss, T., and Dencher, N.A. (2003) Insertion of externally administered amyloid beta peptide 25–35 and perturbation of lipid bilayers. *Biochemistry*, **42**, 13667–13672.

35 Dante, S., Hauss, T., and Dencher, N.A. (2006) Cholesterol inhibits the insertion of the Alzheimer's peptide Abeta(25–35) in lipid bilayers. *Eur. Biophys. J.*, **35**, 523–531.

36 Sasahara, K., Hall, D., and Hamada, D. (2010) Effect of lipid type on the binding of lipid vesicles to islet amyloid polypeptide amyloid fibrils. *Biochemistry*, **49**, 3040–3048.

37 Hertel, C., Terzi, E., Hauser, N., Jakob-Rotne, R., Seelig, J., and Kemp, J.A. (1997) Inhibition of the electrostatic interaction between beta-amyloid peptide and membranes prevents beta-amyloid-induced toxicity. *Proc. Natl. Acad. Sci. USA*, **94**, 9412–9416.

38 Winterhalter, M. (2000) Black lipid membranes. *Curr. Opin. Colloid Interface Sci.*, **5**, 250–255.

39 Arispe, N., Pollard, H.B., and Rojas, E. (1993) Giant multilevel cation channels formed by Alzheimer disease amyloid beta-protein [A beta P-(1–40)] in bilayer membranes. *Proc. Natl. Acad. Sci. USA*, **90**, 10573–10577.

40 Lin, H., Bhatia, R., and Lal, R. (2001) Amyloid beta protein forms ion channels: implications for Alzheimer's disease pathophysiology. *FASEB J.*, **15**, 2433–2444.

41 Micelli, S., Meleleo, D., Picciarelli, V., and Gallucci, E. (2004) Effect of sterols on beta-amyloid peptide (AbetaP 1–40) channel formation and their properties in planar lipid membranes. *Biophys. J.*, **86**, 2231–2237.

42 Lin, M.C., Mirzabekov, T., and Kagan, B.L. (1997) Channel formation by a neurotoxic prion protein fragment. *J. Biol. Chem.*, **272**, 44–47.

43 Stipani, V., Gallucci, E., Micelli, S., Picciarelli, V., and Benz, R. (2001) Channel formation by salmon and human calcitonin in black lipid membranes. *Biophys. J.*, **81**, 3332–3338.

44 Caughey, B. and Lansbury, P.T. (2003) Protofibrils, pores, fibrils, and neurodegeneration: separating the responsible protein aggregates from the innocent bystanders. *Annu. Rev. Neurosci.*, **26**, 267–298.

45 Hartley, D.M., Walsh, D.M., Ye, C.P., Diehl, T., Vasquez, S., Vassilev, P.M., Teplow, D.B., and Selkoe, D.J. (1999) Protofibrillar intermediates of amyloid beta-protein induce acute electrophysiological changes and progressive neurotoxicity in cortical neurons. *J. Neurosci.*, **19**, 8876–8884.

46 Kayed, R., Head, E., Thompson, J.L., McIntire, T.M., Milton, S.C., Cotman, C.W., and Glabe, C.G. (2003) Common structure of soluble amyloid oligomers implies common mechanism of pathogenesis. *Science*, **300**, 486–489.

47 Lesne, S., Koh, M.T., Kotilinek, L., Kayed, R., Glabe, C.G., Yang, A., Gallagher, M., and Ashe, K.H. (2006) A specific amyloid-beta protein assembly in the brain impairs memory. *Nature*, **440**, 352–357.

48 Lue, L.F., Kuo, Y.M., Roher, A.E., Brachova, L., Shen, Y., Sue, L., Beach, T., Kurth, J.H., Rydel, R.E., and Rogers, J. (1999) Soluble amyloid beta peptide concentration as a predictor of synaptic change in Alzheimer's disease. *Am. J. Pathol.*, **155**, 853–862.

49 Walsh, D.M., Klyubin, I., Fadeeva, J.V., Cullen, W.K., Anwyl, R., Wolfe, M.S., Rowan, M.J., and Selkoe, D.J. (2002) Naturally secreted oligomers of amyloid beta protein potently inhibit hippocampal long-term potentiation *in vivo*. *Nature*, **416**, 535–539.

50 Kayed, R., Sokolov, Y., Edmonds, B., McIntire, T.M., Milton, S.C., Hall, J.E., and Glabe, C.G. (2004) Permeabilization of lipid bilayers is a common

conformation-dependent activity of soluble amyloid oligomers in protein misfolding diseases. *J. Biol. Chem.*, **279**, 46363–46366.

51 Friedman, R., Pellarin, R., and Caflisch, A. (2009) Amyloid aggregation on lipid bilayers and its impact on membrane permeability. *J. Mol. Biol.*, **387**, 407–415.

52 Engel, M.F., Khemtemourian, L., Kleijer, C.C., Meeldijk, H.J., Jacobs, J., Verkleij, A.J., de Kruijff, B., Killian, J.A., and Hoppener, J.W. (2008) Membrane damage by human islet amyloid polypeptide through fibril growth at the membrane. *Proc. Natl. Acad. Sci. USA*, **105**, 6033–6038.

53 Keshet, B., Yang, I.H., and Good, T.A. (2010) Can size alone explain some of the differences in toxicity between beta-amyloid oligomers and fibrils? *Biotechnol. Bioeng.*, **106**, 333–337.

54 Friedman, R., Pellarin, R., and Caflisch, A. (2010) Soluble protofibrils as metastable intermediates in simulations of amyloid fibril degradation induced by lipid vesicles. *J. Phys. Chem. Lett.*, **1**, 471–474.

55 Widenbrant, M.J., Rajadas, J., Sutardja, C., and Fuller, G.G. (2006) Lipid-induced beta-amyloid peptide assemblage fragmentation. *Biophys. J.*, **91**, 4071–4080.

56 Martins, I.C., Kuperstein, I., Wilkinson, H., Maes, E., Vanbrabant, M., Jonckheere, W., Van Gelder, P., Hartmann, D., D'Hooge, R., De Strooper, B. *et al.* (2008) Lipids revert inert Abeta amyloid fibrils to neurotoxic protofibrils that affect learning in mice. *EMBO J.*, **27**, 224–233.

57 Arispe, N., Diaz, J.C., and Simakova, O. (2007) Abeta ion channels. Prospects for treating Alzheimer's disease with Abeta channel blockers. *Biochim. Biophys. Acta*, **1768**, 1952–1965.

58 Selkoe, D.J. (2006) Amyloid beta-peptide is produced by cultured cells during normal metabolism: a reprise. *J. Alzheimers Dis.*, **9**, 163–168.

59 Demuro, A., Mina, E., Kayed, R., Milton, S.C., Parker, I., and Glabe, C.G. (2005) Calcium dysregulation and membrane disruption as a ubiquitous neurotoxic mechanism of soluble amyloid oligomers. *J. Biol. Chem.*, **280**, 17294–17300.

60 Kawahara, M., Kuroda, Y., Arispe, N., and Rojas, E. (2000) Alzheimer's beta-amyloid, human islet amylin, and prion protein fragment evoke intracellular free calcium elevations by a common mechanism in a hypothalamic GnRH neuronal cell line. *J. Biol. Chem.*, **275**, 14077–14083.

61 Lashuel, H.A. and Lansbury, P.T. Jr. (2006) Are amyloid diseases caused by protein aggregates that mimic bacterial pore-forming toxins? *Q. Rev. Biophys.*, **39**, 167–201.

62 Kourie, J.I. and Henry, C.L. (2002) Ion channel formation and membrane-linked pathologies of misfolded hydrophobic proteins: the role of dangerous unchaperoned molecules. *Clin. Exp. Pharmacol. Physiol.*, **29**, 741–753.

63 Kourie, J.I. and Shorthouse, A.A. (2000) Properties of cytotoxic peptide-formed ion channels. *Am. J. Physiol. Cell. Physiol.*, **278**, C1063–C1087.

64 de Planque, M.R., Raussens, V., Contera, S.A., Rijkers, D.T., Liskamp, R.M., Ruysschaert, J.M., Ryan, J.F., Separovic, F., and Watts, A. (2007) Beta-sheet structured beta-amyloid(1–40) perturbs phosphatidylcholine model membranes. *J. Mol. Biol.*, **368**, 982–997.

65 Quist, A., Doudevski, I., Lin, H., Azimova, R., Ng, D., Frangione, B., Kagan, B., Ghiso, J., and Lal, R. (2005) Amyloid ion channels: a common structural link for protein-misfolding disease. *Proc. Natl. Acad. Sci. USA*, **102**, 10427–10432.

66 Schauerte, J.A., Wong, P.T., Wisser, K.C., Ding, H., Steel, D.G., and Gafni, A. (2010) Simultaneous single-molecule fluorescence and conductivity studies reveal distinct classes of Abeta species on lipid bilayers. *Biochemistry*, **49**, 3031–3039.

67 Volinsky, R., Kolusheva, S., Berman, A., and Jelinek, R. (2006) Investigations of antimicrobial peptides in planar film systems. *Biochim. Biophys. Acta*, **1758**, 1393–1407.

68 Ravault, S., Flore, C., Saurel, O., Milon, A., Brasseur, R., and Lins, L. (2009) Study of the specific lipid binding properties of Abeta 11–22 fragment at endosomal pH. *Langmuir*, **25**, 10948–10953.

69 Dorosz, J., Volinsky, R., Bazar, E., Kolusheva, S., and Jelinek, R. (2009) Phospholipid-induced fibrillation of a prion amyloidogenic determinant at the air/water interface. *Langmuir*, **25**, 12501–12506.

70 Yahi, N., Aulas, A., and Fantini, J. (2010) How cholesterol constrains glycolipid conformation for optimal recognition of Alzheimer's beta amyloid peptide (Abeta1–40). *PLoS ONE*, **5**, e9079.

71 Kong, J. and Yu, S. (2007) Fourier transform infrared spectroscopic analysis of protein secondary structures. *Acta Biochim. Biophys. Sin.*, **39**, 549–559.

72 Pelton, J.T. and McLean, L.R. (2000) Spectroscopic methods for analysis of protein secondary structure. *Anal. Biochem.*, **277**, 167–176.

73 Chiti, F. and Dobson, C.M. (2006) Protein misfolding, functional amyloid, and human disease. *Annu. Rev. Biochem.*, **75**, 333–366.

74 Maltseva, E., Kerth, A., Blume, A., Mohwald, H., and Brezesinski, G. (2005) Adsorption of amyloid beta (1–40) peptide at phospholipid monolayers. *ChemBioChem*, **6**, 1817–1824.

75 Flach, C.R., Brauner, J.W., Taylor, J.W., Baldwin, R.C., and Mendelsohn, R. (1994) External reflection FTIR of peptide monolayer films *in situ* at the air/water interface: experimental design, spectra–structure correlations, and effects of hydrogen–deuterium exchange. *Biophys. J.*, **67**, 402–410.

76 Xu, Z., Brauner, J.W., Flach, C.R., and Mendelsohn, R. (2004) Orientation of peptides in aqueous monolayer films. Infrared reflection–absorption spectroscopy studies of a synthetic amphipathic beta-sheet. *Langmuir*, **20**, 3730–3733.

77 Lopes, D.H., Meister, A., Gohlke, A., Hauser, A., Blume, A., and Winter, R. (2007) Mechanism of islet amyloid polypeptide fibrillation at lipid interfaces studied by infrared reflection absorption spectroscopy. *Biophys. J.*, **93**, 3132–3141.

78 Koppaka, V. and Axelsen, P.H. (2000) Accelerated accumulation of amyloid beta proteins on oxidatively damaged lipid membranes. *Biochemistry*, **39**, 10011–10016.

79 Liu, L., Komatsu, H., Murray, I.V., and Axelsen, P.H. (2008) Promotion of amyloid beta protein misfolding and fibrillogenesis by a lipid oxidation product. *J. Mol. Biol.*, **377**, 1236–1250.

80 Fu, L., Ma, G., and Yan, E.C. (2010) *In situ* misfolding of human islet amyloid polypeptide at interfaces probed by vibrational sum frequency generation. *J. Am. Chem. Soc.*, **132**, 5405–5412.

81 Homola, J. (2008) Surface plasmon resonance sensors for detection of chemical and biological species. *Chem. Rev.*, **108**, 462–493.

82 Mozsolits, H. and Aguilar, M.I. (2002) Surface plasmon resonance spectroscopy: an emerging tool for the study of peptide–membrane interactions. *Biopolymers*, **66**, 3–18.

83 Subasinghe, S., Unabia, S., Barrow, C.J., Mok, S.S., Aguilar, M.-I., and Small, D.H. (2003) Cholesterol is necessary both for the toxic effect of Aβ peptides on vascular smooth muscle cells and for Ab binding to vascular smooth muscle cell membranes. *J. Neurochem.*, **84**, 471–479.

84 Smith, D.P., Tew, D.J., Hill, A.F., Bottomley, S.P., Masters, C.L., Barnham, K.J., and Cappai, R. (2008) Formation of a high affinity lipid-binding intermediate during the early aggregation phase of alpha-synuclein. *Biochemistry*, **47**, 1425–1434.

85 Devanathan, S., Salamon, Z., Lindblom, G., Grobner, G., and Tollin, G. (2006) Effects of sphingomyelin, cholesterol and zinc ions on the binding, insertion and aggregation of the amyloid Abeta(1–40) peptide in solid-supported lipid bilayers. *FEBS J.*, **273**, 1389–1402.

86 Cross, G.H., Reeves, A.A., Brand, S., Popplewell, J.F., Peel, L.L., Swann, M.J., and Freeman, N.J. (2003) A new quantitative optical biosensor for protein

characterisation. *Biosens. Bioelectron.*, **19**, 383–390.

87 Swann, M.J., Peel, L.L., Carrington, S., and Freeman, N.J. (2004) Dual-polarization interferometry: an analytical technique to measure changes in protein structure in real time, to determine the stoichiometry of binding events, and to differentiate between specific and nonspecific interactions. *Anal. Biochem.*, **329**, 190–198.

88 Sanghera, N., Swann, M.J., Ronan, G., and Pinheiro, T.J. (2009) Insight into early events in the aggregation of the prion protein on lipid membranes. *Biochim. Biophys. Acta*, **1788**, 2245–2251.

89 Valincius, G., Heinrich, F., Budvytyte, R., Vanderah, D.J., McGillivray, D.J., Sokolov, Y., Hall, J.E., and Losche, M. (2008) Soluble amyloid beta-oligomers affect dielectric membrane properties by bilayer insertion and domain formation: implications for cell toxicity. *Biophys. J.*, **95**, 4845–4861.

90 Kotarek, J.A. and Moss, M.A. (2010) Impact of phospholipid bilayer saturation on amyloid-beta protein aggregation intermediate growth: a quartz crystal microbalance analysis. *Anal. Biochem.*, **399**, 30–38.

91 Yip, C.M., Darabie, A.A., and McLaurin, J. (2002) Abeta42-peptide assembly on lipid bilayers. *J. Mol. Biol.*, **318**, 97–107.

92 Kayed, R., Pensalfini, A., Margol, L., Sokolov, Y., Sarsoza, F., Head, E., Hall, J., and Glabe, C. (2009) Annular protofibrils are a structurally and functionally distinct type of amyloid oligomer. *J. Biol. Chem.*, **284**, 4230–4237.

93 Diociaiuti, M., Polzi, L.Z., Valvo, L., Malchiodi-Albedi, F., Bombelli, C., and Gaudiano, M.C. (2006) Calcitonin forms oligomeric pore-like structures in lipid membranes. *Biophys. J.*, **91**, 2275–2281.

94 Jo, E., McLaurin, J., Yip, C.M., St George-Hyslop, P., and Fraser, P.E. (2000) Alpha-synuclein membrane interactions and lipid specificity. *J. Biol. Chem.*, **275**, 34328–34334.

95 Weise, K., Radovan, D., Gohlke, A., Opitz, N., and Winter, R. (2010) Interaction of hIAPP with model raft membranes and pancreatic beta-cells: cytotoxicity of hIAPP oligomers. *ChemBioChem*, **11**, 1280–1290.

96 Haque, F., Pandey, A.P., Cambrea, L.R., Rochet, J.C., and Hovis, J.S. (2010) Adsorption of alpha-synuclein on lipid bilayers: modulating the structure and stability of protein assemblies. *J. Phys. Chem. B*, **114**, 4070–4081.

97 Paquet, M.J., Laviolette, M., Pezolet, M., and Auger, M. (2001) Two-dimensional infrared correlation spectroscopy study of the aggregation of cytochrome $c$ in the presence of dimyristoylphosphatidylglycerol. *Biophys. J.*, **81**, 305–312.

98 Noda, I. (1989) Two-dimensional infrared spectroscopy. *J. Am. Chem. Soc.*, **111**, 8116–8118.

99 Pancoska, P., Kubelka, J., and Keiderling, T.A. (1999) Novel use of a static modification of two-dimensional correlation analysis. Part I. Comparison of the secondary structure sensitivity of electronic circular dichroism, FT-IR, and Raman spectra of proteins. *Appl. Spectrosc.*, **53**, 655–665.

100 Choucair, A., Chakrapani, M., Chakravarthy, B., Katsaras, J., and Johnston, L.J. (2007) Preferential accumulation of Abeta(1–42) on gel phase domains of lipid bilayers: an AFM and fluorescence study. *Biochim. Biophys. Acta*, **1768**, 146–154.

101 Fantini, J., Garmy, N., Mahfoud, R., and Yahi, N. (2002) Lipid rafts: structure, function and role in HIV, Alzheimer's and prion diseases. *Expert Rev. Mol. Med.*, **4**, 1–22.

102 Porat, Y., Kolusheva, S., Jelinek, R., and Gazit, E. (2003) The human islet amyloid polypeptide forms transient membrane-active prefibrillar assemblies. *Biochemistry*, **42**, 10971–10977.

103 Sokolovski, M., Sheynis, T., Kolusheva, S., and Jelinek, R. (2008) Membrane interactions and lipid binding of casein oligomers and early aggregates. *Biochim. Biophys. Acta*, **1778**, 2341–2349.

104 Kolusheva, S., Wachtel, E., and Jelinek, R. (2003) Biomimetic lipid/polymer colorimetric membranes: molecular and

cooperative properties. *J. Lipid Res.*, **44**, 65–71.
105 Jelinek, R. and Kolusheva, S. (2001) Polymerized lipid vesicles as colorimetric biosensors for biotechnological applications. *Biotechnol. Adv.*, **19**, 109–118.
106 Kolusheva, S., Boyer, L., and Jelinek, R. (2000) A colorimetric assay for rapid screening of antimicrobial peptides. *Nat. Biotechnol.*, **18**, 225–227.
107 Giannakis, E., Pacifico, J., Smith, D.P., Hung, L.W., Masters, C.L., Cappai, R., Wade, J.D., and Barnham, K.J. (2008) Dimeric structures of alpha-synuclein bind preferentially to lipid membranes. *Biochim. Biophys. Acta*, **1778**, 1112–1119.

# Index

## a

acetylcholine (ACh)
- nicotinic acetylcholine receptor (nAChR)  162
- α7 nicotinic acetylcholine receptors (α7AchR)  148
- synthesis, suppression of  200

acetylcholinesterase (AChE) inhibitor  200
- activation of ERK by  201
- effect on NMDAR response, impared in  199–201
- enhance cholinergic transmission  200
- NMDAR-EPSC, regulation of  200

acidic fibroblast growth factor (AFGF)  78
acyl CoA binding protein (ACBP)  68, 69
- denaturation, stages of  68
- stages of denaturation  68

AD. see Alzheimer's disease (AD)

aggregation
- in fluorinated organic solvents  74, 75
- on lipids  78–81
- in presence of surfactants  58
- of proteins by SDS  61, 63
- protein structure, effect of surfactants on  60, 61
- protein-surfactant interactions  58–60
- stoichiometry of SDS binding  61

albumin  58, 66
alcadein (Alc)  180, 187, 188
- primary protease  188

alkyltrimethylammonium bromides  65

Alzheimer amyloid peptide interaction
- association with lipid particles,  238
- Aβ catalysis, oligomerization by cell surface  234
- Aβ oligomerization at the cell surface  233, 234
- Aβ type complexes,  234–238
- with cell surfaces and artificial membranes  231
- future directions,  238, 239
- oligomeric and fibrillar Alzheimer amyloid peptides
-- comparison of neurotoxicity  232
- probe membrane interactions of, 246

Alzheimer's disease (AD)  33, 67, 123, 143. see also Alzheimer's disease (AD) brains
- aberrant CaMKII subcellular distribution  198
- amyloid hypothesis in  143
- β-amyloid impairment
-- AMPAR-mediated synaptic transmission and ionic current  195, 196
-- CaMKII involvement in AMPAR trafficking and  196, 197
-- glutamate receptor channels and signaling molecules, impaired regulation  195
-- NMDAR response, AChE inhibitor effect in APP transgenic mice  199–201
-- PIP$_2$ regulation of  197–199
-- PKC-dependent signaling and functions  201, 202
- ApoE, lipoprotein receptors  181–183
- development of  187
- feature of  195
- future directions of  231
- index for  180
- lipid rafts and  183–185
- LRP1 and LRP1B  182
- as membrane-associated enzymopathy  177–180
- neuropathology  144, 168
- pathogenesis of  161
-- cholesterol level  180, 181
- γ-secretase, alcadein processing by  187–190

- SorLA/LR11, 182, 183
- transgenic mice model 181
- type I membrane proteins, intramembrane-cleaving enzyme 186, 187

Alzheimer.s disease (AD) brains
- Cav-1 expression 161
- changes in VDAC levels, in cortex 167
- detection of intracellular Aβ, 144
- ERα–VDAC complex, disruption of 167, 168
- reduced levels of PIP$_2$, 198
- VDAC in lipid rafts 166

amorphous aggregates 65, 125, 210
AMPA receptors (AMPARs) 195
- mediated excitatory postsynaptic current (EPSC) 196
- reduction 196
- synaptic responses and ionic currents, decrease in 197

amyloid β (Aβ) hypothesis 202
amyloid β (Aβ) oligomers 63, 64, 126, 134, 143, 181, 197, 232, 234, 257
amyloid β (Aβ) peptides
- aggregated species of 132
- aggregation 180
-- cascade 57
-- critical concentration 234
-- progression of 239
-- regulation of 81
-- sphingomyelin initiating 255
-- on the surface of neuronal cell lines 233
-- *in vitro* studies of 234
- catalysis 234
- extracellular senile plaques composed primarily of 143
- flotillins appear to upregulated at sites fo 161
- generated through sequential processing of 186
- interactions (*see* Alzheimer amyloid peptide interaction)
- N-terminus 233
- progression 239
- punctate stain of 235

amyloid-derived diffusible ligands (ADDLs) 63, 134, 231, 232
amyloid–lipid binding, schematic depiction 256
amyloid-lipid colocalization complexes 108
amyloid–membrane interactions
- black lipid membranes (BLMs) 252, 253
- challenges and future work 261–263
- experimental approaches and technical challenges 245

- Langmuir monolayers 253–255
- solid-supported bilayers 255–259
- unilamellar vesicles and micelles 248–251

amyloidogenic proteins 3. *see also* α-synuclein
- membrane interactions of 245
- pathological processes involving 4

amyloid oligomers 127, 130, 252, 257
- generation 127–129
- growth 132–135
- interactions 127–129
- polymorphisms of 129
- structural features of 124–127
- structure-toxicity relations of 135

amyloid precursor protein (APP) 145, 161, 195
- APP–X11L–Alc complex 188
- intracellular domain (AICD) 147, 188
- processing 130

amyloid precursor protein intracellular domain (AICD) 147, 186, 188

β-amyloid precursor protein secretases
- Alzheimer's disease as membrane-associated enzymopathy 177–180
- amyloidogenic β-cleavage 179
- ApoE, lipoprotein receptors 181–183
- ApoER2/LRP8, 183
- lipid rafts and 183–185
- LRP1 and LRP1B 182
- metabolism 178
- pathogenesis, cholesterol 180, 181
- γ-secretase, alcadein processing by 187–190
- SorLA/LR11, 182, 183
- type I membrane proteins, intramembrane-cleaving enzyme
-- γ-secretase 186
-- γ-secretase and cholesterol 186, 187

amyloid(s)
- aggregates 2, 33, 124, 126, 128, 130, 133, ???, 256, 259, 262
- assemblies 122
-- cytotoxicity 124
-- morphological modifications 125
-- polymorphisms of 133
- cytotoxicity 129–131
- diseases 121
- fibrils 115, 122, 123, 131
-- assembly 125
-- cytotoxicity of 130
-- formation 96, 103
-- *in vitro* and in tissue 130

– fold 122
– hypothesis 1, 121, 123, 143
– neurotoxicty of intracellular 145, 146
– polymorphisms *in vivo* 133
– pores hypothesis 261
– precursor protein
– – and Alc, co-association 188
– – in context of aging 185
– – interactions of Cav-1 with 161
– – mutant transgenic mice 146, 197, 201
– – processing 130
– property of 126
– toxicity 123, 131
– – data, *in vivo* disease of 124
– vaccine therapies 149
anionic surfactants (ASs) 57
anterior pharynx-defective 1 (APH-1) 179
anti-amyloid
– antibody 149
– design of 115
antibody-$\beta_2$-GPI-lipid complex 67
antigen–antibody complexes 218
antimicrobial peptide novispirin 63
ApoE–cholesterol complexes 182
ApoE receptors 181, 182
– LDLR-related protein 1, 182
APP. *see* amyloid precursor protein
arachidonic acid 3, 21, 64, 67
Arctic mutant 133, 134
assembling oligomers 129
atomic force microscopy (AFM) 72, 125, 257, 258, 262
ATP-binding cassette sub-family A member 1 (ABCA1) 181
– over-expression 182
attenuated total reflection Fourier transform infrared (ATR-FTIR) 95, 99–101, 105, 106, 109, 115

**b**

BACE. *see* β-secretase (BACE)
bacterial hydrogenase maturation factor F (HypF-N)
– N-terminal domain of 132
band-shifting solvatochromic probes 9
β-cell
– apoptosis 93
– exposed to hyperlipidemia 94
– function 93
– main mechanism for hIAPP and 95
black lipid membranes (BLMs) 252, 253
– application 253
– measurements 262

boradiazaindacene (BODIPY) excimer formation 108, 247
bovine spongiform encephalopathy (BSE)
– cellular prion protein (PrP$^c$) and conversion to PrP$^d$ 207–210
– follicular dendritic cells in lymphoid germinal centers 217–219
– membrane changes in 207
– pathological features, and immunohistochemistry 211
– PrP$^d$ aberrant endocytosis, and trafficking in 213–217
– PrP$^d$ accumulation in
– – central nervous system and lymphatic tissues 210–213
– – plasma membranes associated with molecular changes 219, 220
– PrP$^d$ cells transfer 220
– PrP$^d$ extracellular amyloid form 221, 222
– strain-directed effects of prion infection 222
BSE. *see* bovine spongiform encephalopathy (BSE)
β-sheet
– amyloid cross-β-sheet structure 75
– cross-β-sheet conformation for AS 5
– cytotoxicity associated with 34, 81
– formation 75
– formation by Aβ peptide 210
– form serves as seed for formation of amyloid fibrils 80
– intermolecular β-sheet formation 68, 100
– SDS inducing 60, 103

**c**

calcein efflux 40
CaMKII synaptic pool 197
6-carboxyfluorescein (FAM) 234
carboxypeptidase E (CPE) 94
cationic surfactants 65
caveolae/raft signaling hypothesis 159
caveolar scaffolding domain (CSD) 164
caveolin-mediated endocytosis 208
cell-surface sialoglycoprotein 208. *see also* cellular prion protein (PrP$^c$)
cell/tissue impairment, with amyloid oligomers 123
cellular prion protein (PrP$^c$) 160, 207
– glycoforms 209
– trafficking/localization 209
cellular stress 34
central cholinergic system 199
central nervous system (CNS) 180, 182, 207, 210

– abundant PrP$^c$ is detected in association with   209
– ApoE receptors expressed in   182
– endocytosis in   219
– immunohistochemical labeling for PrP$^d$   212
– plasma membrane changes, and immunogold PrP$^d$ labeling   215
– scrapie-and BSE-infected animals, evidence of   220
cholesterol   19, 36, 41, 128, 129, 148, 160, 167, 180
– apoE–cholesterol complexes   182
– homeostasis   181
– level   34
– monomeric A$\beta_{40}$ cause, release of   238
– regulating, binding affinity of A$\beta_{1-40}$ to different glycolipids by   253
– and sphingomyelin in lipid raft models   261
– statins, synthesis inhibitors   186
chronic wasting disease (CWD)   207
α-chymotrypsin   77
circular dichroism (CD) spectroscopy   13, 16, 33, 41, 61, 95, 96, 249, 261
– conformation for proIAPP   96
CNS. see central nervous system (CNS)
complex fluorescence experiments   247
confocal microscopy images   233
Congo Red   64
Creutzfeldt-Jakob disease (CJD)   67, 207
– variant CJD (vCJD)   207
critical micelle concentration (cmc)   102
cross-polarization magic angle spinning (CPMAS)   249, 250
cytotoxic protein-lipid complexes   72–74

### d

dehydroergosterol (DHE)   246
deleted in colorectal carcinoma (DCC)   187
detergent resistant membrane (DRM) domains   179
dialkylaminocoumarin   10
2-(4′-N,N-dialkylaminophenyl)-3-hydroxychromone   11
7-diethylaminocoumarin-3-carbonyl (DAC) fluorophore, applications   246
1,2-dioleoyl-sn-glycero-3-phosphatidylglycerol (DOPG )   16, 41
1,2-dioleoyl-sn-glycero-3-phosphocholine (DOPC)   96
– bilayer   105–107
– DOPC-DOPE membranes   112
– DOPC-DOPG membrane interface   99
– DOPC-DOPG membrane system   101

– DOPC-DPPC-Chol (1:2:1) model raft mixture   115
1,2-dioleoyl-sn-glycero-3-phosphoethanolamine (DOPE)   112
1,2-dioleoyl-sn-glycero-3-phosphoserine (DOPS)   40, 43, 45, 112
dipalmitoylphosphatidylglycerol (DPPG)   18, 37, 79
dipalmitoyl-sn-glycero-3-phosphate (DPPA)   37
disease specific prion protein (PrP$^d$)   208
– aggregation   221
– cellular processing   222
– membrane–molecular interactions   209
– presumptive conversion   213
– trafficking   222
docosahexanoic acid (DHA)   167, 168
double electron electron resonance (DEER)   251
Down's syndrome (DS)   144
Drosophila disc large tumor suppressor (Dlg)   184
Drosophila model   133
dual polarization interferometry (DPI)   256

### e

electrochemical impedance spectroscopy (EIS)   257
electron microscopy (EM)   125
electron paramagnetic resonance (EPR) spectroscopy   5, 250, 251
ELOA complex   74
endoplasmic reticulum (ER)   2, 147, 186
– membranes   128
– organizing membrane protein   217
endosomal/lysosomal system   148
environment-sensitive dyes   9
epi-fluorescence microscopy   258
estrogen receptors (ERs)   161
– estrogen receptor α (ERα),   161, 163
– estrogen receptor β (ERβ),   163
excited state intermolecular proton transfer (ESIPT)   10
– characteristics of ESIPT probes   11
– probes, advantages of   11
extracellular signal-regulated kinase (ERK)   146, 200, 201

### f

familial Alzheimer's disease (FAD)   177. see also Alzheimer's disease (AD)
– causal factor   178
– linked PS1 mutations   189

fibrillar aggregates  33
– β-sheet aggregates  64
fibrillization  33
fibril structural nuclei, conformational features of  126
fibril structure, alteration by TFE  77, 78
flotillins, accumulation at sites of amyloid β peptide (Aβ)  161
fluorescence anisotropy  246, 248
fluorescence microscopy  248, 249
fluorescence resonance energy transfer (FRET)  236, 246
– application  247
– schematic depiction  247
fluorescence spectroscopy  133, 246–248
fluorescence techniques  262
fluorinated organic solvents (FOSs)  57
follicular dendritic cells (FDCs)  209, 217
– maturation cycle  219
– morphological forms  218
– plasma membrane  219
– subcellular changes and ubiquitin and IgG, PrP$^d$ labeling  218
Fourier self-deconvolution (FSD)  100
2-(2-furyl)-3-hydroxychromone (FC)  11

## g

gangliosides  80, 160, 166, 167, 246, 247
Gerstmann–Sträussler–Scheinker disease  213
giant unilamellar vesicles (GUVs)  108, 246
glial cells  145, 168, 209, 210
glial-derived neurotrophic factor (GDNF)- family ligands  161
glutamate receptors  166, 195
glutamatergic transmission, long-term potentiation (LTP)  195
glycogen synthase kinase-3 (GSK-3) activation  165
glycophosphatidylinositol (GPI)  208
glycosaminoglycans (GAGs)  127
glycosylphosphatidylinositol (GPI)-anchored proteins  160
Golgi apparatus  2, 186
– elements  146
GPI-anchored proteins  160, 208, 210, 220
G-protein coupled receptor kinase  5 (GRK5)- deficient mice, selective working memory impairment  146
grazing incidence X-ray diffraction (GIXD)  254
green fluorescent protein (GFP)-RACK1 Sindbis virus  202

## h

hexafluoroisopropanol (HFIP)  64
hIAPP. *see* human islet amyloid polypeptide (hIAPP)
Hippocampal cells 2-D electrophoresis and Western blotting  166
human α-lactalbumin made lethal to tumor cells (HAMLET)  72
human calcitonin (hCT)  250
human cerebral cortex
– association of IGF-1R with caveolin-1 (Cav-1), ERα, and VDAC in lipid rafts  162
human islet amyloid polypeptide (hIAPP)  93
– adsorption kinetics of  108
– AFM image of  99
– amide-I′ bands, time evolution of  99, 106, 107
– amyloid formation, AFM images of  104
– amyloidogenic propensity of  101
– – membrane-mimicking anionic surfactant SDS, effect of  102–105
– anionic DOPC-DOPG lipid interface  98
– ATR-FTIR spectroscopic measurements of  105
– biosynthesis, stages of  101
– conformational transition  96
– cytotoxicity of  95, 112–116
– DOPC-DPPC-Chol (1:2:1) membrane with  110
– fibril formation  111
– fibrillation kinetics, and conformational changes of  96
– hIAPP-proIAPP peptide mixtures
– – ThT binding assay for  101
– interaction of  93
– lipid interactions  94
– membrane interactions  111, 116
– native conformations of  95, 96
– N-terminal region of  96
– oligomers, size of  97
– with raft-like GUVs, interaction of  109
– species, AFM image of  106
– ThT assay monitoring, fibril formation of  113
– zwitterionic lipid raft membranes  109
3-hydroxychromones (3HCs)  10
– chromophores  11
– dyes  11
– membrane probes with different localizations in  12
3-hydroxymethyl 3-glutaryl-coenzymeA (HMG-CoA) reductase  180
hyperpolarization-activated cyclic nucleotide gated (HCN) channel  184

## i

immunocytochemical experiments 197
infrared reflection absorption spectroscopy (IRRAS) 95, 100
– application of 254
infrared reflection adsorption spectroscopy (IRRA) 254
INS-1E cell assay, hIAPP fibrils, least reactive species in 114
insulin growth factor-1 receptor (IGF-1R) 161
insulin-membrane interaction vs. hIAPP-membrane interactions 111, 112
insulin receptor (IR) 161
intensiometric dyes 9
intracellular amyloid. *see also* amyloid(s)
– amyloid β peptide (Aβ), 143, 144
– – origin and source of 147
– detection of 144, 145
– extracelluar, relationship 148, 149
– hypothesis 143
– mechanisms of 146
– neurotoxicty of 145, 146
– sources of 146–148
– toxicity, prevention of 149
intracellular signaling 159
intramolecular charge transfer (ICT) 9
intrinsic fluorescence 247
ion membrane channels, putative formation of 253
islet amyloid polypeptide (IAPP) 78–80, 93, 248
– activity 248
– co-localization 249
– human IAPP (hIAPP) 250
– interaction 258
– lipid interactions of 254
– for membrane interactions and fibrillation 248
– N-terminal part of 80
– PDA–IAPP experiment 260
– stimulate aggregation 78, 79
isothermal titration calorimetry (ITC) 251

## j

JNK interacting proteins (JIPs) 183

## k

K3 aggregation 59
2-keto-3-deoxyoctonoic (KDO) acid 101

## l

α-lactalbumin 58, 73, 75
Langmuir monolayers 253–255

large unilamellar vesicles(LUVs) 246
Lewy body (LBs) 2
Lewy neurites (LNs) 2
lipid rafts 159–162
– and Alzheimer's disease 183–185
– ERα–VDAC interactions, role of 164–167
– integral molecules 160
– neuronal 163
– resident protein 161
– role in 159
– zwitterionic membranes 109
lipids
– function 3
– lipid-water interface 100
– in synucleinopathies 3
long-term potentiation (LTP) formation 143
low-density lipoprotein (LDL) levels 180

## m

MAPK pathways 161, 165
MAPK signaling, mER 165
mass spectrometry (MS) 260
matrix-assisted laser desorption/ionization time-of-flight mass spectrometry (MALDI-TOF-MS) 260
matrix metalloproteases (MMPs) 186
membrane proteins
– proteolytic processing of 128
metabotropic glutamate receptor (mGluR)
– PKC-dependent functions 201
N-methyl-D-aspartic acid receptor (NMDAR) 195
$\beta_2$-microglobulin (β2m) 66, 76, 77
– fibril disassembly 131
– helical-rich conformation 102
mild cognitive impairment (MCI) 144
monoclonal/ polyclonal antibodies 126
motionally restricted lipid component (MRLC) 250
multiparametric ratiometric dyes 10
multiple system atrophy (MSA) 2
multivesicular body (MVBs) 215, 216
– membranes of 145

## n

nerve cells, preservation of 165
neurodegenerative diseases 1
neurofibrillary tangles (NFTs) 143
– formation 149
neuronal apoptosis, pl-VDAC, role of 166
neuronal cell lines 162
neuronal lipid rafts, disruption, schematic hypothesis of 168
neuronal loss 133, 145, 149

neuronal physiological activities
– development of 162
neuronal preservation 159, 164, 169
– intracellular signaling 159
neuronal rafts, estrogen receptors 163, 164
neuroprotection 162
neurotoxic Aβ, production of 179
neurotoxicity 132, 146, 149, 165
neurotoxins 231
nicastrin (NCT) 179
Niemann–Pick type C disease 181
Nile red 10
nitrobenzoxadiazole (NBD) 10
NMDAR channels 198
– $PIP_2$ regulation of 199
NMDAR-EPSC
– physostigmine-induced reduction 201
non-ESIPT probe prodan 10
nuclear magnetic resonance (NMR) 5, 249, 250, 262
– $^{13}C$ NMR spectroscopy 249, 250
– $^{31}P$ magic angle spinning (MAS) 249

**o**

oligomer/fibril polymorphisms 124, 129, 130
oligomer formation in presence of SDS. 65
oligomeric aggregation nuclei, formation of 125

**p**

p3-Alc and Aβ species
– amino acid sequence 190
– profiles 189
1-palmitoyl 2-oleoylphosphatidyletanolamine (POPE) membrane 252
1-palmitoyl-2-oleoyl-sn-glycero-3-phosphocholine (POPC) 40
Parkinson's disease (PD) 1, 132, 164
– drug-based therapies 1
– genetic factors involved in 1, 2
– lipids, central role in 2
– molecular and cellular processes related to 3
– multifactorial nature of 1
pathogenic factor 180
PC12 cells 236
PDA–IAPP experiment, schematic description 260
peptides
– aggregation 127
– fibrillization of 123
peptidyl amidating monooxygenase complex (PAM)
– activation of 94

phenylarsine oxide (PAO) 198
– reducing effect 199
phosphatidylglycerol (PG) 78, 94
phosphatidylserine (PS) 36, 78, 94
phospholipid phosphatidylinositol 4,5-bisphosphate ($PIP_2$) 197
phospho-protein kinase C (p-PKC) 202
photobleaching fluorescence resonance energy transfer (pbFRET) 236
– co-localization 237
PI3K/Akt pathway 165
plasmalemmal voltage-dependent anion channel (pl-VDAC) 164, 165
– pl-VDAC–mERa interactions 168
– regulation 169
plasmon waveguide resonance (PWR) spectroscopy 255
platelet-derived growth factor receptor (PDGFR) 161
point mutations 2, 33
polarized attenuated total internal reflection Fourier transform infrared (PATIR-FTIR) spectroscopy 254
polydiacetylene (PDA) 259
– based sensor 259
– based vesicle assays 259
polymorphic amyloid oligomers
– structural and physicochemical features of 134
polyunsaturated fatty acids (PUFAs) 3, 160
post-synaptic density protein 95 (PSD95) 184
presenilin enhancer 2 (PEN-2) 179
presenilin (PS) genes
– presenilin 1 (PS1) genes 177
– presenilin 2 (PS2) genes 177
primarily biomimetic membrane models 245
prion proteins (PrPs) 67
pro-islet amyloid polypeptide (proIAPP) 93
– aggregation/fibrillation process of 97
– amide-I' band of 107
– amyloid formation 94
– amyloidogenic propensity of
– – membrane-mimicking anionic surfactant SDS, effect of 102–105
– fibrillation 104
– – kinetics and conformational changes of 96
– intracellular processing of 94
protease-resistant prion protein ($PrP^{res}$) 208
– fibrillar amyloid deposits 210
protein-amphiphile complexes 73

protein β$_2$-glycoprotein I, (β$_2$-GPI)   66
protein kinase C (PKC)   201
– mGluR activation   201
– signaling   238
protein-lipid complex   73
protein-oleic acid complexes   58
protein/peptide fibrillization   128
proteins
– aggregation   57, 122, 125, 129, 131
– conformational disorders   1
– degradation   34
– fibrillization of   123
– folding   125
– interactions, lipid rafts   159
– *in vitro*   123
– misfolding   127
– only hypothesis   208
– oxidation   2
– protein L   78
PrP$^d$–ligand–clathrin–ubiquitin array   214
PrP$^d$–ligand–ubiquitin complex   219
PrP$^d$ membrane–protein interactions   219

## q

quartz crystal microbalance (QCM), application   257

## r

Rat IAPP (rIAPP)   93
receptor for activated C-kinase 1(RACK1)   201
RNase Sa protein   78

## s

scrapie prions
– cellular prion protein (PrP$^c$) and conversion to protein (PrP$^d$)   207–210
– central nervous system and lymphatic tissues, PrP$^d$ accumulation in   210–213
– follicular dendritic cells, in lymphoid germinal centers
– – abnormal maturation cycle and immune complex trapping   217–219
– membrane changes in   207
– pathological features, and immunohistochemistry   211
– PrP$^d$ aberrant endocytosis and trafficking in   213–217
– PrP$^d$ accumulation, plasma membranes associated with
– – molecular changes   219, 220
– PrP$^d$ cells transfer   220
– PrP$^d$ extracellular amyloid form   221, 222
– strain-directed effects of prion infection   222

β-secretase (BACE)   179, 182, 183, 186
γ-secretase   186–190
– pathogenic malfunction   190
– role   177
secreted amyloid precursor protein (sAPP)   147
signaling proteins
– lipid modifications of   160
signal transduction cascades   159
small-angle neutron scattering (SANS)   251
small-angle X-ray scattering (SAXS)   126
small unilamellar vesicles (SUVs)   246
SN56 cells   208
– for non-genomic mechanism of estrogen action   165
– recycling endosomal compartment   208
sodium dodecyl sulfate (SDS)   57, 102, 103
– amyloid fibrils   105
– binding, stoichiometry of   61
– cylindrical/spherical   102
– electrostatic repulsion   104
– α-helical conformation   104
– interaction with hIAPP   102
– micelles   250
solid-state nuclear magnetic resonance (ss-NMR) spectroscopy   126
solid-supported bilayers   255–259
sporadic Alzheimer's disease (SAD)   178
– fixed population   180
– genetic causes   181
– risk factor associated with   181, 182
Src tyrosine kinases (STKs) family   159
stomatin/prohibiting/flotillin/HflK/C (SPFH)   161
sum frequency generation (SFG) experiments   254
– advantages   255
surface-enhanced laser desorption/ionization time-of-flight mass spectrometry (SELDI-TOF-MS)   260
surface plasmon resonance (SPR)   255
surfactants   58
– on protein structure   60, 61
α-synuclein (αSN)   2, 69–72
– aggregation and monitored by AS140-MFC   22
– AS-lipid interactions   6, 13
– – AS binds to membranes as α-helix   14
– – calorimetry with DPPC   14

-- CD spectra studies 14
-- conformation(s) of AS on lipid
   membranes 14
-- fluorescence correlation spectroscopy
   (FCS) with POPC vesicles 14
-- fluorescence techniques 6–9
-- higher affinity for raftlike domains 14
-- high-resolution NMR 14
-- ITC measurements 14
-- parameters that determine 13
-- smFRET technology for studies in 15
- binding to charged membranes 21
- determination of oligomer structure by
   Trp fluorescence 44
- diamers 260
- fibrillation 2
- influence of acyl chains
-- aggregates lead to death of dopaminergic
   neurons 33, 34
-- aggregation and effects of fatty acids
   monitored with ESIPT probes
   21–23
-- binding efficiencies of pathogenic protein
   variants 38
-- binding kinetics 19–21
-- binding to giant unilamellar vesicles
   35–39
-- dilauroyl-phosphatidylglycerol
   (DLPG)-rich domains 37
-- E46K and A53T variant 38
-- influence of cholesterol 19
-- lipid phase specificity of lipid binding
   of 39
-- non-Aβ component, amino acids for
   aggregation and 34
-- oligomeric, species inducing
   permeabilization of 35
- molecular and cellular processes related
   to 3
- monomeric, interactions with artificial
   membranes monitored with ESIPT
   probes 15
-- influence of acyl chains 18
-- influence of membrane charge 16
-- influence of membrane curvature
   16, 17
-- influence of membrane phase 17, 18
- oleic acid trigger aggregation *in vitro* 3
- oligomerization facilitating
   oligomerization 3
- oligomers
-- biological significance, interaction
   sites 45–47
-- membrane penetration 47–49

-- model membrane permeabilization,
   by 39–41
-- species of 132
-- structural features of 41–45
- protein 78
- αSN-SDS interactions 70
- structural biology of 4, 5
-- C-terminal region 5
-- NAC regions 5
-- primary and secondary 4
-- variable conformations 5
- toxicity 3

*t*
Tau protein 67
tetramethylrhodamine (TMR)
   234
TGFBIp protein 78
thioflavin S-positive material 134
thioflavin T (ThT) 64, 95
- binding complexes 70
- fluorescence 103
-- assay 111, 112
-- intensity 113
- monitored amyloid fibril formation
   97
tinctorial dyes, Congo Red 212
tingible body macrophages (TBMs)
   213, 216
- cell membrane 216
- primary function 216
- subcellular changes and ubiquitin and
   PrP$^d$ labeling 217
total internal reflection fluorescence
   (TIRF) microscopy 258, 259
trans Golgi network (TGN) 147, 182
transmissible spongiform encephalopathy
   (TSEs) 207
transmission electron microscopy (TEM)
   study 258
trifluoroethanol (TFE) 57
- effects on aggregation 76
- induced formation of fibrils by β2m
   peptide K3, 77
- induce protein aggregation 78
Triton X-100, 184
two-dimensional IR correlation (2D-IR)
   spectroscopy 258
type 2 diabetes mellitus (T2DM) 93
- prediabetic stage of 94

*v*
vesicle-encapsulated fluorescence
   dyes

– leakage 248
voltage-dependent anion channel (VDAC) 161
– inactivation 169
– post-translational regulation 167

## w

water-soluble tetrazolium salt (WST-1) 95
– colorimetric assay 114

## x

X11 family proteins (X11s) 184
– dysfunction 185
– role 185
X-ray reflectivity (XRR) 95

## z

zonula occludens-1 protein (ZO-1) 184
zwitterionic raft membranes, with anionic membranes 109